This book describes the application of new high-temperature superconducting materials to microwave devices and systems. It deals with the fundamentals of the interaction between microwaves and superconductors, and includes a basic description of how microwave devices can be constructed using these materials.

Since the discovery of high-temperature superconductors in 1986 there has been an enormous effort worldwide to develop and characterise these materials. Work on applications has proceeded more slowly, however. Nevertheless, commercial applications are now beginning to be possible, including use in passive microwave devices. The advantages of using high-temperature superconductors in these devices is carefully described by the author, enabling scientists and engineers to form a complete understanding of the subject. The rest of the book is devoted to examples of superconducting microwave filters, antennas and systems. The examples chosen relate not only to what can be achieved at present, but indicate trends for future research and what may be expected for superconducting devices in the future.

This book will be of value to graduate students and researchers, especially electrical engineers interested in new devices that are possible using high-temperature superconductors, as well as physicists, materials scientists and chemists interested in the application of these materials.

PASSIVE MICROWAVE DEVICE APPLICATIONS OF HIGH-TEMPERATURE SUPERCONDUCTORS

PASSIVE MICROWAVE DEVICE APPLICATIONS OF HIGH-TEMPERATURE SUPERCONDUCTORS

M. J. LANCASTER
School of Electronic and Electrical Engineering
University of Birmingham

CAMBRIDGE UNIVERSITY PRESS
Cambridge, New York, Melbourne, Madrid, Cape Town, Singapore, São Paulo

Cambridge University Press
The Edinburgh Building, Cambridge CB2 2RU, UK

Published in the United States of America by Cambridge University Press, New York

www.cambridge.org
Information on this title: www.cambridge.org/9780521480321

First published 1997
This digitally printed first paperback version 2006

A catalogue record for this publication is available from the British Library

Library of Congress Cataloguing in Publication data

Lancaster, M. J.
Passive microwave device applications of high-temperature
superconductors / M. J. Lancaster.
p. cm.
Includes index.
ISBN 0 521 48032 9 (hc)
1. High temperature superconductors. 2. Microwave devices –
Materials. I. Title.
TK7872.S8L357 1996
621.381′3–dc20 96-14062 CIP

ISBN-13 978-0-521-48032-1 hardback
ISBN-10 0-521-48032-9 hardback

ISBN-13 978-0-521-03417-3 paperback
ISBN-10 0-521-03417-5 paperback

To my mother and father

Contents

Preface

A revolution in the field of superconductivity occurred in 1986 with the discovery of superconductors with a transition temperature greater than the boiling point of liquid nitrogen, and many laboratories around the world began the exciting work of developing these materials. This book stems from one such laboratory, which has been looking at the microwave aspects of these materials, not only from a basic science view point, but also from a desire to demonstrate their potential for new applications. The development of microwave applications has proceeded very rapidly, and in less than ten years superconducting communication and signal processing systems are being flown in space, with many other microwave devices and systems to be found in the market place. This book essentially charts this development from the basic fundamental considerations of superconductors in high-frequency fields to the use of superconductors in microwave passive applications.

The book should be suitable as a basic introduction to the microwave applications of superconductors, and can be read independently of any previous knowledge of superconductors. However, the reader is recommended to consult one of the many texts about superconductivity in general in order to obtain a balanced view of the subject. It is expected that a number of different groups will find this book of interest. It could form the text for a specialised undergraduate course or be used in a more general course on microwaves. Examples of both fundamental electromagnetic principles (including plane waves, waveguides and cavities) can be extracted from the first three chapters or, alternatively, examples of microwave devices are given in Chapters 5–8. One of the book's primary uses will be in the introduction of new researchers to the subject, and as a general reference not only for groups working in superconductivity at high frequencies, but also for other groups wanting to know more about the high-frequency applications. By reading this book a new researcher will soon be able to assimilate the basic ideas, concepts and problems of this field. The book should also be suitable for microwave engineers who have heard about superconductivity but want to find out more about the subject.

The book has been written purposely as a mixture of some very detailed mathematical derivations and some sections which are much less detailed. Where detail is included, fundamental concepts relevant to a basic understanding of the entire field are described. Examples include surface impedance, plane waves and Poynting's theorem in Chapter 1, the wide microstrip transmission line in Chapter 2 and the cylindrical cavity in Chapter 3. A fundamental knowledge of this detailed work gives a solid background which can be built upon. The less detailed work in Chapters 5–8 includes a discussion of actual devices and systems, with examples from the literature. Information is given on the basic design principles but with an emphasis on fundamental principles. Some of the examples given here will eventually go out of date but the principles discussed with reference to them will not.

In any book some subjects will have to be omitted for various reasons. Probably the most obvious in this text is any information about cryogenics and information on how to cool the devices described. In parallel with superconductivity research, there has been much research into coolers and a wide range is now available. The problem of cooling is not a technical one, but, rather, one of the extra power consumption and size associated with the requirement. As discussed above, this book is essentially about the development of the field of passive microwave device applications of superconductors since the introduction of HTS, so some of the work on low-temperature superconductors prior to 1986 is not discussed to any great extent. Probably the largest application here is the development of accelerator cavities; however, this is not related directly to the theme of the book.

Acknowledgements

I would like to thank a large number of individuals and organisations for making this book possible. Firstly, my thanks go to all members of the Birmingham University Superconductivity Research Group (UBSRG), which began with the advent of high temperature superconductors and has grown into one of the largest UK groups working on all aspects of superconductivity. The many discussions with the members of this interdisciplinary group have been of enormous benefit over the last few years. Specifically in the microwave group in the School of Electronic and Electrical Engineering, I would like to thank Dr Adrian Porch, not only for reading this text but for the many stimulating discussions we have had and his contribution to much of the work described. I would also like to thank other lecturing members of the group including Professor Tom Maclean, Dr Fred Huang and Dr Zipeng Wu, the former two having worked principally on antennas and delay line filters respectively. Much of the work described has been done by the staff and students of the group, including Bea Avenhaus, Jia Sheng Hong Hong, Leo Ivrissimtzis, Alan Elston, Jeff Powell, Nick Exon, Mazlina Esa, Marcos De Melo, Dung Shing Hung, Martin Holyroyd, Graeme McCaffery, Philip Woodall, Henry Cheung and Janet Li.

I would also like to acknowledge specific members of the UBSRG, including Professor Colin Gough its director, Dr Stuart Abell, Professor Peter Edwards, Dr Chris Muirhead and Dr Colin Greaves for much advice and help on many aspects of superconductivity. This should also include Dr Fee Wellhofer, who has provided many of the superconducting thin films used for the devices and measurements described.

Our external collaborators have also been a source of inspiration and these include the members of the former ICI group, in particular Professor Neil Alford, Dr Tim Button and Dr Paul Smith. I not only need to thank them for supplying samples but also for discussions on some of the microwave aspects of their materials work. I would also like to thank Professor Richard Humphreys and members of his group at DRA Malvern for supplying thin films and the much valued discussions on many aspects of HTS. Our European collaborative work has also proved invaluable in the writing of this book; specifically I would like to thank Dr Günter Müller, Professor Dr Bernd Stritzker, Dr Jean Dumas, Professor Dr J. Senateur, Professor Dr Olivier Thomas, Professor Dr Claire Schlenker, Dr Rui Henriques and all members of their groups. There are also a number of visiting research fellows to the school including Dr Y. Huang, Dr A. Gorur and Professor Alan Portis whose contributions have proved valuable and whom I wish to acknowledge. Thanks are also given to other collaborators, too numerous to mention here, in the UK and around the world.

Finally I wish to acknowledge the all important financial contributions to this work which include the UK EPSRC, the European Union through an ESPRIT award (No. 6113) and the UK DTI. I would like to thank the University of Birmingham and the School of Electronic and Electrical Engineering for financial contributions and permission for my sabbatical leave.

1
Superconductivity at microwave frequencies

1.1 Introduction

This first chapter deals with some fundamental aspects of how superconductors interact with high-frequency fields, and discusses the theoretical tools available for the solution of problems. Although superconductors were discovered in 1911 by H. Kamerlingh Onnes,[1] it was not until the early 1930s that significant consideration was given to high-frequency effects. The thermal properties of superconductors were investigated by Gorter and Casimir in 1934,[2] and they predicted a temperature dependence of superconducting carriers by minimising the Helmholtz free energy. To do this the carriers within a superconductor were assumed to consist of both superconducting and normal carriers whose relative densities changed as a function of temperature. This two-fluid model was taken further by the London brothers in 1934 to account for the high-frequency properties of superconductors.[3–5] Their contribution is outlined in Section 1.2. The London equations can be used in conjunction with Maxwell's equations in order to allow them to be applicable to superconductors. Complex conductivity also follows from the two-fluid model. This simplifies the problem in that the normal conductivity σ can be replaced by a complex conductivity $\sigma_1 - j\sigma_2$, which accounts for superconducting phenomena. Heinz London also produced some early measurements on the surface resistance of tin, during which he discovered the anomalous skin effect, and Fritz London was the first to suggest that flux in a superconductor is quantised. However, it was not until a number of years later that the significance of these observations was recognised. The backbone of the major theories of superconductivity were accomplished in the 1950s, with Pippard's discussion of non-local effects in 1953,[6] the Ginzburg–Landau phenomenological theory in 1950[7] and, finally, the theory by Bardeen, Cooper and Schrieffer (BCS theory) in 1957.[8] The BCS theory was elucidated for high-frequency fields by Mattis and Bardeen in 1958.[9] Further discussion of the important aspects of these works is given in Section 1.9. The discovery of high-temperature superconductors (HTSs) by Bednorz and Müller[10] in 1986 has fundamentally changed the outlook for the applications of superconductors and has led to an enormous expansion of work in theoretical, experimental and application areas.

Two fundamental concepts are described in this chapter: the complex conductivity and the surface impedance of a superconductor. Complex conductivity can be used in conjunction with Maxwell's equations to predict the effects of high-frequency fields. It can be measured just like the normal conductivity of metals or predicted by the various theories. The theoretical approach is, of course, of paramount importance in yielding a fundamental understanding of the materials. Complex conductivity is introduced in Section 1.2 and its use is discussed with Maxwell's equations in Section 1.3. Following this, plane wave propagation in superconductors is discussed in Section 1.4, leading on to the concept of surface impedance. Surface impedance is important because it simplifies the solution of boundary value problems. The real part, the surface resistance, is discussed extensively and is an important quantity in the prediction of the performance of a microwave device. Complex conductivity is directly related to surface impedance through Maxwell's equations; this is discussed extensively in Sections 1.6 and 1.8. The remainder of the chapter discusses how both complex conductivity and surface impedance can be calculated from fundamental concepts as functions of frequency, temperature, microwave power and steady magnetic fields.

1.2 London equations and complex conductivity

Superconductivity is a consequence of paired and unpaired electrons travelling within the lattice of a solid. The paired electrons travel, under the influence of an electric field, without resistive loss. In addition, due to the thermal energy present in the solid, some of the electron pairs are split, so that some normal electrons are always present at temperatures above absolute zero. It is therefore possible to model the superconductor in terms of a complex conductivity $\sigma_1 - j\sigma_2$. This section looks into the basis of electron transport in superconductors and shows that superconductors can be represented by a complex conductivity. Although a very simple classical model is assumed, it gives a basis for the understanding of the microscopic processes in superconductors. This is called the 'two-fluid model'.

Consider the force exerted on an electron pair, which can be written

$$2m\frac{d\mathbf{v}_s}{dt} = -2e\mathbf{E} \tag{1.2.1}$$

where $\mathbf{v}_s$ is the velocity of the electron pair, e is the charge on an electron, m is the mass of an electron and $\mathbf{E}$ is the applied electric field. A similar equation can be written for the normal electrons in the solid, travelling at velocity $\mathbf{v}_n$:

$$m\frac{d\mathbf{v}_n}{dt} + m\frac{\mathbf{v}_n}{\tau} = -e\mathbf{E} \tag{1.2.2}$$

where τ is the momentum relaxation time. The extra term which appears in this equation is due to the scattering of the normal electrons with the lattice. If the electric field were switched off ($\mathbf{E} = 0$), then the velocity would decay with a characteristic time τ. This can be seen by the solution of Equation (1.2.2) in this case. The current densities for the normal electrons ($\mathbf{J}_n$) and paired electrons ($\mathbf{J}_s$) are

$$\mathbf{J}_s = -n_s e\mathbf{v}_s \tag{1.2.3}$$

$$\mathbf{J}_n = -n_n e\mathbf{v}_n \tag{1.2.4}$$

where n_n and n_s are the normal and paired electron densities respectively.

These general equations have two simpler cases, both of which have been tremendously useful in the study of superconductors. Firstly, consider the case when the superconductor is composed only of superconducting electron pairs. Combining Equations (1.2.1) and (1.2.3) gives

$$\Lambda\frac{\partial \mathbf{J}_s}{\partial t} = \mathbf{E} \tag{1.2.5}$$

where $\Lambda = m/n_s e^2$ is the London parameter, which is related to the superconducting penetration depth, as will be seen later. This equation has come to be known as the first London equation, as it was first introduced by the London brothers in 1934. Now take the curl of Equation (1.2.5) to give

$$\Lambda\frac{\partial}{\partial t}(\nabla \times \mathbf{J}_s) = \nabla \times \mathbf{E} \tag{1.2.6}$$

and using Faraday's law

$$\nabla \times \mathbf{E} = -\frac{\partial \mathbf{B}}{\partial t} \tag{1.2.7}$$

gives

$$\Lambda\nabla \times \mathbf{J}_s = -\mathbf{B} \tag{1.2.8}$$

The constant of integration with respect to time to obtain Equation (1.2.8) has been put to zero; this has been found to be correct for superconductors through experimental observation. Equation (1.2.8) is known as the second London equation. Equations (1.2.5) and (1.2.8) are important because they represent a simple method of introducing superconductivity to Maxwell's equations that enables conventional solution methods to be used. However, these equations are not useful for high-frequency fields in themselves, as there may be a significant number of normal electrons present which contribute to the electrodynamics. This brings us to the second case; here all the time dependencies are taken as sinusoidal, that is,

$$\mathbf{J}_s = \mathbf{J}_{s0}e^{j\omega t} \quad \mathbf{J}_n = \mathbf{J}_{n0}e^{j\omega t} \text{ and } \mathbf{E} = \mathbf{E}_0 e^{j\omega t} \tag{1.2.9}$$

The relationship with the current density and the electric field is given by

$$\mathbf{J}_0 = \mathbf{J}_{n0} + \mathbf{J}_{s0} = \sigma_{TF}\mathbf{E}_0 \tag{1.2.10}$$

In keeping with convention, we will drop the zero subscript on the variables and assume that the time dependence will carry through the equations below, that is,

$$\mathbf{J} = \mathbf{J}_n + \mathbf{J}_s = \sigma_{TF}\mathbf{E} \tag{1.2.11}$$

Here σ_{TF} is the two-fluid model conductivity. Now Equation (1.2.3) can be substituted into Equation (1.2.1) in a similar fashion to that above, and Equation (1.2.4) can be substituted into Equation (1.2.2). Then sinusoidal time dependence is assumed for both (Equations (1.2.9)) and a relationship between $\mathbf{J}_s$, $\mathbf{J}_n$ and $\mathbf{E}$ results. Using Equation (1.2.11) this results in

$$\mathbf{J} = \mathbf{J}_n + \mathbf{J}_s = (\sigma_{1TF} - j\sigma_{2TF})\mathbf{E} \tag{1.2.12}$$

where σ_{1TF} and σ_{2TF} are

$$\sigma_{1TF} = \frac{n_n e^2 \tau}{m(1+\omega^2\tau^2)} \tag{1.2.13}$$

$$\sigma_{2TF} = \frac{n_s e^2}{\omega m} + \frac{\omega n_n e^2 \tau^2}{m(1+\omega^2\tau^2)} \tag{1.2.14}$$

For most practical situations $(\omega\tau)^2 \ll 1$ and relaxation effects can be ignored. This is a very crude derivation but it allows us to reach Equation (1.2.12) in quite a simple manner. The concept of complex conductivity is fundamental to the high-frequency calculations in which we are interested.

A simple equivalent circuit is shown in Figure 1.2.1, which describes complex conductivity. The total current in the circuit is split between the reactive inductance and the resistance which represents dissipation in the

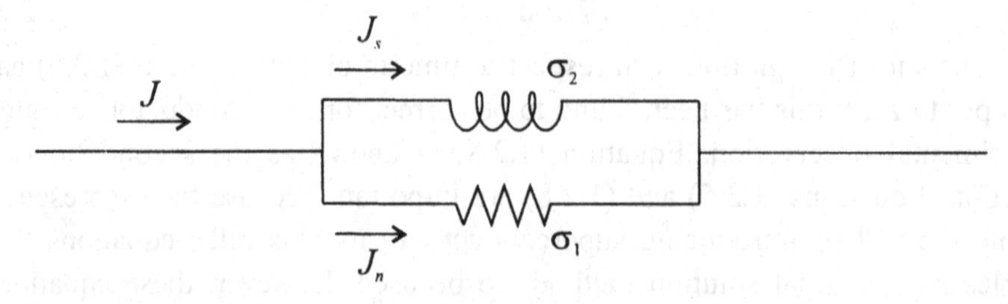

Figure 1.2.1. Equivalent circuit depicting complex conductivity.

system. As the frequency decreases, the reactance becomes lower and more of the current flows through the inductance. When the current is constant this inductance completely shorts the resistance, allowing resistance-free current flow. This section has not considered many of the important quantum mechanical effects in superconductors; this is left until Section 1.9. However, the concept of using complex conductivity does not change; it is only the calculation of σ_1 and σ_2 from the fundamental physical properties of the materials which is substantially modified. In fact, it is possible to measure the complex conductivity of a superconductor in the same way as measuring the conductivity of a normal conductor, and this measured value can be used for calculations of the performance of a microwave device. The exception is of course if the device under consideration is based upon quantum mechanical tunnelling processes. The following sections now deal with the effect of using complex conductivity firstly in Maxwell's equations and then in the wave equation; the discussion then moves to the propagation of plane waves in superconductors. Discussion of how σ_1 and σ_2 vary as a function frequency, temperature and magnetic fields, in addition to the physical properties of the materials, is left to Section 1.9 and the sections that follow. An alternative approach to using a complex conductivity is to consider a complex permittivity for the medium, introducing a negative dielectric constant.[11] However, although this leads to correct expressions, it gives a non-physical model which is much more difficult to interpret.

1.3 Maxwell's equations and superconductors

Maxwell's equations form the basis of electromagnetic theory and can easily be adopted for use with superconductors for many practical problems. The four fundamental relationships are

$$\nabla \times \mathbf{H} = \mathbf{J} + \frac{\partial \mathbf{D}}{\partial t} \tag{1.3.1}$$

$$\nabla \times \mathbf{E} = -\frac{\partial \mathbf{B}}{\partial t} \tag{1.3.2}$$

$$\nabla \cdot \mathbf{B} = 0 \tag{1.3.3}$$

$$\nabla \cdot \mathbf{D} = \rho \tag{1.3.4}$$

where the displacement current is related to the electric field by $\mathbf{D} = \varepsilon \mathbf{E}$ and the magnetic flux density is related to the magnetic field strength by $\mathbf{B} = \mu \mathbf{H}$. The wave equation for superconductors, useful for many microwave calculations in this book, can be derived from Maxwell's equations as follows.

Take the curl of Faraday's law (Equation (1.3.2)) and the time derivative of

Ampère's law (Equation (1.3.1)) and substitute for the time derivative of curl **H**, resulting in

$$\nabla \times \nabla \times \mathbf{E} = -\mu \frac{\partial \mathbf{J}}{\partial t} - \mu\varepsilon \frac{\partial^2 \mathbf{E}}{\partial t^2} \tag{1.3.5}$$

Now using the vector identity

$$\nabla \times \nabla \times \mathbf{E} = \nabla(\nabla \cdot \mathbf{E}) - \nabla^2 \mathbf{E} \tag{1.3.6}$$

with Equation (1.3.4) and the assumption that the regions of interest have no free charge ($\rho = 0$) gives

$$\nabla^2 \mathbf{E} = \mu \frac{\partial \mathbf{J}}{\partial t} + \mu\varepsilon \frac{\partial^2 \mathbf{E}}{\partial t^2} \tag{1.3.7}$$

This is a very general form of the wave equation which can apply to both superconductors and normal conductors. At present, no assumptions have been made about the relationship between current density and electric field. In order to make this equation useful, the connection between these two quantities must be included. To do this first substitute $\mathbf{J} = \mathbf{J}_n + \mathbf{J}_s$, then use London's first equation (Equation (1.2.5)) to connect the supercurrent density with the electric field. For the normal current density use the conventional relation $\mathbf{J}_n = \sigma_{1TF}\mathbf{E}$. Here, non-local effects are ignored, and this is appropriate for HTS where the mean free path is short compared with the penetration depth. These substitutions result in

$$\nabla^2 \mathbf{E} = \mu\sigma_{1TF} \frac{\partial \mathbf{E}}{\partial t} + \frac{\mu}{\Lambda}\mathbf{E} = \mu\varepsilon \frac{\partial^2 \mathbf{E}}{\partial t^2} \tag{1.3.8}$$

This is the wave equation for superconductors using the two-fluid approximation. A similar expression can be obtained for the magnetic field. Two cases will now be looked at; first the case where the fields are not time varying and, second, when the fields have sinusoidal time variations. For the first case, Equation (1.3.8) becomes

$$\nabla^2 \mathbf{E} = \frac{\mu}{\Lambda}\mathbf{E} \tag{1.3.9}$$

For a plane geometry with no field variation in the x and y directions the solution to this equation is given by

$$\mathbf{E} = E_x e^{-z/\lambda_L} \mathbf{a}_x \tag{1.3.10}$$

Here $\mathbf{a}_x$ is a unit vector in the x direction and λ_L is a depth where the field decays by an amount e^{-1} compared to the magnitude of **E** at the surface ($\mathbf{E}_x$). The depth of penetration of the field into the superconductor is called the London penetration depth (λ_L). This is, of course, quite unlike normal conductors where there is full penetration of the field into the material in this case. It can also be deduced by substitution of Equation (1.3.10) into Equation (1.3.9), giving

$$\lambda_L^2 = \frac{\Lambda}{\mu} \tag{1.3.11}$$

Thus London's first equation can be written

$$\frac{\partial \mathbf{J}_s}{\partial t} = \frac{1}{\mu \lambda_L^2} \mathbf{E} \tag{1.3.12}$$

and London's second equation becomes

$$\nabla \times \mathbf{J}_s = -\frac{\mathbf{H}}{\lambda_L^2} \tag{1.3.13}$$

Now consider the case of sinusoidal time variations. Equation (1.3.8) becomes

$$\nabla^2 \mathbf{E} = j\omega\mu\left(\sigma_{1TF} - j\frac{1}{\omega\Lambda} + j\omega\varepsilon\right)\mathbf{E} \tag{1.3.14}$$

Again, this leads to the concept of complex conductivity and this equation can be written

$$\nabla^2 \mathbf{E} = j\omega\mu(\sigma_{TF} + j\omega\varepsilon)\mathbf{E} \tag{1.3.15}$$

where

$$\sigma_{TF} = \sigma_{1TF} - j\sigma_{2TF} \tag{1.3.16}$$

and

$$\sigma_{2TF} = \frac{1}{\omega\Lambda} = \frac{1}{\omega\mu\lambda_L^2} \tag{1.3.17}$$

At this point a generalisation will be made so that the expressions developed do not necessarily apply to the two-fluid model only. For the following sections $\sigma_{TF} = \sigma_{1TF} - j\sigma_{2TF}$ will be replaced by $\sigma = \sigma_1 - j\sigma_2$, so that all the expressions developed apply not only to the two-fluid model but also to any material which can be shown to have complex conductivity. The usefulness of this will be seen in the later sections where, for example, type II superconductors will be found to have a complex conductivity different from σ_{TF}.

1.4 Plane waves

A plane wave has a constant amplitude and phase over an infinite single plane (the *xy* plane in the derivation given below). The wave propagates normal to the plane (*z* direction). It is useful to consider plane waves because the solution to Maxwell's equations is relatively straightforward and a great deal of understanding of the problem can be gained. Expressions for the complex propagation constant, giving the attenuation and velocity of electromagnetic waves in a superconductor, can be derived. In addition, the intrinsic impedance of a superconducting material can then be found, from which the surface impedance is derived.

First, consider a plane electromagnetic wave travelling in the z direction in a material filling the half-space $z \geqslant 0$. The solution to the wave equation (Equation (1.3.15)) for a forward-propagating wave is

$$\mathbf{E} = E_x e^{-\gamma z}\mathbf{a}_x \quad \text{and} \quad \mathbf{H} = H_y e^{-\gamma z}\mathbf{a}_y \tag{1.4.1}$$

This can be checked by back substitution of this expression into Equation (1.3.15), resulting in a value of the propagation constant γ of

$$\gamma = \surd(j\omega\mu(\sigma + j\omega\varepsilon)) = \alpha + j\beta \tag{1.4.2}$$

The real and imaginary parts of the propagation constant (termed α and β) represent the attenuation and the phase, respectively, of the travelling wave. Also, $\mathbf{E}$ and $\mathbf{H}$ are not independent of each other but are related through Faraday's law. The relationship can be found if Equation (1.4.1) is substituted into Faraday's law (Equation (1.3.2)), with sinusoidal time dependence, giving

$$-j\omega\mu H_y = \frac{\partial E_x}{\partial z} \tag{1.4.3}$$

Equation (1.4.3) shows that $\mathbf{E}$ and $\mathbf{H}$ are at right angles and defines the impedance for the medium. If Equation (1.4.1) is now substituted, then

$$Z_s = \frac{E_x}{H_y} = \frac{j\omega\mu}{\gamma} = \sqrt{\left(\frac{j\omega\mu}{\sigma + j\omega\varepsilon}\right)} \tag{1.4.4}$$

This intrinsic or wave impedance is used extensively in many calculations, for example, in the reflection of waves between different media. The intrinsic impedance of a medium is equal to the surface impedance, as discussed in Section 1.6.

1.4.1 Plane waves in superconductors

The propagation constant and impedance of a superconductor can be deduced by replacing σ by $\sigma_1 - j\sigma_2$ in Equations (1.4.2) and (1.4.4). In both cases, the usual assumption that the conduction current is much larger than any displacement current ($\sigma \gg \omega\varepsilon$) can be made, giving

$$\gamma = \surd(j\omega\mu(\sigma_1 - j\sigma_2)) \tag{1.4.5}$$

and

$$Z_s = \sqrt{\left(\frac{j\omega\mu}{\sigma_1 - j\sigma_2}\right)} \tag{1.4.6}$$

The real and imaginary parts of both these equations can be separated, resulting in

$$\gamma = \frac{1}{2}\sqrt{(\omega\mu)}f_1(\sigma_1, \sigma_2) \tag{1.4.7}$$

and

$$Z_s = \frac{1}{2}\sqrt{(\omega\mu)}f_2(\sigma_1, \sigma_2) \tag{1.4.8}$$

where

$$f_1(\sigma_1, \sigma_2) = (\sqrt{(p+\sigma_1)} + \sqrt{(p-\sigma_1)}) - j(\sqrt{(p-\sigma_1)} - \sqrt{(p+\sigma_1)}) \tag{1.4.9}$$

and

$$f_2(\sigma_1, \sigma_2) = \frac{1}{p}(\sqrt{(p+\sigma_1)} - \sqrt{(p-\sigma_1)}) + j\frac{1}{p}(\sqrt{(p-\sigma_1)} + \sqrt{(p+\sigma_1)}) \tag{1.4.10}$$

where

$$p = \sqrt{(\sigma_1^2 + \sigma_2^2)} \tag{1.4.11}$$

These equations are rather unwieldy and in order to simplify the general equations for the propagation constant and the intrinsic impedance of the medium the useful approximation $\sigma_2 \gg \sigma_1$ can be made. This is the case provided that the temperature is not too close to the transition temperature, where more normal carriers are present. Making the approximation $\sigma_2 \gg \sigma_1$ and expanding the square root in Equation (1.4.5) to first order in σ_1/σ_2 gives an approximate expression for γ:

$$\gamma \approx \sqrt{(\omega\mu\sigma_2)}\left(1 + j\frac{\sigma_1}{2\sigma_2}\right) \tag{1.4.12}$$

The real part of this is the attenuation coefficient, α, and is given by

$$\alpha = \mathrm{Re}\,(\gamma) = \sqrt{(\omega\mu\sigma_2)} \tag{1.4.13}$$

This attenuation coefficient represents an exponential decay of the electric and magnetic fields as the wave propagates in the z direction. The amplitude of the magnetic and electric fields can be represented by

$$A = A_0 \exp\left(-\sqrt{(\omega\mu\sigma)}z\right) \tag{1.4.14}$$

A characteristic depth (λ) can be defined such that the wave is attenuated by e^{-1} of its initial value, which may be at the surface of the superconductor

$$\lambda = \frac{1}{\mathrm{Re}\,(\gamma)} = \frac{1}{\sqrt{(\omega\mu\sigma_2)}} \tag{1.4.15}$$

The characteristic depth λ is the same as the London penetration depth if $\sigma_2 = \sigma_{2TF}$, and is governed mainly by the properties of the electron pairs, that is, the value of σ_2 rather than σ_1. It represents a depth to which electromagnetic fields penetrate superconductors and defines the extent of a region near the surface of a superconductor in which currents can be induced. A distinction must be made between the a.c. penetration depth and the d.c. penetration depth at this point. The depth to which the alternating fields penetrate the surface of the superconductor is always $1/\mathrm{Re}(\gamma)$, and using Equation (1.4.5) gives the actual depth irrespective of the values of σ_1 and σ_2. This depth will alter from the value given by Equation (1.4.15) when $\sigma_1 \ll \sigma_2$ (low temperature) to the normal state skin depth when $\sigma_2 = 0$ (at and above T_c). The d.c. superconducting penetration depth, the London depth, varies from a low-temperature value to an infinitely deep value as the superconductor becomes normal. In this case the normal carriers do not need to be considered as they have no effect on field penetration at d.c. As will be seen in Section 1.11, the a.c. penetration depth, as with all the other calculations in this section using complex conductivity, can also apply in the case of type II superconductors provided that σ_1 and σ_2 are defined appropriately. In the two-fluid model ($\sigma = \sigma_{TF}$) outlined in Section 1.2, this penetration depth is independent of frequency when $\sigma_1 \ll \sigma_2$. This can also be seen by using the approximations $n_s \gg n_n$ and $\omega\tau \ll 1$, where, from Equation (1.2.15), it can be seen that λ will be independent of frequency.

The imaginary part of the propagation constant is

$$\beta = \sqrt{\left(\frac{\omega\mu\sigma_1^2}{4\sigma_2}\right)} = \frac{\omega}{c} \tag{1.4.16}$$

where c is the velocity of the electromagnetic wave in the superconductor, which, by rearranging Equation (1.4.16), is

$$c = \sqrt{\left(\frac{4\sigma_2\omega}{\mu\sigma_1^2}\right)} \tag{1.4.17}$$

Now consider the intrinsic impedance, using the same approximations as given above ($\sigma_1 \ll \sigma_2$). Equation (1.4.6) becomes

$$Z_s = \sqrt{\left(\frac{\omega\mu}{\sigma_2}\right)}\left(\frac{\sigma_1}{2\sigma_2} + j\right) \tag{1.4.18}$$

which can again be split into real and imaginary parts giving

$$R_s = \frac{\sigma_1}{2\sigma_2}\sqrt{\left(\frac{\omega\mu}{\sigma_2}\right)} \quad \text{and} \quad X_s = \sqrt{\left(\frac{\omega\mu}{\sigma_2}\right)} \tag{1.4.19}$$

As will be seen later, these are referred to as the surface resistance and surface reactance of the superconductor. If $\sigma_1 \ll \sigma_2$, which, as mentioned above, is a good approximation at low temperatures, then $R_s \ll X_s$. Additionally, if $\sigma_2 \gg \omega\mu$, which is valid in all practical cases, then the values of R_s and X_s are much less than 1 ohm. Continuing with the derivation, σ_2 as defined in Equation (1.4.15) can be substituted into Equation (1.4.19), giving

$$R_s = \frac{\omega^2\mu^2\sigma_1\lambda^3}{2} \quad \text{and} \quad X_s = \omega\mu\lambda \tag{1.4.20}$$

These are well-known equations and are useful in all aspects of the work undertaken in later sections. For the two-fluid model ($\sigma = \sigma_{TF}$), provided σ_{1TF} and λ are independent of frequency, then R_s will increase as ω^2. This is of practical significance and will be discussed later when we compare superconductors with normal conductors. Also, an inductance can be defined using Equation (1.4.20):

$$L_I = \mu\lambda \tag{1.4.21}$$

This inductance is called the internal inductance and is discussed later in Section 1.7.

If both R_s and X_s have been determined by measurement it is often useful to determine σ_1 and σ_2 from their values. Equation (1.4.6) can be rearranged to give

$$\sigma_1 - j\sigma_2 = \frac{\omega\mu}{X_s^4 + 2R_s^2X_s^2 + R_s^4}(2R_sX_s - j(X_s^2 - R_s^2)) \tag{1.4.22}$$

If required, this can be simplified further using $R_s \ll X_s$.

1.5 Comparison of normal conductors with superconductors

It is interesting at this point to look at the comparison of the preceding derivations of the propagation characteristics of plane waves in superconductors with that of the well-known expressions for normal conductors. The approximation for a good conductor will be made in the following, that is:

$$\frac{\sigma}{\omega\varepsilon} \gg 1 \tag{1.5.1}$$

The derivation for the equations for normal conductors follows exactly the same steps as described above for superconductors by simply substituting $\sigma_2 = 0$ in Equations (1.4.7) and (1.4.8). In this case both the imaginary and real parts of the propagation constant are equal, that is,

$$\alpha = \beta = \sqrt{\left(\frac{\omega\mu\sigma_n}{2}\right)} \tag{1.5.2}$$

where σ_n is now the conductivity of a normal conductor and is purely real. This is not the same as superconductors where the real and imaginary parts differ by a factor of approximately $\sigma_1/2\sigma_2$ at low temperatures. We can also define a characteristic depth, known as the skin depth, which is given by

$$\delta = \sqrt{\left(\frac{2}{\omega\mu\sigma_n}\right)} \tag{1.5.3}$$

Provided we are in the limit where the conductivity is independent of frequency, then this skin depth is a function of frequency ($\propto \omega^{-0.5}$), where, as for the superconductor, it is independent of frequency (when $\sigma_1 \ll \sigma_2$). Moving on to the surface impedance, for a normal conductor the surface resistance and surface reactance are equal and are given by

$$R_s = X_s = \sqrt{\left(\frac{\omega\mu}{2\sigma_n}\right)} \tag{1.5.4}$$

Both are proportional to the square root of frequency, which has particular significance when comparing the applicability of superconductors to microwave devices. Because the surface resistance of a superconductor increases more rapidly (as frequency squared) there is a frequency where the surface resistance of normal metals actually becomes lower than that of superconductors. This has become known as the 'crossover frequency'. Appendix 3 summarises these expressions for normal conductors, superconductors and dielectrics. Appendix 1 also places some numerical values on the parameters discussed in this and the preceding section, giving the values of the surface impedance of various materials and their crossover frequencies.

1.6 Surface impedance

The surface impedance of a material has been introduced rather arbitrarily in the preceding sections by saying it is equal to the intrinsic or wave impedance of a material. Under this assumption the surface impedance, including the surface resistance and reactance, has been discussed in some detail, with Equations (1.4.6) and (1.4.8) giving the surface impedance of a superconductor in full, which in the limit $\sigma_1 \ll \sigma_2$ has been shown to reduce to

$$Z_s = \sqrt{\left(\frac{\omega\mu}{\sigma_2}\right)}\left(\frac{\sigma_1}{2\sigma_2} + j\right) \tag{1.6.1}$$

The assumption that the intrinsic and surface impedances are equal, which has been made in the preceding sections, requires some justification and this is done here.

The concept of surface impedance is used in solving electromagnetic problems when complicated geometry can be reduced to a single boundary with a (usually frequency dependent) boundary impedance value. Problems can be solved using this surface impedance value without further reference to the geometry beyond the boundary. The impedance value which is required for the solution of this type of problem is the ratio of the tangential electric field E_t and magnetic field H_t at a point on the boundary, and gives the definition of surface impedance:

$$Z_s = \frac{E_t}{H_t} \tag{1.6.2}$$

The concept can be discussed with reference to an infinite half-plane. In order to check that the intrinsic impedance is equal to the surface impedance consider a plane wave incident upon a boundary where the intrinsic impedances are Z_1 to the left of the boundary, and Z_2 to the right, as shown in Figure 1.6.1. Ignoring the time dependence, the incident field is given by

$$E_i = E_{0i}e^{-\gamma_1 z} \quad H_i = \frac{E_{0i}}{Z_1}e^{-\gamma_1 z} \tag{1.6.3}$$

Part of this wave is transmitted and can be represented by

$$E_t = E_{0t}e^{-\gamma_2 z} \quad H_t = \frac{E_{0t}}{Z_2}e^{-\gamma_2 z} \tag{1.6.4}$$

and the part that is reflected is given by

$$E_r = E_{0r}e^{\gamma_1 z} \quad H_r = -\frac{E_{0r}}{Z_1}e^{\gamma_1 z} \tag{1.6.5}$$

The exponent is positive because the reflected wave is travelling in the opposite direction to the incident wave. This also gives rise to the negative sign on the magnetic field reflected amplitude, and can be proved by consideration of the Poynting vector, which must be directed in the negative z direction.

On the boundary the incident and reflected fields must sum to the transmitted field in the right-hand medium of Figure 1.6.1, that is,

$$\begin{aligned} E_i + E_r &= E_t \\ H_i + H_r &= H_t \end{aligned} \tag{1.6.6}$$

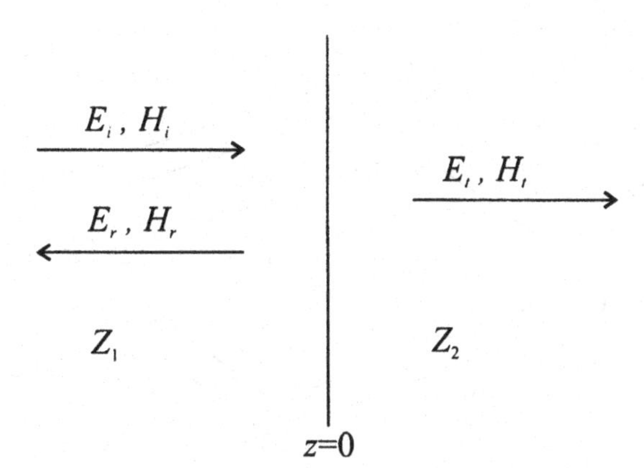

Figure 1.6.1. Reflection and transmission of a plane wave at a boundary.

These are also equal to the tangential electric and magnetic fields on the right-hand side of the boundary, so, with the definition of surface impedance,

$$Z_s = \frac{E_t}{H_t} = \frac{E_i + E_r}{H_i + H_r} \tag{1.6.7}$$

Substituting for the fields using Equations (1.6.3), (1.6.4), and (1.6.5) gives, at $z = 0$,

$$Z_s = Z_2 = Z_1 \left(\frac{E_{0i} + E_{0r}}{E_{0i} - E_{0r}} \right) \tag{1.6.8}$$

This gives the intrinsic impedance; Z_2 is equal to the surface impedance Z_s.

Another definition quoted in the literature for surface impedance is that it is equal to the impedance of any square of material measured from one end to the opposite end, as depicted in Figure 1.6.2. This can be shown to be simply equivalent to the definition given above by considering the voltage across, and the current through, the material:

$$Z_s = \frac{V}{I} = \frac{El}{Kw} = \frac{E_t}{H_t} \tag{1.6.9}$$

where $l = w$, the length and width of the sample respectively; K is the surface current density, which is equal to the surface tangential magnetic field H_t. The impedance of a square connects the field derivations to a circuit idea, which is also useful in some of the work to be considered later.

A further concept sometimes used when considering surface impedance and connecting it with the penetration depth is to write the surface impedance as

$$Z_s = \frac{E_t}{\displaystyle\int_{-\infty}^{0} J_x(z)\, dz} \tag{1.6.10}$$

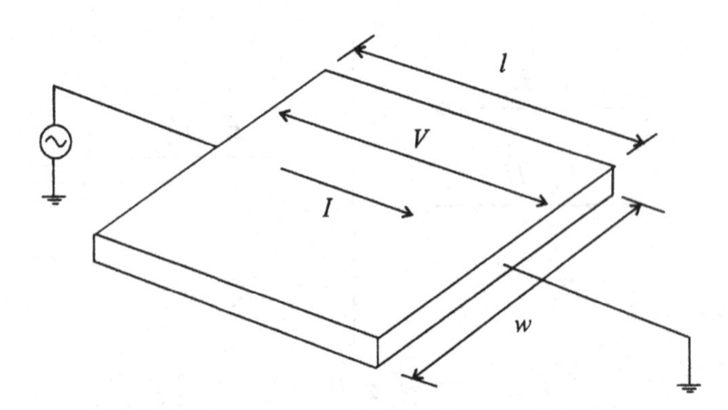

Figure 1.6.2. Voltage across and current through a section of material.

Here E_t is the tangential electric field at the surface and $J_x(z)$ can be, for example, the exponentially decaying current into the material. This definition gives a physical insight into how the decaying current is taken into account in the value of surface impedance and how a three-dimensional field can be reduced to a surface impedance value at the boundary.

1.7 Poynting's theorem and superconductors

Poynting's theorem tells us about energy and power flow in an electromagnetic field and is conventionally given by

$$-\oint_{surface} \mathbf{E} \times \mathbf{H} \cdot \mathbf{ds} = \int_{volume} \mathbf{E} \cdot \mathbf{J}\, dv + \frac{\partial}{\partial t} \int_{volume} \left(\frac{\varepsilon \mathbf{E}^2}{2} + \frac{\mu \mathbf{H}^2}{2} \right) dv \quad (1.7.1)$$

For normal conductors the first term on the right-hand side represents the ohmic power dissipated, but, as seen below, this will also contain an energy storage term when applied to superconductors. The second term on the right-hand side represents the rate of increase of energy storage in both the electric and magnetic fields. The term on the left-hand side of the equation is the total power flowing into the volume of interest. The Poynting vector is $\mathbf{P} = \mathbf{E} \times \mathbf{H}$. Assuming that the current density can be split into the sum of a super current and a normal current, that is, $\mathbf{J} = \mathbf{J}_s + \mathbf{J}_n$ and that the supercurrent is governed by London's first equation (Equation (1.3.12)), then Poynting's theorem for superconductors can be written

$$-\oint_{surface} \mathbf{E} \times \mathbf{H} \cdot \mathbf{ds} = \int_{volume} \mathbf{E} \cdot \mathbf{J}_n\, dv + \frac{\partial}{\partial t} \int_{volume} \left(\frac{\varepsilon \mathbf{E}^2}{2} + \frac{\mu \mathbf{H}^2}{2} + \frac{\mu \lambda_L^2 \mathbf{J}_s^2}{2} \right) dv \quad (1.7.2)$$

An extra term has appeared, which is due to energy storage in the motion of the superconducting electrons. This can easily be demonstrated by substitution of Equations (1.2.14) and (1.2.3) into $\mu\lambda^2\mathbf{J}_s^2/2$ using Equation (1.3.11) under the approximation $n_s \gg n_n$, resulting in a kinetic energy density of $n_s m \mathbf{v}_s^2/2$.

An alternative approach will now be pursued which considers Poynting's theorem under the approximation of sinusoidal field variations. Although Equation (1.7.2) can be reduced to a similar expression, the approach given below gives a slightly different, more informative, interpretation of the problem.

Ampère's and Faraday's laws under the approximation of sinusoidal time variations are

$$\nabla \times \mathbf{H} = \mathbf{J} + j\omega\varepsilon\mathbf{E} \quad (1.7.3)$$

and

$$\nabla \times \mathbf{E} = -j\omega\mu\mathbf{H} \tag{1.7.4}$$

These expressions can be substituted into the right-hand side of the vector identity:

$$\nabla \cdot (\mathbf{E} \times \mathbf{H}^*) = \mathbf{H}^* \cdot (\nabla \times \mathbf{E}) - \mathbf{E} \cdot (\nabla \times \mathbf{H}^*) \tag{1.7.5}$$

The result can be integrated over a volume and the divergence theorem applied to the volume integral of $\nabla \cdot (\mathbf{E} \times \mathbf{H}^*)$, resulting in[12]

$$-\oint_{surface} \mathbf{E} \times \mathbf{H}^* \cdot d\mathbf{s} = \int_{volume} \mathbf{E} \cdot \mathbf{J}^* \, dv + j\omega \int_{volume} (\mu|\mathbf{H}|^2 - \varepsilon|\mathbf{E}|^2) \, dv \tag{1.7.6}$$

Equation (1.7.6) is different from Equation (1.7.1), not only in the fact that sinusoidal time dependence has been introduced but also the complex conjugate (denoted by *) of some of the fields has now been introduced. We can now substitute for the complex conductivity and the term $\mathbf{E} \cdot \mathbf{J}^*$ becomes

$$\mathbf{E} \cdot \mathbf{J}^* = (\sigma_1 + j\sigma_2)|\mathbf{E}|^2 \tag{1.7.7}$$

Thus Equation (1.7.6) can be written

$$-\frac{1}{2}\oint_{surface} \mathbf{E} \times \mathbf{H}^* \cdot d\mathbf{s} = \frac{1}{2}\int_{volume} \sigma_1 |\mathbf{E}|^2 \, dv + j\omega \int_{volume} \left(\frac{\mu|\mathbf{H}|^2}{2} + \frac{\sigma_2|\mathbf{E}|^2}{2\omega} - \frac{\varepsilon|\mathbf{E}|^2}{2} \right) dv \tag{1.7.8}$$

A factor of $1/2$ has been introduced into all terms so the interpretation of the result in terms of energy and power is straightforward. The first term on the right-hand side of Equation (1.7.8) represents the time average power loss due to ohmic dissipation. The volume integral on the right-hand side represents the difference between the time average inductively stored energy and capacitively stored energy; it includes the inductive term containing the kinetic energy of the electron pairs.

Now consider the dissipative term in Equation (1.7.8), which gives the time average power dissipated. Take the volume under consideration to be an infinite half-plane of superconducting material with the origin of the z coordinate at the surface and increasing z into the superconductor, as shown in Figure 1.7.1. In this case the volume integral can be reduced to a surface integration by knowing that the field decreases exponentially into the superconductor, that is,

$$P_{av} = \frac{1}{2}\sigma_1 |Z_2|^2 \int_{surface} \int_0^{\infty} |\mathbf{H}_s|^2 e^{-2z/\lambda} dz \, ds \tag{1.7.9}$$

Here the magnetic field within the superconductor is given by $\mathbf{H} = \mathbf{H}_s e^{-z/\lambda}$ and the magnitude of the electric field is related to the magnetic field through

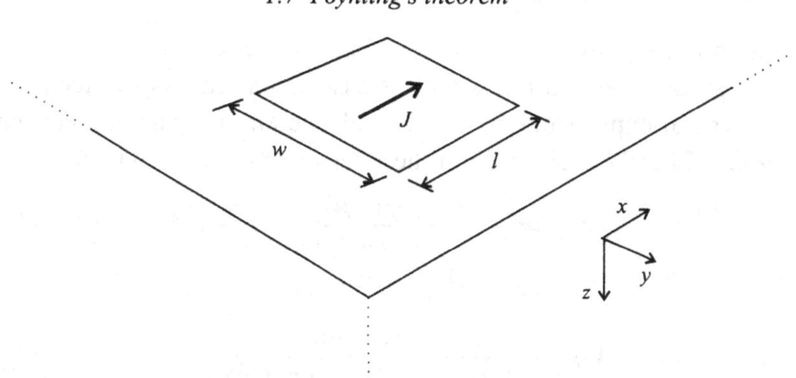

Figure 1.7.1. Plane surface showing coordinate directions.

the intrinsic impedance Z_s of the superconductor, that is, $|\mathbf{E}|^2/|\mathbf{H}|^2 = |Z_s|^2$. Integration is straightforward and results in

$$P_{av} = \frac{1}{4}\sigma_1 |Z_2|^2 \lambda \int_{surface} |\mathbf{H}_s|^2 ds \tag{1.7.10}$$

Using the general expressions for the surface impedance and propagation constant, Equations (1.4.7) and (1.4.8), it can be shown that Equation (1.7.10) reduces to

$$P_{av} = \frac{1}{2} R_s \int_{surface} |\mathbf{H}_s|^2 ds \tag{1.7.11}$$

This is a well-known expression and has now been shown to be correct for all values of σ_1 and σ_2. It is also interesting to look at the inductive term in the same manner. The average rate of change of energy after integration into the superconductor is given by

$$U'_{av} = \frac{j\omega\mu\lambda}{4}\left(1 + \frac{\sigma_2 |Z_s|^2}{\omega\mu}\right) \int_{surface} |\mathbf{H}_s|^2 ds \tag{1.7.12}$$

In a similar manner to the proof of Equation (1.7.11), using general expressions for the surface impedance and propagation constant, this can be shown to be equal to

$$U'_{av} = j\frac{1}{2} X_s \int_{surface} |\mathbf{H}_s|^2 ds \tag{1.7.13}$$

It can be seen that there are two terms associated with the surface reactance (X_s): a term associated with the kinetic energy of the electrons (X_{sk}), as described above, and a term associated with the penetration of the magnetic field (X_{sH}), that is,

$$X_s = X_{sk} + X_{sH} \tag{1.7.14}$$

The first term in Equation (1.7.12) corresponds to the term which is associated with the magnetic field and the second is associated with super electron flow. Using the general expressions for the impedance and propagation constant of Equations (1.4.7) and (1.4.8), each of these terms can be shown to be

$$X_{sH} = \frac{\sqrt{(\omega\mu)}}{\sqrt{(p-\sigma_1)} + \sqrt{(p+\sigma_1)}} \tag{1.7.15}$$

and

$$X_{sk} = \frac{\sigma_2\sqrt{(\omega\mu)}}{p(\sqrt{(p-\sigma_1)} + \sqrt{(p+\sigma_1)})} \tag{1.7.16}$$

where

$$p = \sqrt{(\sigma_1^2 + \sigma_2^2)} \tag{1.7.17}$$

If the approximation is now made that $\sigma_2 \gg \sigma_1$ in both equations for X_{sk} and X_{sH}, then the result is the same in both cases, giving

$$\frac{1}{2}X_s = X_{sk} = X_{sH} = \frac{1}{2}\sqrt{\left(\frac{\omega\mu}{\sigma_2}\right)} = \frac{1}{2}\omega\mu\lambda \tag{1.7.18}$$

However, for the case where $\sigma_1 \gg \sigma_2$ when the superconductor approaches the normal state, then the result is different, with

$$X_s \approx X_{sH} = \sqrt{\left(\frac{\omega\mu}{2\sigma_1}\right)} \tag{1.7.19}$$

and

$$X_{sk} = \frac{\sigma_2}{\sigma_1}\sqrt{\left(\frac{\omega\mu}{2\sigma_1}\right)} \tag{1.7.20}$$

Hence, it can be seen that energy storage in the surface of a superconductor takes place via two mechanisms: kinetic energy storage and the storage due to the penetration of the field. At low temperatures, when the transport is dominated by superconducting carriers, there is an equal storage of energy in each mechanism. However, as the temperature increases towards a point very close to T_c, then the normal carriers dominate the transport mechanism. At this point the energy storage due to kinetic energy reduces and the storage in the internal magnetic field dominates.

Internal or surface inductance can be defined at this point using Equation (1.7.18) such that

$$L_I = L_m + L_k = \mu\lambda \tag{1.7.21}$$

Here the internal inductance is made up of two terms: a magnetic and a kinetic inductance, which are of course equal at low temperatures. It should be noted that some authors sometimes call the total inductance, L_I, the 'kinetic inductance'; in this text the term will be referred to as the internal inductance, leaving the term kinetic inductance for L_k in the expression given above.

1.8 Surface impedance of thin films

Films of superconducting material are the main constituent of many of the device applications described later, and it is crucial for these applications that a good understanding of the properties of these films is obtained. The concept of surface impedance can be applied to films, that is, the surface impedance described in Section 1.6 for an infinitely thick film can be modified in order to take the finite thickness of the film into account. This concept is particularly useful since this new, modified surface impedance can be used in calculations involving boundary impedance values. In order to calculate the surface impedance of a film consider a plane wave incident upon the film shown in Figure 1.8.1.

If a film of superconducting material with surface or intrinsic impedance Z_s and propagation constant γ_s is grown on a substrate material with intrinsic impedance Z_d and propagation constant γ_d, then the impedance, looking from the upper surface of the film towards the substrate, is given by

$$Z_f = Z_s \frac{Z_d + Z_s \tanh(\gamma_s t)}{Z_s + Z_d \tanh(\gamma_s t)} \tag{1.8.1}$$

where t is the thickness of the thin film. This is a well-known expression for calculating the impedance for a block of material, and is discussed in any good book on electromagnetic theory.[13] The following approximation will now be made:

$$Z_d \tanh(\gamma_s t) \gg Z_s \tag{1.8.2}$$

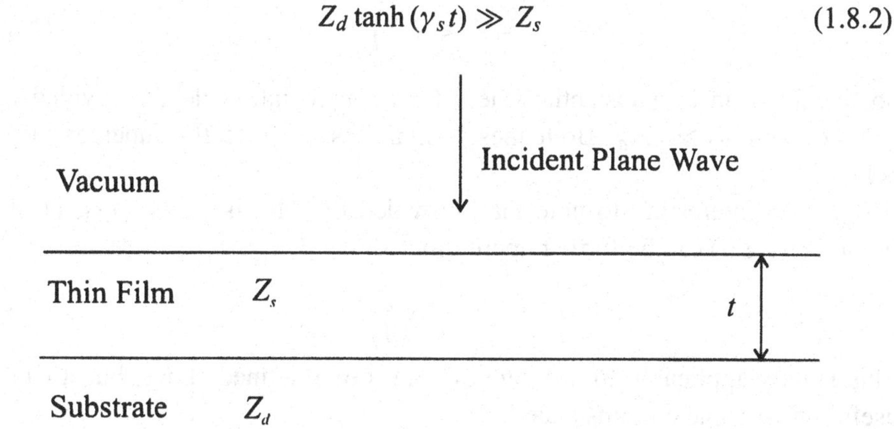

Figure 1.8.1. Film of material on the surface of a substrate showing an incident plane wave.

If this approximation holds, then it also implies that $Z_d \gg Z_s \tanh(\gamma_s t)$ provided $\tanh(\gamma_s t)$ is not too large. Under this approximation Equation (1.8.1) is reduced to

$$Z_f = Z_s \coth(\gamma_s t) \tag{1.8.3}$$

This expression holds for both normal conductors and superconductors and for most practical thicknesses of film. However, for ultra-thin films, when $\gamma_s t < Z_s/Z_d$ it is no longer valid; this is discussed further below. The expressions given above can be used directly in numerical calculations, but for further insight into how films behave, Equation (1.8.3) needs to be split into its real and imaginary parts; this will be done in the next section.

Before proceeding, the approximations made to derive Equation (1.8.3) will be looked at in more detail. The approximation assumes that both the real and imaginary parts of the superconductor surface impedance are less than the substrate surface impedance and that the film is not too thin. If the substrate is lossless, then its impedance is given by

$$Z_d = \sqrt{\left(\frac{\mu}{\varepsilon_d}\right)} \tag{1.8.4}$$

The imaginary part of the surface impedance of the substrate is zero in this case. Using the superconductor surface impedance with the assumption that $\sigma_2 \gg \sigma_1$, the approximation implies

$$1 \ll \frac{\sigma_2}{\omega\varepsilon_d} \tag{1.8.5}$$

with $\gamma_s t$ not too small. The limit of the approximation for small t can be seen if the term $\tanh(\gamma_s t)$ is replaced by $\gamma_s t$ and $\mathrm{Re}(\gamma_s) = 1/\lambda$, that is,

$$\frac{\omega\varepsilon_d}{\sigma_2} \ll \left(\frac{t}{\lambda}\right)^2 \tag{1.8.6}$$

So the film can be substantially less than a penetration depth provided that $\sigma_2 \gg \sigma_1$ and $\sigma_2 \gg \omega\varepsilon_d$. Both these conditions are true for superconductors below T_c.

It is also interesting to note that, provided the film is thin ($\gamma_s t \ll 1$), then Equation (1.8.3) can be further approximated by

$$Z_f = \frac{Z_s}{\gamma_s t} \tag{1.8.7}$$

This is only applicable for a range of values of t defined above, but it can be useful under some circumstances.

Before these equations for the surface impedance of thin films can be used, the approximations need to be considered carefully. Obviously, the parameters

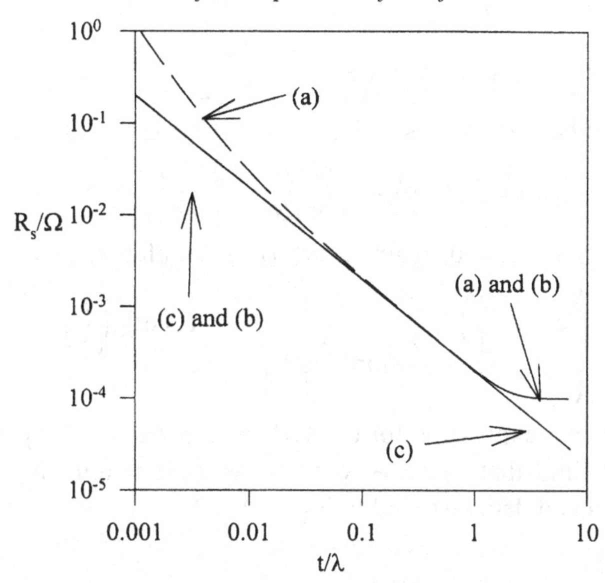

Figure 1.8.2. The surface impedance of a HTS thin film with $R_s = 100\ \mu\Omega$, $\lambda = 0.14\ \mu\text{m}$, at 10 GHz on a substrate with relative permittivity 10. (*a*) The full solution using Equation (1.8.1). (*b*) The approximate solution using Equation (1.8.3). (*c*) The thin film approximate solution using Equation (1.8.7).

of the particular material under consideration and the thickness of the film need to be looked at for each application. As an example, Figure 1.8.2 shows how the surface resistance of a typical HTS thin film varies with the three different expressions.

The limits for the approximations can be clearly seen for this particular thin film, and Equation (1.8.3) is a good approximation for films of thickness just less than a penetration depth, as well as those greater than a penetration depth. In Figure 1.8.2 it cannot be distinguished from the accurate expression for film thickness around and above a penetration depth. It can also be seen that Equation (1.8.7) is not accurate for film thicknesses above a penetration depth.

1.8.1 Surface resistance and reactance of superconducting films

It is instructive to look in more detail at Equation (1.8.3) and split it into its real and imaginary parts, deriving expressions for the surface resistance and reactance. For superconducting films with the approximation $\sigma_1 \ll \sigma_2$, the expression in Equation (1.4.12) for the complex propagation constant can be substituted into Equation (1.8.3). In addition, using $\sigma_2 = 1/\omega\mu\lambda^2$, the equation is reduced to

$$Z_f = Z_s \coth\left(\frac{t}{\lambda}\left(1 + j\frac{\sigma_1 \omega \mu \lambda^2}{2}\right)\right) \quad (1.8.8)$$

This equation can be split into its real and imaginary parts by using[14]

$$\coth(a + jb) = \frac{\coth(a)\coth(jb) + 1}{\coth(a) + \coth(jb)} \quad (1.8.9)$$

and using $Z_s = R_s + jX_s$ and, again, $\sigma_1 \ll \sigma_2$, gives, after some manipulation,

$$Z_f = R_s\left\{\coth\left(\frac{t}{\lambda}\right) + \frac{t}{\lambda}\frac{1}{\sinh^2\left(\frac{t}{\lambda}\right)}\right\} - jX_s \coth\left(\frac{t}{\lambda}\right) \quad (1.8.10)$$

This is a well-known expression for the surface impedance of superconducting thin films in the limit that $\sigma_1 \ll \sigma_2$, with the approximations discussed in the preceding section and demonstrated in Figure 1.8.2.

1.8.2 Surface impedance of thin superconducting films close to T_c

Another approximation sometimes quoted in the literature is derived using Equation (1.8.7) but not the approximation $\sigma_1 \ll \sigma_2$. It is therefore more useful near T_c. In this case the film is also assumed to be thin, so the approximation $\gamma_s t \ll 1$ discussed above applies. However, the approximation $\gamma_s t \gg Z_s/Z_d$ still applies, giving only a region of thickness where the approximation is valid. This region must be carefully deduced when the approximation is used with known values of parameters, although it is generally good for most practical cases. In this case, Equation (1.8.7) becomes

$$Z_f = \frac{1}{t}\frac{\sigma_1 + j\sigma_2}{\sigma_1^2 + \sigma_2^2} \quad (1.8.11)$$

Also, replacing σ_2 by $1/\omega\mu\lambda_L^2$, so that the two-fluid model applies ($\sigma = \sigma_{TF}$), gives

$$Z_f = \frac{1}{t}\frac{\omega^2\mu^2\lambda_L^4}{1 + \sigma_1^2\omega^2\mu^2\lambda_L^4}(\sigma_1 + j\sigma_2) \quad (1.8.12)$$

This, of course, is not a good approximation for films of the order of one penetration depth and above.

1.8.3 Superconducting thin films on lossy substrates

The above derivation assumed that the substrate has no effect on the surface impedance of the thin film. This is not always the case, and lossy substrates or substrates with a high dielectric constant can affect the surface impedance significantly if the films are of the appropriate thickness. An expression for the surface impedance of a superconducting thin film on a lossy dielectric is now derived.

Equation (1.8.1) can be written

$$Z_f = \frac{Z_s}{\tanh(\gamma_s t)} \left\{ \frac{1 + \dfrac{Z_s}{Z_d} \tanh(\gamma_s t)}{\dfrac{Z_s}{Z_d} \dfrac{1}{\tanh(\gamma_s t)} + 1} \right\} \tag{1.8.13}$$

This can be expanded assuming that $Z_d > Z_s$, and that $\gamma_s t$ is not too small. This is usually a good approximation provided the superconducting film is still the dominant influence on the surface impedance. In general it assumes that

$$\sqrt{\left(\frac{\omega\mu}{\sigma_2}\right)\left(\frac{\sigma_1}{2\sigma_2} + j\right)} \gg \sqrt{\left(\frac{\mu}{\varepsilon_d}\right)\left(1 + j\frac{\sigma_d}{2\omega\varepsilon_d}\right)} \tag{1.8.14}$$

It has of course already been assumed that $\sigma_2 \gg \sigma_1$, although it does not necessarily have to be true for the approximation to be valid. Here ε_d and σ_d are the permittivity and conductivity, respectively, of the substrate. The expansion of Equation (1.8.14) gives

$$Z_f = Z_s\left\{\coth(\gamma_s t) - \frac{Z_s}{Z_d}\frac{1}{\sinh^2(\gamma_s t)}\right\} \tag{1.8.15}$$

This is the same Equation (1.8.3) except that an additional term has appeared which contains the parameters associated with the substrate. As in the preceding section this can be split into its real and imaginary parts; after some manipulation and assuming that $\sigma_2 \gg \sigma_1$ this gives

$$Z_f = R_s\left\{\coth\left(\frac{t}{\lambda}\right) + \frac{t}{\lambda}\frac{1}{\sinh^2\left(\dfrac{t}{\lambda}\right)}\right\} + X_s^2\sqrt{\left(\frac{\varepsilon_d}{\mu}\right)}\left(1 - j\frac{\sigma_d}{2\omega\varepsilon_d}\right)\frac{1}{\sinh^2\left(\dfrac{t}{\lambda}\right)} - jX_s\coth\left(\frac{t}{\lambda}\right) \tag{1.8.16}$$

A similar equation has been discussed in the literature.[15] The loss tangent of the substrate is given by

$$\tan(\delta) = \frac{\sigma_d}{\omega\varepsilon_d} \tag{1.8.17}$$

The main effect of the substrate is to add an extra term in the surface resistance which represents the penetration of energy through the film into the substrate. This equation has been used extensively in the assessment of superconducting thin films. When a film is measured in a cylindrical cavity, for example, it is the effective surface resistance of the film, R_f, which is measured and Equation (1.8.16) can then be used to extract the intrinsic value of R_s. This is discussed further in Chapter 3.

Surface roughness is another factor which can affect the surface impedance of a superconductor. As with normal metals, the effect is only apparent when the depth of field penetration is of the order of, or less than, the roughness of the surface. In normal metals the problem was discussed as early as 1949[16] for square, rectangular and triangular grooves, and these results are still used widely today in curve fitted form. For superconductors the problem of square grooves in the surface of a superconductor has been discussed by Wu.[17] The results of this work are given in Equations (1.8.18) and (1.8.19). These are curve fitted expressions to numerical evaluations of the exact solution for an rms groove depth of Δ ($\Delta = d/4$, d being the actual groove depth). They are accurate to within 4% of the numerical results, and because an rms roughness is used here, they are applicable for an estimate of the effect of surface roughness on any surface. The expressions give the ratio of the actual surface resistance of the rough surface to the surface resistance of the same material if perfectly flat. The approximation $\sigma_2 \gg \sigma_1$ has been used:

$$\frac{R_s}{R_{flat}} = 1 + \frac{2}{\pi}\operatorname{atan}\left[\left(\frac{\Delta}{\lambda}\right)^2\left[0.35 + 1.05\exp\left[-0.5\left(\frac{2\sigma_2}{\sigma_1}\right)\right]\right]\right] \tag{1.8.18}$$

$$\frac{X_s}{X_{flat}} = 2 - \exp\left[-\frac{\Delta}{\lambda}\left[1 + 2.5\exp\left[-0.5\left(\frac{2\sigma_2}{\sigma_1}\right)\right]\right]\right] \tag{1.8.19}$$

These results show an increase in the surface resistance and surface reactance of a superconductor as the surface roughness approaches the penetration depth. However, the effect is only small and gives a maximum deterioration of a factor of about 2 when the rms surface roughness increases to several λ. In practice, superconducting thin films are fairly smooth on the scale of a penetration depth and the effect is small. The same may not necessarily apply to bulk polycrystalline or thick film materials.

1.9 Temperature and frequency dependence of complex conductivity

So far in this chapter complex conductivity and the resulting parameters of the surface impedance and propagation constant have been discussed in detail with little reference to the actual numerical values of these parameters for different materials or how the parameters vary as a function of conditions. In this section and the rest of the chapter the variation of the complex conductivity with a number of parameters will be discussed. During the discussion some reference to the numerical values of the parameters is given. However, for a more comprehensive look at the actual values of complex conductivity, penetration depth and surface impedance the reader is referred to Appendix 1, where a set of experimentally measured values with emphasis on high-temperature superconductors is given.

The variation of complex conductivity with temperature, frequency, microwave power and applied d.c. magnetic field is given in the following sections using various approaches. The complex conductivity can then be used with all the expressions in the preceding sections to determine the surface impedance and propagation constant under the various approximations given.

1.9.1 Temperature dependence of classical complex conductivity

In 1934 Gorter and Casimir[18] realised that a model in which the charge carriers consist of two types, superconducting and normal, could account for the thermal properties of superconductors. The consequences of this in terms of complex conductivity have been discussed in Section 1.2. They also deduced the temperature dependence of the density of normal and superconducting carriers as

$$n_n = n\left(\frac{T}{T_c}\right)^4 \tag{1.9.1}$$

$$n_s = n\left(1 - \left(\frac{T}{T_c}\right)^4\right) \tag{1.9.2}$$

where $n = n_n + n_s$ is the total number of carriers. Using Equation (1.2.13), this immediately gives a value of σ_1 of

$$\sigma_1 = \sigma_n\left(\frac{T}{T_c}\right)^4 \tag{1.9.3}$$

where σ_n is the normal state conductivity at T_c. The imaginary part of the complex conductivity is less straightforward. Using Equation (1.2.14) and assuming that $\omega\tau \ll 1$ and the number of normal carriers is not too large, then σ_2 can be approximated as

$$\sigma_2 = \frac{1}{\omega\mu\lambda_0^2}\left(1 - \left(\frac{T}{T_c}\right)^4\right) \tag{1.9.4}$$

The constant $1/\omega\mu\lambda_0^2$ is found by setting $T = 0$ and $n = n_s$ and using Equation (1.4.15). Here λ_0 is the penetration depth as the temperature approaches zero Kelvin and is the same as the London penetration depth in that limit. The calculation of Equation (1.9.4) makes it not strictly valid close to T_c.

1.9.2 Non-local electrodynamics

Early measurements by London[19,20] showed the low-temperature surface resistance of tin to be much higher than expected. This was attributed to the fact that the mean free path, the average distance that electrons travel between scattering, became large compared with the skin depth. When this is the case the scattering process changes from that due to conventional scattering centres in the metal to surface scattering. Under normal conditions, when the mean free path is small ($l \ll \delta$), the electric field decays into the surface a distance of around a skin depth and can be assumed to be constant for the distance which the electrons travel between scattering. Under this condition the current density is connected by the normal relation $\mathbf{J} = \sigma\mathbf{E}$. However, when the mean free path is large ($l \gg \delta$) the electrons can travel through a varying electric field before they are scattered, and within a penetration depth the relation $\mathbf{J} = \sigma\mathbf{E}$ no longer holds. The electrodynamics are said to be non-local in this case and an averaging process must take place in order to relate $\mathbf{J}$ to $\mathbf{E}$. This case is also called the anomalous limit and the surface resistance of a normal conductor for diffuse surface scattering in this case is given by[21]

$$R_s = \left(\frac{3^{1/2}\omega^2\mu_0^2 l}{16\pi\sigma_n} \right)^{1/3} \tag{1.9.5}$$

This relationship gives a frequency dependence of the surface resistance of the normal metal of $\omega^{2/3}$ in the anomalous limit, compared with $\omega^{1/2}$ under normal circumstances when the mean free path is short. This means that for a fixed temperature the frequency dependence will gradually change from $\omega^{1/2}$ at low frequencies when the skin depth is large to $\omega^{2/3}$ at higher frequencies when the skin depth is small. The change of frequency will be lower at lower temperatures because of the increased mean free path.

Experimental observation of the surface resistance of tin doped with indium lead prompted Pippard to suggest that a similar non-local effect also occurs in superconductors.[22–27] Pippard suggested that the electrons in a superconductor may act coherently over a distance ξ, which he called the coherence length. In the anomalous limit this coherence becomes greater than the superconducting penetration depth and the electrodynamics become non-local. The relation between coherence length and mean free path proposed by Pippard is

$$\frac{1}{\xi} = \frac{1}{\xi_0} - \frac{1}{l} \tag{1.9.6}$$

In a normal conductor, as the mean free path becomes very large it becomes

Table 1.9.1. *Some superconducting materials*[28]

Superconductor	T_c/K	Δ/k_BT_c	ξ_0/nm	λ_0/nm	μ_0H_c/T
Nb	9.2	1.9	39	32	0.2
Pb	7.2	2.0	110	28	0.08
Sn	3.7	1.8	300	28	0.03
Nb_3N	17	2.1	6	50	—
$YBa_2Cu_3O_{7-x}$	92	~ 2.2	$ab < 2$ $c < 0.4$	ab 140 c 770	$\mu_0H_{c1} \sim 0.02$ $\mu_0H_{c2} \sim 100$

closer to a perfect conductor. Some coherence lengths for a selection of superconductors are shown in Table 1.9.1.

1.9.3 Microscopic complex conductivity

Bardeen, Cooper and Schrieffer's (BCS) theory,[29] published in 1957, provides a quantum mechanical description of the superconducting state. It predicts the properties of a large number of superconducting materials, including high-frequency microwave properties. Expressions for these high-frequency properties were developed, using the BCS theory, by Mattis and Bardeen in 1958.[30]

As the temperature of a superconductor drops below its transition temperature the normal electrons form into pairs; this condensation is due to the formation of an energy gap $\Delta(T)$ in the electron density of states which is temperature dependent. The temperature dependence of this gap is predicted by BCS, and a good fit to it is given by the empirical expression[31]

$$\Delta(T) = \Delta_0 \left(\cos\left(\frac{\pi}{2} \left(\frac{T}{T_c} \right)^2 \right) \right)^{1/2} \tag{1.9.7}$$

Here Δ_0 is the energy gap as the temperature approaches absolute zero, and in weakly coupled BCS theory it is given by

$$\Delta_0 = 1.764 k_B T_c \tag{1.9.8}$$

This parameter is found experimentally for different superconductors, and increases to 2 and above for superconductors which are strongly coupled. The Heisenberg uncertainty principle predicts how the coherence length is related to the energy gap, that is,

$$\xi(T) = \frac{v_F \hbar}{\pi \Delta(T)} \tag{1.9.9}$$

where v_F is the Fermi velocity. Table 1.9.1 shows some parameters discussed above for a few superconducting materials.

The Mattis and Bardeen[30] formulation for superconductors is not analytically soluble and will not be reproduced here. However, a number of approximate expressions for the various parameters have been deduced and are given below. An approximation for the surface resistance for temperatures less than about half the transition temperature is given by[32,33]

$$R_s \approx A\frac{\Delta}{T}\left(\frac{\omega}{\Delta}\right)^2 \ln\left(\frac{\Delta}{\hbar\omega}\right) e^{-\Delta/k_B T} \qquad (1.9.10)$$

Here A is a numerical factor dependent upon the properties in the normal state and not strongly dependent on frequency or temperature. This expression has been used widely in order to determine the energy gap of superconducting materials by looking at an exponential temperature dependence of the surface resistance at low temperatures. For practical situations a residual surface resistance must be added to Equation (1.9.10) so that the exponential decrease of surface resistance does not continue to very low temperatures. The value of this residual surface resistance depends on the type and quality of the superconductor and arises from the effects of impurities, trapped flux or weak links at grain boundaries. It may be increased in high temperature superconductors, perhaps due to their short coherence length.

Although the prediction of high-frequency surface impedance using the BCS theory is not straightforward, a number of numerical calculations have been made which can cover the parameter range of interest. In addition, it has been possible to fit expressions to these numerical results, enabling BCS predictions to be made simply and quickly. An expression, due to Linden,[34] for the complex conductivity of a superconductor which is derived using a fitting technique to numerical solutions, is given in Equation (1.9.11). Curve fitting provides an excellent way of obtaining fast numerical predictions of complex conductivity and hence surface impedance, although careful consideration must be taken of the accuracy of these methods in the context in which the results are to be used. The expression (Equation (1.9.11)) discussed below represents a speed increase of between one and five orders of magnitude over numerical computations, which are not easy to implement.

Complex conductivity as a function of temperature and frequency is defined by Linden as[34]

$$\sigma_{1MTF} - j\sigma_{2MTF} = \frac{1}{j\omega\mu\lambda_d^2(T)} + \frac{\sigma_n}{1 + j\dfrac{\hbar\omega}{\pi\Delta_0}\dfrac{l}{\xi_0}}\eta(\omega, T) \quad \omega < \omega_s(T) \qquad (1.9.11)$$

This expression is a modified two-fluid (MTF) model which can be seen in comparison with Equation (1.2.12) in Section 1.2. Here σ_n is the normal state conductivity at T_c, l is the mean free path, ξ_0 is the low-temperature coherence length and Δ_0 is the low-temperature energy gap is described above. The depth $\lambda_d(T)$ is similar to the London penetration depth except that it is modified to take non-local effects into account, that is, $\lambda_d^2(0) \approx \lambda_L^2(0)(1 + \xi_0/l)$. The scattering time has been replaced by an equivalent BCS parameter, given approximately by[34]

$$\tau = \frac{\hbar}{\pi \Delta_0} \frac{l}{\xi} \tag{1.9.12}$$

The frequency limit ω_s in Equation (1.9.11) represents the gap frequency, above which the energy of the photons in the applied field is large enough to split the electron pairs, and therefore the carriers become normal electrons. Since the energy gap decreases as the temperature increases, this gap frequency also decreases and is given by

$$\omega_s(T) = \frac{2\Delta(T)}{\hbar} \tag{1.9.13}$$

where the temperature dependence of the gap $\Delta(T)$ is given by Equation (1.9.7). The temperature dependence of the penetration depth, $\lambda_d(T)$, for a clean, weakly coupled superconductor can be approximated by[34,35]

$$\lambda_{BCS}^2(T) = \frac{\lambda_0^2}{1 - (T/T_c)^{3-T/T_c}} \tag{1.9.14}$$

This expression has been observed to fit experimental results over a wide temperature range (but not necessarily at very low temperatures). This can be used in Equation (1.9.11) in the clean weak coupling case ($\xi_0/l \ll 1$ and $\Delta(T)/k_B T_c \sim 1.76$). For strong coupling ($\Delta(T)/k_B T_c > 2$) the conventional two-fluid penetration depth is a good approximation of the penetration depth in this case, that is,

$$\lambda_L^2(T) = \frac{\lambda_0^2}{1 - (T/T_c)^4} \tag{1.9.15}$$

The choice of the particular function for the penetration does not affect the value of the surface resistance appreciably. The function $\eta(\omega, T)$ in Equation (1.9.11) is the ratio of normal electrons to the total number of electrons and is given by

$$\eta(\omega, T) = \frac{2\Delta(T)}{k_B T} \exp\left(\frac{\Delta(T)}{k_B T}\right) \ln\left(\frac{\Delta(T)}{\hbar\omega_1}\right) F(\omega) \tag{1.9.16}$$

The value of $F(\omega)$ is given by Linden[34] as

$$F(\omega) = \frac{a}{1+(\omega/\omega_0)^b} + c \tag{1.9.17}$$

where

$$a = 0.16 + 0.17\begin{cases} 1 & l/\xi_0 < 1 \text{ and } (\omega/\omega_0)^b < e^3 \\ 1-(1/3)\log(l/\xi_0) & 1 < l/\xi_0 < 1000 \text{ and } (\omega/\omega_0)^b < e^3 \\ 0 & \text{otherwise} \end{cases} \tag{1.9.18}$$

$$b = \frac{1}{\ln(10)}\begin{cases} 1+0.225\log(l/\xi_0) & l/\xi_0 > 1 \\ 1 & \text{otherwise} \end{cases} \tag{1.9.19}$$

$$c = \begin{cases} \dfrac{a((\omega/\omega_0)^b + e^3)}{1+e^3} & \left(\dfrac{\omega}{\omega_0}\right)^b > e^3 \\ 0 & \text{otherwise} \end{cases} \tag{1.9.20}$$

$$\omega_0 = 2\pi\begin{cases} 3.9811 \times 10^9 & \xi_0/l < 10 \\ (\xi_0/l)^{2/3}(1.9953 \times 10^{10}) & \text{otherwise} \end{cases} \tag{1.9.21}$$

with $\omega_1 = 1$. This is very similar to an expression given by Hinken;[36] in this case $F(\omega) = 1$ and $\omega_1 = \omega$. Hinken's approximation is good over a limited parameter range, that is, $\lambda \gg \xi_0$, $\omega \ll \omega_s$ and $k_B T \ll \Delta$.

The expression for $\eta(\omega, T)$ has been found by curve fitting to the value of σ_1 from the BCS results calculated numerically by Zimmerman,[37] and fine tuning accomplished with the use of Halbritter's[38] program. The fitting is optimised for frequencies at or below the microwave regime. Figure 1.9.1 shows the value of complex conductivity for niobium using this MTF model and the parameters are given in [39]. The value of σ_1 increases below T_c due to the coherence of the BCS pairing and has a maximum value that is quite different from the two-fluid model, which doesn't have a maximum.

The surface resistance of niobium can also be calculated by using Equation (1.4.8) and this is shown in Figure 1.9.2. In this diagram there is a comparison with the full numerical BCS calculation and some experimental data, and all agree very closely. Equation (1.9.15) is used for the penetration depth in this case.

The value of the surface resistance using the MTF model is accurate to about a factor of two over the entire parameter range investigated (dirty to clean $0.01 < \xi_0/l < 100$, low to high temperature $0.022 < T/T_c < 0.94$, weak to strong coupling $1.75 < \Delta_0/k_B T < 2.75$), and significantly better over a large range of parameters especially in the weak coupling limit. Proper specification of the penetration depth variation gives improved accuracy.

BCS theory is an isotropic s-wave theory which does not fully explain the properties of high-temperature superconductors. In order to explain these

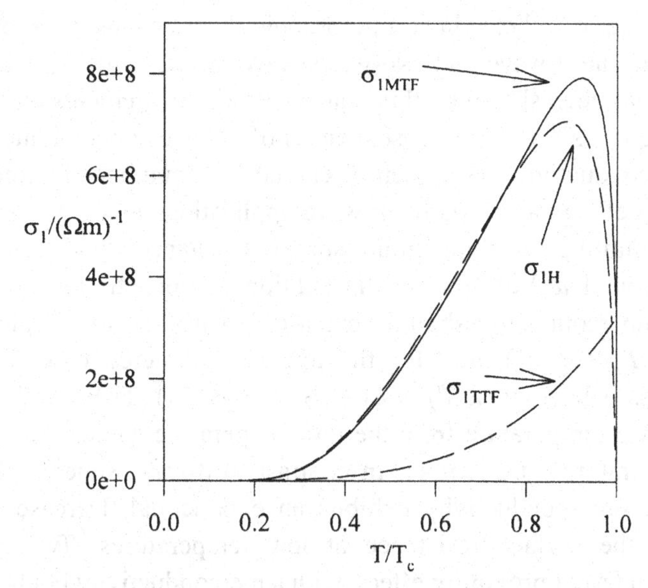

Figure 1.9.1. The temperature variation of σ_l for niobium for the modified two-fluid σ_{1MTF}, the traditional two-fluid σ_{1TTF} and the Hinken σ_{1H} approximations. The parameters used are $\Delta_0/k_BT_c = 1.97$, $T_c = 9.2$ K, $\lambda_0 = 34.1$ nm, $\rho_n = 1/\sigma_n = 0.31\ \mu\Omega$ cm, $l = 100$ nm and $\xi_0 = 39$ nm.

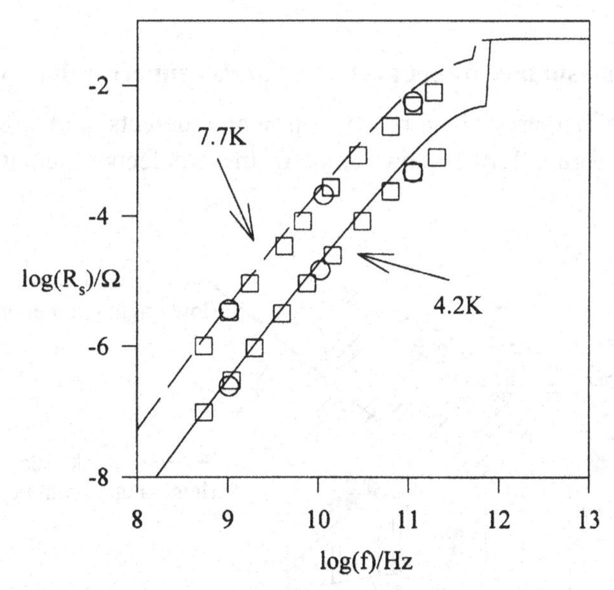

Figure 1.9.2. The frequency variation of the surface resistance of niobium using the parameters in Figure 1.9.1. The circles represent the full BCS numerical calculation, the squares represent experimental results[39] and the lines represent the modified two-fluid model. (After Linden[34].)

properties a number of other related microscopic theories have been proposed, among which are the d-wave and two-gap s-wave models. Strong evidence of d-wave pairing has been shown by flux quantisation measurements on tricrystal superconducting rings[40] and the consequences of d-wave superconductivity on the microwave conductivity have been discussed by a number of authors.[41–43] One of the main effects as far as microwave applications are concerned is the prediction of a limiting low-temperature surface resistance which has the very low value of about 1 $\mu\Omega$ at 10 GHz. In addition, the penetration depth for a pure d-wave superconductor should increase linearly at low temperatures ($\lambda(T) = \lambda_0 + c_1 T,\ T \ll T_c$), and for the impure disordered case there is a quadratic gapless behaviour ($\lambda(T) = \lambda_0 + c_2 T^2,\ T \ll T_c$). There is the prediction of a crossover temperature from the low-temperature gapless behaviour to the higher-temperature pure regime. In contrast to d-wave superconductivity, two-gap s-wave superconductivity exhibits an exponential decrease with the temperature of the surface resistance at low temperatures. Two gaps are predicted by an internal proximity effect, with superconductivity being induced in the chains from the CuO_2 planes. The induced gap is predicted to be smaller in the chains than in the planes.[44,45] A comparison of experimental evidence for the two models has been given by Klein,[46] with some experimental data being given in Appendix 1.

1.10 The surface impedance of granular superconductors

High-temperature superconductors contain many defects and are far from perfect crystals. Figure 1.10.1 shows some of these defects schematically. The

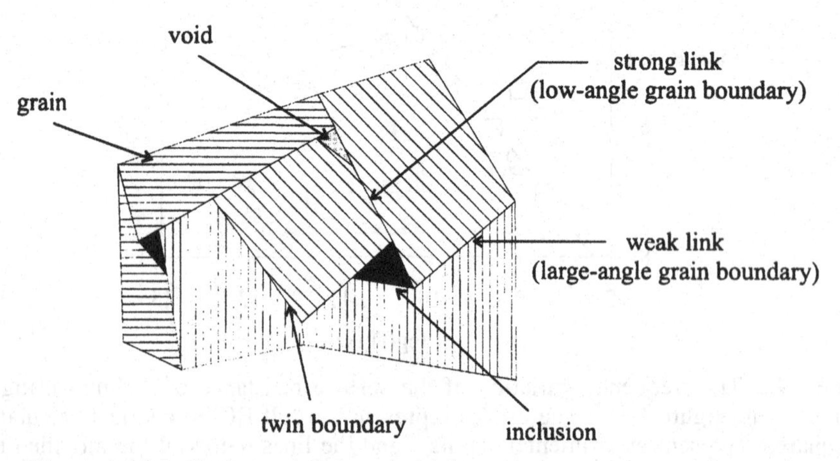

Figure 1.10.1. Grains and defects in high-temperature superconductors.

superconductor is composed of inclusions of contaminating material, voids in the structure, and large and small grains of material connected by regions of low critical current density which may be of varying magnitude. There are regions where twinning occurs, with boundaries between also producing weak links. The density of these defects is vastly different dependent upon the type of material. Polycrystalline material is dominated by grain boundary weak links, whereas in single-crystal materials there are relatively fewer defects. Thin films lie in between the two. The quality of the manufacture of the material obviously has a significant effect upon the defects present. The scale of the defects ranges from the atomic scale to almost the size of the actual piece of a sample. With such a wide variety of defects it is difficult to analyse analytically, although a number of attempts have been made, with some success.

Effective medium theory can be used to tackle the problem to a certain extent. The simplest approach is to ignore the effect of the grain boundaries and assume that the material is composed of superconducting material and inclusions of insulating or poorly conducting material. This has been discussed by Gittleman and Matey[47] using the Bruggeman effective medium approximation.[48] This approach clearly explains higher than expected values of surface resistance and the non-ω^2 frequency dependence of a superconducting sample as the volume fraction of the inclusion increases. However, in the majority of polycrystalline materials the fraction of non-superconducting material is small. Some polycrystalline materials have intentional metallic inclusions in the structure. This has been found to improve their properties and the effective medium theory may then help to explain their behaviour.[49] The inclusion of silver improves the surface resistance by increasing the critical current between the grains, and will obviously only ultimately provide superior polycrystalline material if the critical current is not improved by alternative material processing techniques. Effective medium theory can also be used to study superconductor/insulator composites.[50,51]

A more useful model is to consider superconducting material as an array of coupled superconducting grains.[52] The coupling between the grains is characterised by a reduced critical current and can be modelled by a resistively shunted Josephson junction, as depicted in Figure 1.10.2.

First consider the current through a Josephson junction shunted by a material with conductivity σ_j, given by[53–55]

$$J(t) = J_0 \sin(\delta(t)) + \sigma_j E(t) \tag{1.10.1}$$

where the electric field is

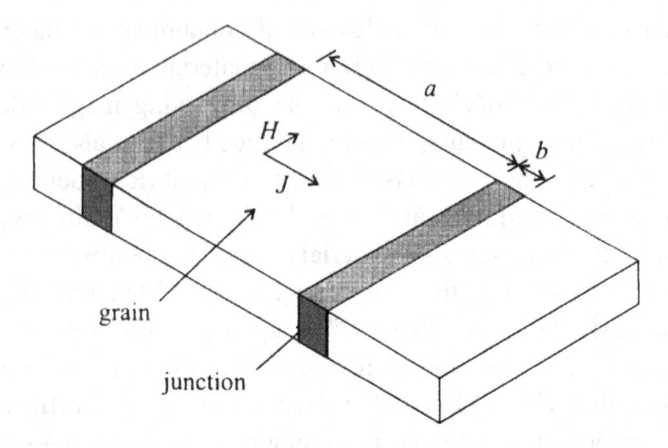

Figure 1.10.2. Three grains connected by two weak links.

$$E(t) = \frac{\Phi_0}{2\pi b_{eff}} \frac{\partial \delta}{\partial t} \tag{1.10.2}$$

where Φ_0 is the flux quantum

$$\Phi_0 = \frac{h}{2e} = 2.0678 \times 10^{-15} \text{ Wb} \tag{1.10.3}$$

and $\delta(t)$ is the phase difference of the macroscopic wavefunction across the junction

$$\delta(t) = \delta_0 + \frac{2\pi}{\Phi_0} Vt \tag{1.10.4}$$

This latter expression leads to an oscillating current at frequency $\omega = 2 \text{ eV}/\hbar$ for an applied constant voltage V across the junction. Also in Equation (1.10.2) the distance b_{eff} is given by $b_{eff} = b + 2\lambda_g$, where b is the distance across the junction and λ_g is the penetration depth in the material at either side of the junction. The expressions given above assume that the microwave current density is small compared with the critical current density of the junction. By differentiating Equation (1.10.1) with respect to time and then substituting into Equation (1.10.2), one finds that

$$\frac{\partial J}{\partial t} = \frac{1}{l_j} E + \sigma_j E \tag{1.10.5}$$

with l_j being the Josephson inductivity given by

$$l_j = \frac{\Phi_0}{2\pi b_{eff} J_0 \cos \delta} \tag{1.10.6}$$

Now assuming a sinusoidal time dependence in Equation (1.10.5), the effective conductivity of a shunted junction can be written as

$$\sigma_{sj} = \frac{1}{j\omega l_j} + \sigma_j \tag{1.10.7}$$

It is now possible to use a simple effective medium approach to consider the effect of grains of superconducting material in series with the junctions. If all the current is forced to go through the junctions and the grains in series, then their effective resistivities can be added with appropriate weighting for the size of the grains and junctions. This results in an expression for complex conductivity for such a medium of

$$\sigma = \frac{\sigma_g(1 + j\omega l_j\sigma_j)(a + b)}{a(1 + j\omega l_j\sigma_j) + j\omega l_j\sigma_g b} \tag{1.10.8}$$

where σ_g is the complex conductivity of the superconducting grain and a and b represent the size of the grain and junction as depicted in Figure 1.10.2. Assuming that the conductance of the junction is dominated by the inductance ($\omega l_j \gg 1/\sigma_j$) and that the grain is dominated by superconducting carriers ($\sigma_{1g} \ll \sigma_{2g}$), the complex conductivity of material with small junctions ($b < a$) is given approximately by[52]

$$\sigma = \left[j\omega\mu_0\lambda_{eff}^2 + \omega^2\mu_0\lambda_j^2\frac{\hbar}{2eI_cR}\right]^{-1} \tag{1.10.9}$$

Here I_c is the critical current of the junction and R is its shunt resistance. The effective penetration depth is given by[52]

$$\lambda_{eff}^2 = \lambda_g^2 + \lambda_j^2 \tag{1.10.10}$$

Here λ_g is the London penetration depth of the grain and λ_j is the effective continuum penetration depth due to the network of grain boundary junctions alone. Although similar to the Josephson penetration depth, it is not the same in this context and is given by

$$\lambda_j^2 = \frac{l_j}{\mu}\frac{b}{a} = \frac{\hbar}{a2eJ_0\mu_0} \tag{1.10.11}$$

Using Equations (1.4.18) and (1.10.9) the surface impedance of granular material with the above approximations is given by

$$Z_s = \omega\mu_0\lambda_{eff}\left[\frac{1}{2}\left(\frac{\lambda_j}{\lambda_{eff}}\right)^2\frac{\hbar\omega}{2eI_cR} + j\right] \tag{1.10.12}$$

The surface impedance is dominated by the grain boundaries if the size of the grains is small or if the junction critical current is small, making λ_{eff} approach λ_j. When the critical current is high, then λ_{eff} approaches λ_g and the surface impedance is dominated by the grains. The magnetic field dependence of the conductivity and surface impedance of this composite structure can be studied by noting that there is a decrease in the critical current with increasing field. A discussion of experimental work related to the surface impedance as a function of field is given in references [55, 56] and some results are given in Appendix 1.

Clearly, if the surface impedance is dominated by weak links with reduced critical currents, then the microwave power dependence, in addition to the d.c. magnetic field dependence,[57,58] will also depend critically upon the granular structure. For the microwave power dependence the junction has three separate regimes dependent upon the level of the microwave field: (i) At low fields, pair breaking in the junction region, which can be described by the Ginzburg–Landau theory, accounts for additional losses. (ii) At higher fields, when the H_{cl} of the junction is exceeded, flux penetrates the junction and increases the losses further. (iii) At the higher power levels, flux flow occurs within the junction. Analytical expressions are developed for these separate regimes as well as comparison with experimental results in references [55, 59–65]. The results will not be covered further here, although some measurements are given and discussed in Appendix 1.

1.11 The surface impedance of type II superconductors

For type II superconductors, when an applied steady magnetic field is increased above the critical field H_{cl}, flux penetrates the superconductor in the form of current vortices forming a lattice of flux lines. The lowest energy state is a triangular array of flux lines, and the separation is given approximately by $\sqrt{\Phi_0/B}$. Here Φ_0 is the flux quantum and B is the applied static field. Movement of flux lines through the material produces a dissipation resulting in a small resistance. For example, in a wire the flux lines are circumferential and the application of a current down the wire causes their diameters to shrink towards the centre, where they are annihilated. However, defects are always present which can cause flux lines to become trapped or pinned at particular positions. Energy is required to move the flux lines from the pinning centre. If pinning dominates, then the movement of flux lines disappears and the dissipation approaches the lossless case again. The inclusion of pinning centres in materials has been widely studied in order to increase their critical currents. The first case given above is called flux flow, where the movement is unhindered by any pinning. Flux creep is where pinning is present and flux lines are able to move by jumping between these centres; this occurs with the assistance of thermal energy. If an alternating field is applied, then this will move the flux lines whether they are pinned or not, and then there is additional dissipation due to this motion. The dynamics of this motion are discussed below. Much of the following discussion is taken from the theory of Coffey and Clem,[66–69] although there has been a substantial amount of work undertaken previously on the subject.[70–74]

The London equations can still be a good approximation in this case if an additional source term is included:

$$\nabla \times \mathbf{J}_s = -\frac{1}{\mu_0 \lambda_L^2}(\mathbf{B} - n(x,t)\Phi_0 \mathbf{a}_\Phi) \tag{1.11.1}$$

where $n(x, t)$ is the density of the vortices, which may be assumed to vary spatially, and $\mathbf{a}_\Phi$ is a unit vector directed along the vortex cores. These vortices are assumed to provide flux of the magnitude of a single flux quantum Φ_0 and of zero radius. Additionally, the microwave magnetic field is assumed to be small compared with the applied static magnetic field. The usual approximations of the London model still apply.

Taking the time derivative of Equation (1.11.1), substituting $\mathbf{J}_s = \mathbf{J} - \mathbf{J}_n$ and $\mathbf{J}_n = \sigma_{1TF}\mathbf{E}$, and using Faraday's law results in

$$\nabla \times \frac{\partial \mathbf{J}}{\partial t} - \frac{\Phi_0}{\lambda_L^2 \mu_0}\frac{\partial n}{\partial t}\mathbf{a}_\Phi = \frac{1}{\lambda_L^2 \mu_0}\nabla \times \mathbf{E} + \sigma_{1TF}\nabla \times \frac{\partial \mathbf{E}}{\partial t} \tag{1.11.2}$$

Here σ_{1TF} is the real part of the two-fluid model complex conductivity given by Equation (1.2.13). Equation (1.11.2) gives the basic relation between **J** and **E** which is required to calculate the effective conductivity of the superconductor. However, a relationship is required between the change in flux line density and the current in the superconductor.

Each of the flux lines can be pinned, and application of an alternating field causes them to move from equilibrium. With this movement a viscous drag is opposing the motion and a force is trying to restore the flux line into a position at the centre of the potential well. The equation of motion for the flux lines in this situation can be written as

$$\eta \frac{\partial \mathbf{u}}{\partial t} + \kappa_p \mathbf{u} = \mathbf{J} \times (\Phi_0 \mathbf{a}_\Phi) \tag{1.11.3}$$

Here **u** is the flux line displacement, η is the viscous drag coefficient and κ_p is the restoring force constant in the pinning potential well (Labusch parameter). Any effects due to the mass of the vortex are small and are ignored. Solving this equation, assuming that the current is directed in the x direction only, the flux in the y direction and sinusoidal time dependencies, results in

$$u_z = \frac{J_x \Phi_0 \hat{\mu}}{j\omega} \tag{1.11.4}$$

where $\hat{\mu}$ is termed the dynamic mobility given by

$$\hat{\mu} = \frac{1}{\eta\left(1 - j\frac{\kappa_p}{\omega\eta}\right)} \tag{1.11.5}$$

This mobility is modified in reference [67] to account for the thermal motion of vortices. It can be seen from Equation (1.11.5) that there is a crossover frequency $\omega_c = \kappa_p/\eta$ when this mobility changes over from dynamics mainly associated with viscous drag (flux flow) to dynamics associated with pinning.

$\partial n/\partial t$ can be related to the flux line velocity by considering a small area $\delta z\,\delta x$ in the plane normal to $\mathbf{a}_x$. Using conservation of n (since $\nabla \cdot \mathbf{B} = 0$), the number of flux lines leaving this small area per second is $n\,(\partial v/\partial z)\,\delta z\,\delta x$, which can be equated to $(\partial n/\partial t)\,\delta z\,\delta x$. Using Equation (1.11.4) gives

$$\frac{\partial n}{\partial t} = n\frac{\partial u^2}{\partial z\,\partial t} = n\Phi_0\hat{\mu}\frac{\partial J_x}{\partial z} \tag{1.11.6}$$

where the flux is in the $\mathbf{a}_y$ direction. This expression can now be substituted into Equation (1.11.2). During this substitution it is also assumed that the plane wave is of the form

$$\mathbf{J} = J_x e^{j\omega t} e^{-z/\lambda}\mathbf{a}_x \quad \text{and} \quad \mathbf{E} = E_x e^{j\omega t} e^{-z/\lambda}\mathbf{a}_x \tag{1.11.7}$$

Here λ represents the distance that the incident wave penetrates the superconductor as depicted in Figure 1.11.1; it is a real quantity. A complex penetration depth is used in Coffey and Clem's theory. This is not included here as the main interest is the effective complex conductivity to make the result consistent with the rest of the book.

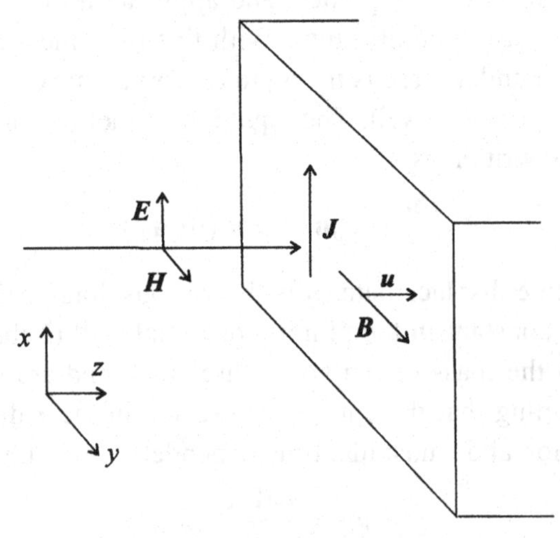

Figure 1.11.1. Plane wave incident on a slab of superconductor showing the directions of the fields.

The result after substitution is

$$J_x = \sigma E_x \tag{1.11.8}$$

where

$$\sigma = \frac{\sigma_{1TF} - j\dfrac{1}{\omega\mu_0\lambda_L^2}}{1 - j\dfrac{B\Phi_0\hat{\mu}}{\omega\mu_0\lambda_L^2}} \tag{1.11.9}$$

Equations (1.11.9) and (1.11.8) define complex conductivity for type II superconductors in the mixed state under the conditions discussed above. The real and imaginary parts of Equation (1.11.9) can be substituted in all the derivations in Sections 1.4–1.8. For example, R_s or X_s can now be calculated. These are plotted in Figure 1.11.2.

Although not shown, the temperature dependence of the surface impedance also shows a change from flux pinning to flux flow. At low temperatures, the pinning is strong and thermal agitation is low. As the temperature increases, the pinning constant reduces and thermal agitation first induces a region where flux creep is dominant, which then gives way to flux flow. A similar change from one regime to another is obtained as a function of the applied magnetic field.

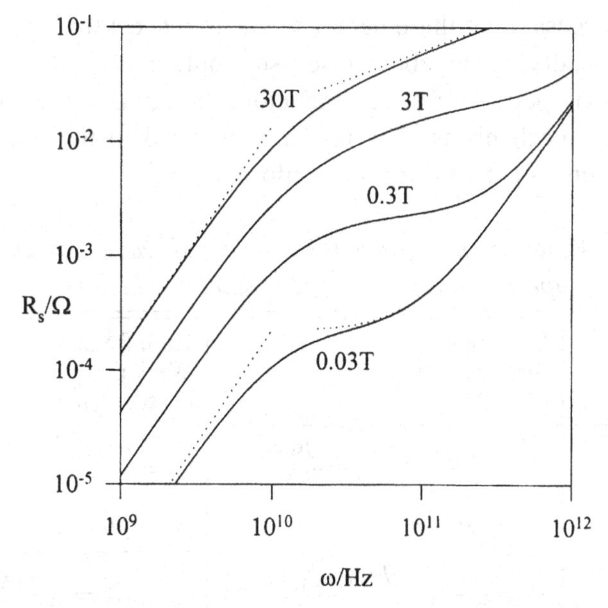

Figure 1.11.2. Surface resistance of a type II superconductor for various applied fields. The dotted lines represent the approximations in Table 1.11.1. Parameters are $\lambda_0 = 0.14\ \mu\text{m}$, $T_c = 93$ K, $B_{c2} = 100T$, $\sigma_{1TF} = 10^7\ (\Omega\text{m})^{-1}$, $\eta = 10^{-6}\ \text{N}\,\text{sm}^{-2}$ and $\kappa_p = 10^4\ \text{Nm}^{-2}$ ($\omega_c = 10^{10}$ Hz). Also, $\lambda_L^2 = \lambda_0^2/(1-(T/T_c)^4)(1-(B/B_{c2}))$.

Both of these changes from one regime to another depend of course upon the frequency of the applied field and give rise to frequency dependent crossover fields and temperatures.

Because of the complicated form of Equation (1.11.9), it is rather difficult to see clearly how the changing parameters affect the complex conductivity or the surface impedance. It is therefore useful to consider approximations to this equation in the low- and high-field limits and the low- and high-frequency limits. Table 1.11.1 shows the results of these limiting cases. These approximations not only show how R_s and X_s vary in the limiting cases, but indicate which regime is applicable from observation of experimental results. It is also interesting to note that the Campbell penetration depth, λ_C, can be extracted in the low-frequency, high-field region where the pinning dominates. Using $\lambda = 1/\mathrm{Re}\,(j\omega\mu_0\sigma)$ and Equation (1.11.9) gives

$$\lambda_C = \sqrt{\left(\frac{B\Phi_0}{\mu_0\kappa_p}\right)} \tag{1.11.10}$$

Similarly, the flux flow penetration depth can be calculated in the limit of high-frequency and strong fields as

$$\lambda_{FF} = \sqrt{\left(\frac{2B\Phi_0}{\omega\mu_0\eta}\right)} \tag{1.11.11}$$

Some further discussion of these parameters with respect to experimental work is given in Appendix 1. The above discussion only outlines the subject of the high-frequency properties of superconductors, bringing out some of the important physical mechanisms. The reader is referred to the references given above plus references [75–83] for more information.

Table 1.11.1. *Values of the surface resistance and reactance under various approximations for type II superconductors*

	Low frequency Strong pinning $\omega \ll \kappa_p/\eta$	High frequency Weak pinning $\omega \gg \kappa_p/\eta$
Low field $B\Phi_0\hat{\mu} \ll \omega\mu_0\lambda_L^2$	$R_s = \dfrac{\sigma_{1TF}\omega^2\mu_0^2\lambda_L^3}{2} + \dfrac{B\Phi_0\eta\omega^2}{2\lambda_L\kappa_p^2}$ $X_s = \omega\mu_0\lambda_L$	$R_s = \dfrac{\sigma_{1TF}\omega^2\mu_0^2\lambda_L^3}{2} + \dfrac{B\Phi_0}{2\lambda_L\eta}$ $X_s = \omega\mu_0\lambda_L$
High field $B\Phi_0\hat{\mu} \gg \omega\mu_0\lambda_L^2$	$R_s = \dfrac{\eta\omega^2}{2\kappa_p}\sqrt{\left(\dfrac{B\Phi_0\mu_0}{\kappa_p}\right)}$ $X_s = \omega\sqrt{\left(\dfrac{B\Phi_0\mu_0}{\kappa_p}\right)}$	$R_s = X_s = \sqrt{\left(\dfrac{\omega\mu_0\Phi_0 B}{2\eta}\right)}$

1.12 References

1 Onnes H. K. *Commun. Phys. Lab. Uni. Leiden*, **120b**, 3, 1911
2 Gorter C. J. and Casimir H. B. G. *Phys. Z.*, **35**, 963, 1934
3 London F. and London H. *Physica*, **2**, 341, 1935
4 London F. and London H. *Proc. Roy. Soc. (Lond.)*, **A149**, 71, 1935
5 London H. Production of heat in superconductors by alternating currents, *Nature*, **133**, 497, 1934
6 Pippard A. B. The surface impedance of superconductors and normal metals at high frequencies Parts I, II and III, *Proc. Roy. Soc.*, **A191**, 370–415, 1947
7 Ginzburg V. L. and Landau L. D. *Zh. Eksp. Theo. Fiz.*, **20**, 1064, 1950. English translation contained in *Men of Physics*, V1: L. D. Landau ed. D. ter Haar Pergamon, pp. 138–67, New York 1965
8 Bardeen J., Cooper L. N. and Schrieffer J. R. Theory of superconductivity, *Phys. Rev.*, **108**, 1175–1204, 1957
9 Mattis D. C. and Bardeen J. Theory of the anomalous skin effect in normal and superconducting metals, *Phys. Rev.*, **111**, 412–17, 1958
10 Bednorz J. G. and Müller K. A. Possible high T_c superconductivity in the Ba–La–Cu–O system *Z. Phys.*, B**64**, 189, 1986
11 Mei K. K. and Liang G.-C. Electromagnetics of superconductors, *IEEE Trans. on Microwave Theory and Techniques*, **39**(9), 1545–52, 1991
12 Ramo S., Whinnery J. R. and van Duzer T. *Fields and Waves in Communication Electronics*, John Wiley and Sons, 1984
13 Kraus J. D. *Electromagnetics*, McGraw-Hill, 1984
14 Abramowitz M. and Stegun I. A. *Handbook of Mathematical Functions*, Dover Publications, New York, 1972
15 Klein N., Chaloupka H., Müller G., Orbach S., Piel H., Roas B., Schultz H., Klein U. and Peiniger M. The effective microwave surface impedance of high T_c thin films, *J. Appl. Phys.*, **67**(11), 6940–45, June 1990
16 Morgan S. P. Jr, *J. Appl. Phys.*, **20**, 352, 1949
17 Wu Z. and Davis L. E. Surface roughness effect on surface impedance of superconductors, *Appl. Phys. Lett.*, **76**(6), 3669–72, 1994
18 Gorter C. J. and Casimir H. B. G. *Phys. Z.*, **35**, 963, 1934
19 London F. *Superfluids*, Vols. 1 and 2, Dover, New York, 1954
20 London F. and London H. *Proc. Roy. Soc. (Lond.)*, A**149**, 71–85, 1935
21 Newhouse V. L. *Applied Superconductivity VII*, Academic Press, 1975
22 Pippard A. B. The surface impedance of superconductors and normal metals at high frequencies I, *Proc. Roy. Soc.*, **191**, 385, 1947
23 Pippard A. B. The surface impedance of superconductors and normal metals at high frequencies II, *Proc. Roy. Soc.*, **191**, 370, 1947
24 Pippard A. B. The surface impedance of superconductors and normal metals at high frequencies III. The relation between impedance and superconducting penetration depth, *Proc. Roy. Soc.*, **191**, 399–415, 1947
25 Pippard A. B. *Proc. Roy. Soc.*, A**203**, 98, 1950
26 Pippard A. B. *Proc. Roy. Soc.*, A**203**, 195, 1950
27 Pippard A. B. *Proc. Roy. Soc.*, A**216**, 547, 1953
28 Turneaure J. P., Halbritter J. and Schwettman H. A. *J. Superconductivity*, **4**, 341, 1991
29 Bardeen J., Cooper L. N. and Schrieffer J. R. Theory of superconductivity, *Phys. Rev.*, **108**, 1175–1204, 1957

30 Mattis D. C. and Bardeen J. Theory of the anomalous skin effect in normal and superconducting metals, *Phys. Rev.*, **111**(2), 412–17, 1958
31 Sheahen T. P. Rules for energy gap and critical field and superconductors, *Phys. Rev.*, **149**, 368, 1966
32 Cooke D. W., Gray E. R., Javadi H. H. S., Houlton R. J., Rusnak B., Meyer E. A., Arendt P. N., Klein N., Müller G., Orbach S., Piel H., Drabeck L. and Grüner G. Frequency dependence of the surface resistance in high temperature superconductors, *Solid State Communications*, **73**(4), 297–300, 1990
33 Turneaure J. P., Halbritter J. and Schwettman H. A. *Supercond.*, **4**, 341, 1991
34 Linden D. S., Orlando T. P. and Lyons W. G. Modified two-fluid model for superconductor surface impedance calculation, *IEEE Trans. on Applied Superconductivity*, **4**(3), 136–42, 1994
35 Bonn D. A., Dosanja P., Liang R. and Hardy W. H. Evidence for rapid suppression of quasiparticle scattering below T_c in $YBa_2Cu_3O_{7-\delta}$, *Phys. Rev. Lett.*, **68**, 2390–3, 1992
36 Hinken J. H. *Superconductor Electronics: Fundamentals and microwave applications*, Springer-Verlag, Berlin, 1988.
37 Zimmerman W., Brandt E. H., Bauer M., Seider E. and Genzel L. Optical conductivity of superconductors with arbitrary purity, *Physica*, C**183**, 99–104, 1991
38 Halbritter J. Kernforschungszentrum Karlsruhe Externer Bericht 3/70–6 Karlsruhe: Institut für Experimentelle Kernphysik. June, 1970
39 Piel H. and Müller G. The microwave surface impedance of high T_c superconductors, *IEEE Trans. on Magnetics*, **27**, 854–62, 1991
40 Tsuei C. C. *et al.* Pairing symmetry and flux quantisation in a tricrystal superconducting ring of $YBa_2Cu_3O_{7-\delta}$, *Phys. Rev. Lett.*, **73**, 595–6, 1994
41 Hirschfeld P. J., Putikka W. O. and Scalapino D. J. Microwave conductivity of d-wave superconductors, *Phys. Rev. Lett.*, **71**(22), 3705–8, 1993
42 Mao J., Anlage S. M., Peng J. L. and Greene R. L. Consequences of d-wave superconductivity for high frequency applications of cuprate superconductors, *IEEE Trans. on Applied Superconductivity*, **5**(2), 1997–2000, 1995
43 Lee P. A. Localised states in a d-wave superconductor *Phys. Rev. Lett.*, **71**, 1887–90, 1993
44 Wolf S. A., Morawitz H. and Kresin V. Z. *Mechanism of Conventional and High-T_c Superconductivity*, Oxford University Press, 1993
45 Klein N., Tellmann N., Schultz H., Urban K., Wolf S. A. and Kresin V. Z. Evidence of two gap s-wave superconductivity in $YBa_2Cu_3O_{7-x}$ from microwave surface impedance measurements, *Phys. Rev. Lett.*, **71**(20), 3355–8, 1993
46 Klein N., Tellmann N., Wolf S. A. and Kresin V. Z. Microwave surface impedance of $YBa_2Cu_3O_{7-x}$: s-wave versus d-wave pairing, *J. of Superconductivity*, **7**(2), 459–61, 1994
47 Gittleman J. I. and Matey J. R. Modelling the microwave properties of $YBa_2Cu_3O_{7-x}$ superconductors, *J. Appl. Phys.*, **65**(2), 688–91, 1989
48 Landauer R. Electrical transport and optical properties of inhomogeneous media, eds. J. C. Garland and D. B. Tanner. *AIP Conf. Proc. No. 40 American Institute of Physics, New York*, 1978
49 Ishii O., Konaka T., Sato M. and Koshimoto Y. Reduction of the surface resistance of $YBa_2Cu_3O_{7-x}$ pellets and thick films by adding Ag, *Japanese J. of App. Phys.*, **29**(7), L1075–L1078, 1990
50 Golosovsky M., Tsindlekht M., Davidov D. and Sarychev A. K. Effective medium

approach to the microwave properties of the high-T_c superconductor insulator composites, *Physica*, C**209**, 337–40, 1993
51 Waldram J. R., Porch A. and Cheah H. M. Microwave surface impedance of patterned $YBa_2Cu_3O_7$ thin films, *Physica*, B**194–6**, 1607–8, 1994
52 Hylton T. L., Kapitulnik A., Beasley M. R., Carini J. P., Drabeck L. and Grüner G. Weakly coupled grain model of high-frequency losses in high T_c superconducting thin films, *Appl. Phys. Lett.*, **53**(14), 1343–5, 1988
53 Franson J. D. and Mercereau J. E. *J. Appl. Phys.*, **47**, 3261, 1976
54 Portis A. M. and Cooke D. W. *Superconductor Science and Technology*, **5**, 395, 1992
55 Portis A. M. Electrodynamics of high temperature superconductors, *Lecture Notes in Physics*, **48**, World Scientific, 1992
56 Portis A. M., and Cooke D. W. Effect of magnetic fields and variable power on the microwave properties of granular superconductors, *Materials Science Forum*, **130–2**, 315–48, 1993
57 Wosik J., Xie L. M., Chau R., Samaan A., Wolfe J. C., Selvamanickam V. and Salama K. Surface resistance of grain aligned $YBa_2Cu_3O_x$ bulk: Evidence for two kinds of weak links, *Phys. Rev.*, B**47**(14), 8969–77, 1993
58 Wosik J., Xie L. M., Halbritter J., Chau R., Samaan A., Wolfe J. C., Selvamanickam V. and Salama K. Effect of the weak links on the surface resistance of $YBa_2Cu_3O_x$ bulk material, *MRS Spring Meeting, Symposium S, San Francisco CA*, April, 1992
59 Attanasio C., Mariato L. and Vaglio X. Residual surface resistance of polycrystalline superconductors, *Phys. Rev.* B**43**, 6128–32, 1991
60 Clem J. R. *Physica*, C**153–5**, 50, 1988
61 Halbritter J. RF residual losses, surface impedance, and granularity in superconducting cuprates, *J. Appl. Phys.*, **68**(12), 6315–26, 1990
62 Halbritter J. Weak link effects in the surface impedance of cuprate superconductors, *J. of Alloys and Compounds*, **195**, 579–82, 1993
63 Herd J. S., Halbritter J. and Herd K. G. Microwave power dependence of HTS thin film transmission lines, *IEEE Trans. on Applied Superconductivity*, **5**(2), 1991–3, 1995
64 Halbritter J. On extrinsic effects in the surface impedance of cuprate superconductors by weak links, *J. Appl. Phys.*, **71**(1), 339–43, 1992
65 Fagerberg R. and Grepstad J. K. Temperature dependant kinetic inductance of $YBa_2Cu_3O_{7-\delta}$ thin films predicted by the coupled grain model in the strong coupling limit, *J. Appl. Phys.*, **75**(11), 7408–13, 1994
66 Coffey M. W. and Clem J. R. *IEEE Trans. on Magnetics*, **27**, 2136, 1991
67 Coffey M. W. and Clem R. Unified theory of effects of vortex pinning and flux creep upon the rf surface impedance of type II superconductors, *Phys. Rev. Lett.*, **67**(3), 386–9, 1991
68 Coffey M. W. and Clem J. R. Theory of rf magnetic permeability of type II superconductors in slab geometry with an oblique applied static magnetic field, *Phys. Rev.*, B**45**(18), 10527–35, 1992
69 Coffey M. W. and Clem J. R. Theory of microwave transmission and reflection in type II superconductors in the mixed state, *Phys. Rev.* B**48**(1), 342–50, 1993
70 Hebard A. F. *et al. Phys. Rev.*, B**40**, 5243, 1989
71 Portis A. M. *et al. Europhys. Lett.*, **5**, 467, 1988
72 Koshelev A. E. and Vinokur V. M. *Physica*, C**173**(5–6), 469, 1991
73 Yeh N.-C. *Phys. Rev.*, B**43**, 3748, 1991

74 Geshkenbein V. B., Vinokur V. M. and Fehrenbacher R. *Phys. Rev.*, B**43**, 3748, 1991
75 Gittleman J. J. and Rosenblum B. Radio frequency resistance in the mixed state for subcritical currents, *Phys. Rev. Lett.*, **16**(17), 734, 1966
76 Brandt E. H. Penetration of magnetic fields into type II superconductors, *Phys. Rev. Lett.*, **67**(16), 2219, 1991
77 Portis A. M. Electrodynamics of high temperature superconductors, *Lecture Notes in Physics*, **48**, World Scientific, 1992
78 Brandt E. M. *Physica*, C**185–9**, 270, 1991
79 Dulcic A. and Pozek M. Microwave surface impedance in the mixed state of type II superconductors, *Physica*, C**218**, 449–56, 1993
80 Yeh N.-C. High frequency vortex dynamics and dissipation of high-temperature superconductors, *Phys. Rev.*, B**43**(1), 523–31, 1991
81 Chen J. L. and Yang T. J. Flux flow of Abrikosov vortices in type II superconductors, *Phys. Rev.*, B**50**(1), 319–323, 1994
82 Owliaei J., Sridhar S. and Talvacchio J. Field dependent crossover in the vortex response at microwave frequencies in $YBa_2Cu_3O_{7-\delta}$ films, *Phys. Rev. Lett.*, **69**(23), 3366–9, 1992
83 Mahel M. Electrodynamics of type II superconductors. Theory of microwave losses, *Solid State Communications*, **90**(7), 447–50, 1994

2
Superconducting transmission lines

2.1 Introduction

In 1947 Pippard[1] pointed out that a wave would be slowed when it propagated along a superconducting transmission line. This effect is due to the increase in inductance of the transmission line because of the penetration of the external magnetic field into the superconductor. The effect increases as the proportion of the magnetic field inside the superconductor increases relative to the proportion of the external magnetic field. This is not the only effect of using superconductors in transmission lines. Provided the transmission line propagates in a TEM mode, a superconducting transmission line is dispersionless, due to the penetration depth not varying with frequency. This is in contrast with a normal conductor where the skin depth is a function of frequency, and increasing the frequency has the effect of reducing the skin depth and hence increasing the velocity due to the decrease of internal inductance. However, for application purposes the most important effect of using superconductors is the very low loss of the transmission line.

This chapter looks at superconducting transmission lines in some detail. Section 2.2 considers the wide microstrip or parallel plate superconducting transmission line. This transmission line is one of the simplest and is close to the type of transmission line used in many applications. Because of its simple nature, the wide microstrip can be analysed and considerable understanding of the effects of using superconductors can be gained. Although the wide microstrip can be used for some application purposes, other types of transmission line are sometimes more appropriate and some are discussed in Section 2.4. Following this, a discussion of some approximate methods and numerical methods for the evaluation of superconducting transmission line performance is given. A description is given of a numerical algorithm to deduce the parameters of an arbitrary-shaped planar superconducting transmission line.

The final section of this chapter discusses the propagation of very narrow pulses on superconducting transmission lines.

2.2 Wide microstrip with complex impedance boundaries

The wide microstrip line is shown in Figure 2.2.1. It consists of two infinitely long plates of conducting or superconducting material separated by a distance h. The width of the plates is w such that $w \gg h$, implying that approximately all the field is contained within the region between the plates and that no fringe fields exist outside the plates.

It can easily be shown than when the plates are perfect conductors the capacitance and inductance per unit length are[2]

$$C = \frac{\varepsilon w}{h} \quad \text{and} \quad L = \frac{\mu h}{w} \tag{2.2.1}$$

resulting in a velocity of propagation and a characteristic impedance of

$$c = \frac{1}{\sqrt{(LC)}} = \frac{c_0}{\sqrt{\varepsilon_r}} \quad \text{and} \quad Z_0 = \sqrt{\left(\frac{L}{C}\right)} = \frac{h}{w}\sqrt{\left(\frac{\mu}{\varepsilon}\right)} \tag{2.2.2}$$

Here c_0 is the velocity of light in a vacuum. These expressions are accurate for a wide microstrip in the perfectly conducting case, and are a good approximation for a superconducting wide microstrip if $h \gg \lambda$ and $t \gg \lambda$. However, if field penetration into the superconductor is important, then the analysis needs to be modified.

The first theoretical consideration of a superconducting wide microstrip was by Swihart[3] in 1961. The solution obtained by Swihart is for a wide microstrip bounded by two finite thickness superconducting films. In this case the geometry is such that Maxwell's equations are directly soluble for the fields within the dielectric and superconductor and the use of the six boundary conditions completes the solution. The velocity and attenuation on the trans-

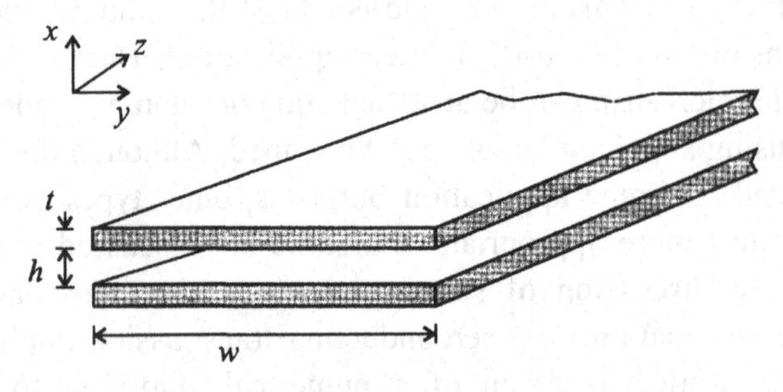

Figure 2.2.1. The wide microstrip line.

mission line can then be calculated. This method can easily be extended for a larger number of layers, for example to consider the effect of buffer layers[4,5] on the propagation. However, the expressions become increasingly more complex and unwieldy as the number of layers increases. An alternative method is described in this section which takes into account any number of layers of superconductor and/or dielectric.

A good understanding of the wide microstrip is important before discussion of more complex transmission lines, and so the wide microstrip is discussed here in some detail. The formulation not only allows the effects of superconductors to be included but also allows for the inclusion of the effects of buffer layers with lossy dielectrics, as well as including any number of superconducting or dielectric layers. The formulation calculates the complex propagation constant and characteristic impedance of a wide microstrip, but instead of having perfectly conducting walls, as discussed above, this derivation assumes that the walls have a complex surface impedance Z_p. Substitution for Z_p can then be made for the surface impedance of the superconductor, the surface impedance of thin films or other surface impedances, depending upon the problem under investigation. This method is extremely powerful, as will be seen later.

The most general solution to the wave equation given the upper and lower infinitely extending boundaries of the wide microstrip is

$$\mathbf{E} = E_x(x,z)\mathbf{a}_x + E_z(x,z)\mathbf{a}_z \tag{2.2.3}$$

$$\mathbf{H} = H_y(x,z)\mathbf{a}_y \tag{2.2.4}$$

This is a transverse magnetic (TM) field, where the magnetic field only has a component directed between the plates normal to the direction of propagation. The electric field also has a component down the transmission line, usually representing the loss down the waveguide.

In the following analysis it is instructive to start from Maxwell's equations. Sinusoidal time dependence is assumed and, following convention, will not be explicitly written into each equation. Also, it is assumed that there is no free charge between the plates. Maxwell's equations are first simplified by substituting Equations (2.2.3) and (2.2.4) in turn, as outlined below.

Substituting Equation (2.2.3) into

$$\nabla \cdot \mathbf{E} = 0 \tag{2.2.5}$$

gives

$$\frac{\partial E_z}{\partial z} = -\frac{\partial E_x}{\partial x} \tag{2.2.6}$$

Substituting Equation (2.2.4) into

$$\nabla \cdot \mathbf{B} = 0 \tag{2.2.7}$$

gives

$$\frac{\partial H_y}{\partial y} = 0 \tag{2.2.8}$$

Substituting Equations (2.2.3) and (2.2.4) into

$$\nabla \times \mathbf{E} = -\frac{\partial \mathbf{B}}{\partial t} \tag{2.2.9}$$

gives

$$H_y = \frac{1}{j\omega\mu}\left(\frac{\partial E_z}{\partial x} - \frac{\partial E_x}{\partial z}\right) \tag{2.2.10}$$

Substituting Equations (2.2.3) and (2.2.4) into

$$\nabla \times \mathbf{H} = \mathbf{J} + \frac{\partial \mathbf{D}}{\partial t} \tag{2.2.11}$$

gives

$$E_x = -\frac{1}{\sigma + j\omega\varepsilon}\frac{\partial H_y}{\partial z} \tag{2.2.12}$$

Equations (2.2.6), (2.2.8), (2.2.10) and (2.2.12) can be modified further. Equation (2.2.10) can be differentiated with respect to z, and Equation (2.2.6) with respect to x. Then $\partial^2 E_z/\partial x\,\partial z$ can be eliminated between the resulting equations, and then using Equation (2.2.12) gives

$$E_x = \frac{1}{\gamma^2}\left(\frac{\partial^2 E_x}{\partial x^2} + \frac{\partial^2 E_x}{\partial z^2}\right) \tag{2.2.13}$$

where the propagation constant of the medium between the plates is

$$\gamma^2 = j\omega\mu(\sigma + j\omega\varepsilon) \tag{2.2.14}$$

The procedure for the solution to the problem is to first use Equation (2.2.13) in order to find E_x. Once E_x is known, Equation (2.2.6) can be used to find E_z and then Equation (2.2.10) to find H_y, thus finding all the electromagnetic field components. The solution of Equation (2.2.13) for the field between the plates of the wide microstrip can now be written as

$$E_x = (C^+ e^{\gamma_x x} + C^- e^{-\gamma_x x})e^{-\gamma_z z} \tag{2.2.15}$$

The first term represents waves travelling in the positive x direction and the second term represents waves travelling in the negative x direction. C^+ and C^- are undetermined constants so far. The propagation down the microstrip in the z-direction is also assumed to be sinusoidal. Substituting Equation (2.2.15) into Equation (2.2.6) results in

$$E_z = \frac{\gamma_x}{\gamma_z}(C^+ e^{\gamma_x x} - C^- e^{-\gamma_x x})e^{-\gamma_z z} \tag{2.2.16}$$

Finally, substitution of Equations (2.2.15) and (2.2.16) into Equation (2.2.10) yields

$$H_y = \frac{\gamma_x^2 + \gamma_z^2}{j\omega\mu\gamma_z}(C^+ e^{\gamma_x x} + C^- e^{-\gamma_x x})e^{-\gamma_z z} \tag{2.2.17}$$

Equations (2.2.15), (2.2.16) and (2.2.17) represent the solution to the problem. All that remains to be done is to find how the different values of the propagation constants are connected and then use the boundary conditions to determine the values of the constants C. Substituting the assumed value of E_x, Equation (2.2.15), into Equation (2.2.13) leads to

$$\gamma^2 = \gamma_x^2 + \gamma_z^2 \tag{2.2.18}$$

The boundary conditions are now applied. The field components normal to the surface of the plates at both the upper and lower boundaries must be equal. This implies that, using Equation (2.2.15), $C^+ = C^-$. With this, Equations (2.2.15)–(2.2.17) can now be summarised as

$$E_x = 2C \cosh(\gamma_x x)e^{-\gamma_z z} \tag{2.2.19}$$

$$E_z = 2C\frac{\gamma_x}{\gamma_z} \sinh(\gamma_x x)e^{-\gamma_z z} \tag{2.2.20}$$

$$H_y = 2C\frac{\gamma_x^2 + \gamma_z^2}{j\omega\mu\gamma_z} \cosh(\gamma_x x)e^{-\gamma_z z} \tag{2.2.21}$$

In order to keep the solution as general as possible, the most general boundary condition will be used. This is when the upper and lower strip surfaces have the same surface impedance as Z_p. This condition implies that

$$\frac{E_z}{H_y} = -Z_p \text{ at } \pm h/2 \tag{2.2.22}$$

The negative sign is because the propagation is in the $-\mathbf{a}_x$ direction, which can be verified by calculation of the Poynting vector. Using this condition and by dividing Equation (2.2.20) by Equation (2.2.21) and using Equation (2.2.18) on the microstrip boundary at $x = \pm h/2$, one finds

$$Z_p = -\frac{j\omega\mu\sqrt{(\gamma^2 - \gamma_z^2)}}{\gamma^2} \tanh(\sqrt{(\gamma^2 - \gamma_z^2)}h/2) \tag{2.2.23}$$

Provided the surface impedance is known, the propagation constant in the z direction down the microstrip waveguide (γ_z) can be found. Of course, the height of the guide h needs to be known, as do the properties of the material filling the gap in order to determine γ. It should be noted that this expression reduces to the preceding expression for a microstrip with perfectly conducting walls if Z_p is made zero, that is $\gamma_z = \gamma$. This leads to a useful approximation by assuming that $\gamma^2 \approx \gamma_z^2$, and is an excellent approximation, provided that the microstrip is not too lossy or too wide. Equation (2.2.23) reduces to

$$\gamma_z = \gamma\sqrt{\left(1 - j\frac{2Z_p}{\omega\mu h}\right)} \tag{2.2.24}$$

We also need to find the characteristic impedance of the transmission line; this is given by

$$Z_0 = \frac{V}{I} = \frac{\int \mathbf{E}\cdot d\mathbf{l}_E}{\oint \mathbf{H}\cdot d\mathbf{l}_H} \tag{2.2.25}$$

Here the integration for the electric field is just between the plates and the integration for the magnetic field is around one of the plates. Using the equations for the electric and magnetic fields (Equations (2.2.19) and (2.2.21) respectively), and the approximation that $\gamma_z \approx \gamma$, this results in

$$Z_0 = \frac{j\omega\mu\gamma_z}{\gamma^2}\frac{h}{w} \tag{2.2.26}$$

Again, this reduces to the previously derived expression for a microstrip with perfectly conducting walls if $Z_p = 0$, implying that $\gamma_z = \gamma$, as discussed above.

This general problem is now solved using Equations (2.2.19), (2.2.20), and (2.2.21), giving the electromagnetic field distribution, Equation (2.2.24), the guide propagation constant, and Equation (2.2.26), the microstrip characteristic impedance. A selection of interesting problems related to superconducting microstrip circuits can be looked at simply by substituting for Z_p in the above expressions.

2.3 Superconducting wide microstrip

In order to find the impedance and propagation constant of a superconducting wide microstrip the surface impedance of the superconductor Z_s, discussed in Section 1.4, needs to be substituted for Z_p into Equations (2.2.24) and (2.2.26). This is trivial and will not be discussed further. However, a more interesting problem, and one of practical significance, is the problem when the microstrip is formed of thin film superconductors. For this case the surface impedance of the film, discussed in Section 1.8, needs to be substituted for Z_p. Equation (1.8.10) in Section 1.8.1 can be written

$$Z_f = R_f + jX_f \tag{2.3.1}$$

Substituting this into Equation (2.2.24) gives the complex propagation constant of the superconducting transmission line:

$$\gamma_z = \gamma\sqrt{\left(1 + \frac{2X_f}{\omega\mu h} - j\frac{2R_f}{\omega\mu h}\right)} \tag{2.3.2}$$

If $R_f \ll X_f$, then this is given approximately by

$$\gamma_z = \gamma \left\{ \sqrt{\left(1 + \frac{2X_f}{\omega\mu h}\right)} - j\frac{R_f}{\omega\mu h}\frac{1}{\sqrt{\left(1 + \frac{2X_f}{\omega\mu h}\right)}} \right\} \tag{2.3.3}$$

This approximation assumes that $\sigma_1 \ll \sigma_2$ and that the film is not too thin; it is discussed in detail in Section 1.8. Substitutions can now be made for R_f and X_f from Section 1.8.1, that is,

$$R_f = R_s \left\{ \coth\left(\frac{t}{\lambda}\right) + \frac{t}{\lambda}\frac{1}{\sinh^2\left(\frac{t}{\lambda}\right)} \right\} \quad \text{and} \quad X_f = X_s \coth\left(\frac{t}{\lambda}\right) \tag{2.3.4}$$

The propagation constant of the dielectric between the two plates ($\gamma = \alpha + j\beta$) is now assumed lossless ($\alpha = 0$) and the velocity of propagation of a wave travelling down the transmission line ($c_z = \omega/\beta_z$) can be calculated:

$$c_z = \frac{1}{\sqrt{(\mu\varepsilon)}}\frac{1}{\sqrt{\left(1 + \frac{2\lambda}{h}\coth\left(\frac{t}{\lambda}\right)\right)}} \tag{2.3.5}$$

An expression for the loss per unit length (α_z) down the guide can be calculated similarly:

$$\alpha_z = \sqrt{\left(\frac{\varepsilon}{\mu}\right)}\frac{R_s}{h}\frac{\left\{ \coth\left(\frac{t}{\lambda}\right) + \frac{t}{\lambda}\frac{1}{\sinh^2\left(\frac{t}{\lambda}\right)} \right\}}{\sqrt{\left(1 + \frac{2\lambda}{h}\coth\left(\frac{t}{\lambda}\right)\right)}} \tag{2.3.6}$$

To complete the picture, the impedance of the microstrip can be calculated using Equation (2.2.26) and the same approximations, but with $\alpha \to 0$, resulting in

$$Z_0 = \sqrt{\left(\frac{\mu}{\varepsilon}\right)}\frac{h}{w}\sqrt{\left(1 + \frac{2\lambda}{h}\coth\left(\frac{t}{\lambda}\right)\right)} \tag{2.3.7}$$

These equations are well known[3] but the method used to calculate them here is rather general. The approximations made in the derivation above have been discussed in detail. For example, it is now quite straightforward to see how similar expressions can be calculated if the dielectric inside the wide microstrip is lossy; the approximation of $\alpha = 0$, just before Equation (2.3.5), need not have been made. Alternatively, for example, if the dielectric material above and below the superconducting thin film boundaries of the microstrip contributed to the loss, then Z_f for the thin film at the start of this section could have been replaced by Equation (1.8.16) in Section 1.8.2.

Figures 2.3.1 and 2.3.2 show the effect of varying the distance h between the plates for a number of different thickness superconducting films. Figure 2.3.1 uses Equation (2.3.5) to calculate the reduced velocity c_z/c_0 and Figure 2.3.2 uses Equation (2.3.6) to calculate the attenuation in the microstrip line. It can

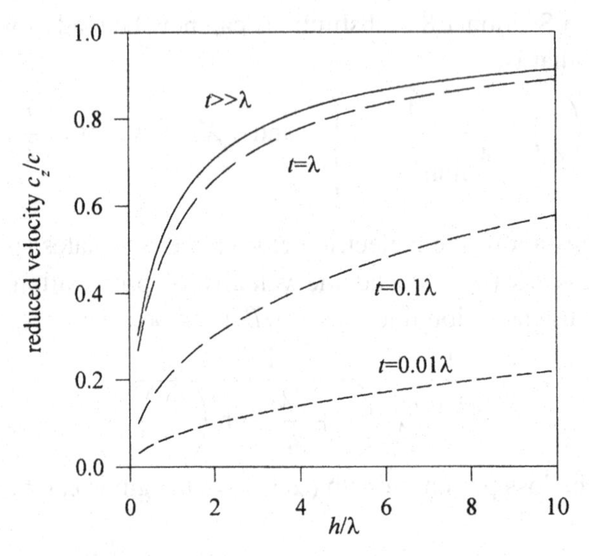

Figure 2.3.1. The change in velocity (c_z/c_0) as a function of h, the spacing between the plates, for a number of different superconductor thicknesses.

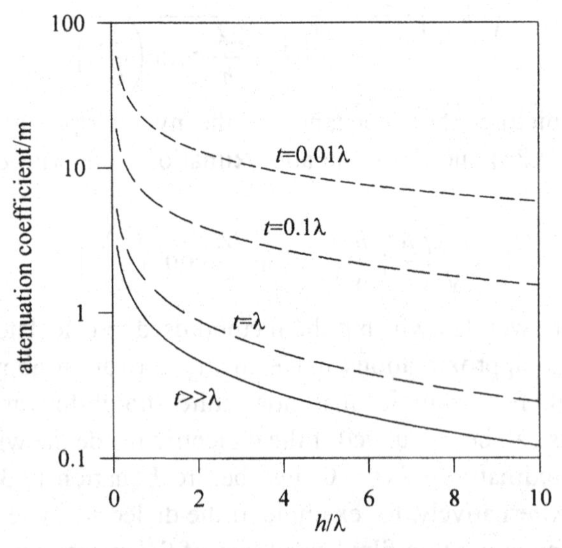

Figure 2.3.2. The attenuation in a superconducting wide microstrip as a function of spacing between the plates for a number of different superconductor thicknesses (assuming $R_s = 100\ \mu\Omega$).

be seen from Figure 2.3.1 that as the spacing h decreases below a penetration depth the velocity can be substantially reduced. This occurs for both thick and thin superconducting films, with the thinner films having a lower velocity than the thicker films. As discussed above, this is due to relative increased penetration of the magnetic field into the film, increasing the inductance without affecting the capacitance of the microstrip. The reduction in velocity for the thinner films is due to the enhancement of the external fields outside the microstrip due to complete penetration through the thin film. However, accompanying this reduction in velocity is an increase in attenuation, as shown in Figure 2.3.2. As the thickness of the films falls below a penetration depth the attenuation increases. In addition, for thinner films, leakage through the film also gives an increase in attenuation. It is clear that if this technique is being used to reduce the velocity and hence miniaturise a device, then careful consideration must also be given to the losses. This problem is discussed further in Chapter 5, along with examples of filter designs which use kinetic inductance for miniaturisation.

2.4 Other planar transmission lines

The calculations on the wide microstrip discussed in Section 2.3 show in detail the predominant effects of using superconductors in microwave transmission lines, and the results are very useful for producing estimates of these effects on the transmission of energy. The equations can in fact be used directly in designing filters when the thickness of the substrate is small compared with the width of the line. This is the case in a new type of superconducting microwave filter discussed in Section 5.6. However, transmission lines do not normally fall into the wide microstrip category, and more accurate estimates of their properties are usually required. Although microstrip is widely used in the microwave industry it is not the only interesting transmission line; some other examples are shown in Figure 2.4.1.

Each of the transmission line types shown in Figure 2.4.1 has specific advantages in certain application areas, and most have been investigated for application with superconductors. Microstrip is the industry standard and a large amount of information about it is available in the form of publications and CAD. However, it requires double-sided deposition of the superconductor. Also, for dense circuits it requires thin substrates to maintain reasonable impedances and to reduce the coupling between different parts of the circuit. The coplanar line consists of two ground planes and a central signal line in between. It has the advantage that only one-sided deposition of the superconductor is required and crosstalk between lines can be reduced by narrowing

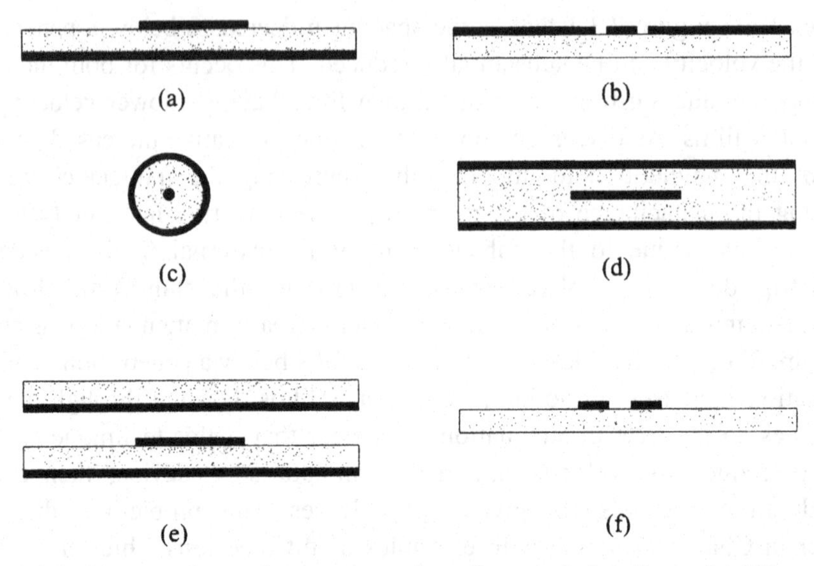

Figure 2.4.1. Cross-sections of some examples of transmission lines. (*a*) Microstrip. (*b*) Coplanar. (*c*) Coaxial. (*d*) Stripline or tri-plate. (*e*) Inverted microstrip. (*f*) Coplanar strips.

the ground plane to signal line spacing. The problem with the coplanar line is the imbalance in the ground planes; although they are connected together at some points in the circuit, the potential of the ground planes can become different on each side, resulting in unwanted modes and interference. These problems can be overcome to a certain extent by placing bonding wires between the ground planes, but this increases the cost and complexity. The stripline shown in Figure 2.4.1 (d) is similar to a microstrip; only an extra ground plane is placed above the signal line. This has an advantage over the microstrip in that radiation losses are eliminated. The propagation is pure TEM and the velocity of transmission is just $c_0/\sqrt{\varepsilon_r}$, whereas for the coplanar line it is of the order $c_0/\sqrt{((\varepsilon_r + 1)/2)}$. This velocity reduction makes smaller circuits possible. The problem with stripline is the difficulty of construction. Two substrates are required to be sandwiched together; small air gaps cause perturbations of the impedance and velocity and it is difficult to form contacts to the external electronics. Two other transmission lines are shown in Figure 2.4.1: the inverted microstrip and the parallel coplanar strip. The inverted microstrip is useful if the dielectric substrate has high loss, and the electric field is now mainly in the air gap between the substrates, thus reducing the dielectric losses. Like the stripline this circuit is difficult to fabricate, requiring tight mechanical tolerances. The parallel coplanar strip line removes the problem of the imbalanced ground planes of the coplanar lines but it is difficult to produce with reasonable impedance values.

All these transmission line types have been studied widely using normal conductors and design equations are available from as far back as 1942.[6] Since then, many books and a considerable amount of CAD have become available for their design. These design tools have good accuracy and can be used for estimates of superconducting transmission line performance without modification. Whether design equations for use with normal conductors or specific design techniques for superconductors are used depends upon the application and the accuracy required. For example, for a transmission line with large effects due to kinetic inductance, conventional design synthesis is useless. Fine tuning when accurate responses are required has to be done by numerical techniques and/or experimental evaluation. Some approximate methods are now discussed which help bridge the gap between the conventional methods of transmission line analysis for normal conductors and numerical analysis for superconductors.

2.5 Incremental inductance rule for superconductors

The incremental inductance rule is used to calculate the losses in a transmission line once the inductance of the transmission line is known. This eliminates separate calculations for the series resistance or the attenuation constant. From first principles, the calculation for the inductance of a transmission line is usually more straightforward than the calculation for the losses. The calculation for the inductance can be analytic or numerical for a transmission line; either way the incremental inductance rule can be used to calculate the losses.

Equation (1.4.18) gives the surface impedance of a superconductor when $\sigma_2 \gg \sigma_1$. For any transmission line where the penetration of the field is small compared with all the conductor dimensions, the internal impedance can be written as

$$Z_i = GZ_s = G\sqrt{\left(\frac{\omega\mu}{\sigma_2}\right)}\left(\frac{\sigma_1}{2\sigma_2}+j\right) \tag{2.5.1}$$

G is a geometrical factor depending upon the effective area over which the current is distributed. From Equation (2.5.1) the internal resistance, reactance and inductance are

$$R_i = X_i\frac{\sigma_1}{2\sigma_2} = \omega L_i\frac{\sigma_1}{2\sigma_2} \tag{2.5.2}$$

Substitution for the penetration depth and the surface resistance using Equations (1.4.15) and (1.4.20) gives

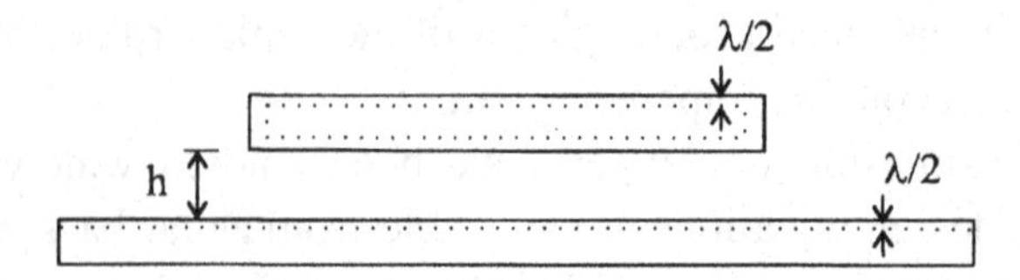

Figure 2.5.1. The microstrip used in the incremental inductance calculation.

$$R_i = R_s \frac{L_i}{\mu\lambda} \tag{2.5.3}$$

Hence, given the surface inductance of the transmission line and knowing the penetration depth and the surface resistance, the internal resistance R_i of the transmission line can be calculated. The internal resistance is, of course, identical to the series resistance of the transmission line. This method assumes that all the loss is due to current flow within the superconductor and does not arise from radiation or losses in the dielectric. Equation (3.2.25) can be used to calculate the attenuation constant. Equation (2.5.3) can be rewritten in a more general sense as

$$Z_i = Z_s \frac{L_i}{\mu\lambda} \tag{2.5.4}$$

When Wheeler first discussed the incremental inductance rule,[7] in terms of normal conductors, he used the expression for the external inductance in order to calculate L_i and hence the losses. Figure 2.5.1 shows how this is achieved with the example of a superconducting microstrip. The external inductance is calculated for the microstrip, then an approximation for the internal inductance can be obtained by using the same calculation but reducing the conductor size by $\lambda/2$; the difference between the two is the internal inductance. Equation (2.5.3) can then be used to calculate the resistance, and hence deduce the attenuation.

This method of calculating the internal inductance is only useful when the thickness of the superconductor is much greater than the penetration depth, and may not be appropriate for transmission lines of very thin film superconductors. However, the next section describes a method which can be used for thin superconducting films.

2.6 Phenomenological loss equivalent method

The problem with the incremental inductance rule described in the preceding section is that if the penetration depth is of the order of the film thickness or larger, then it is no longer applicable. For superconducting thin films it is quite often the case that the films are thin and the incremental inductance rule can no longer be applied. However, another approximate method exists, that is, the phenomenological loss equivalent method (PEM),[8] which takes the current distribution into account. The method is based on finding an equivalent strip which has the same internal impedance as the more complex transmission line of interest, as depicted in Figure 2.6.1.

The PEM method allows the loss, velocity and impedance of any transmission line to be approximated if the inductance and capacitance are known. The resultant parameters include the effects of the internal current distribution, and the film thickness may be less than a penetration depth. The inductance and capacitance can be calculated by conventional means, using the many equations in the literature, commercial CAD programs or numerical programs.[9]

The calculation of the equivalent strip is in fact quite straightforward. The internal impedance of both conductors is defined to be the same. This is the same no matter what the penetration depth, so the width of the conductor can be calculated assuming that the penetration depth is small (the thickness of the strip will then be used to take the field penetration into account). Equation (2.5.4) gives the internal impedance of the transmission line. The internal impedance of the strip is

$$Z_i = \frac{Z_s}{W_e} \tag{2.6.1}$$

Equating Equations (2.5.4) and (2.6.1) gives the equivalent width as

$$W_e = \frac{\mu\lambda'}{L_i} = \frac{1}{G} \tag{2.6.2}$$

where G is the geometric factor given in Equation (2.5.1); λ' rather than λ has been used here because in order to calculate W_e the penetration depth is assumed to be small compared with the other dimensions and need not necessarily relate to the actual penetration depth. The incremental inductance rule can be used to calculate L_i. A better way to calculate G is to take the limit as λ' goes to zero, giving[8]

$$G = \frac{1}{\mu}\sum_j \frac{\partial L}{\partial n} \tag{2.6.3}$$

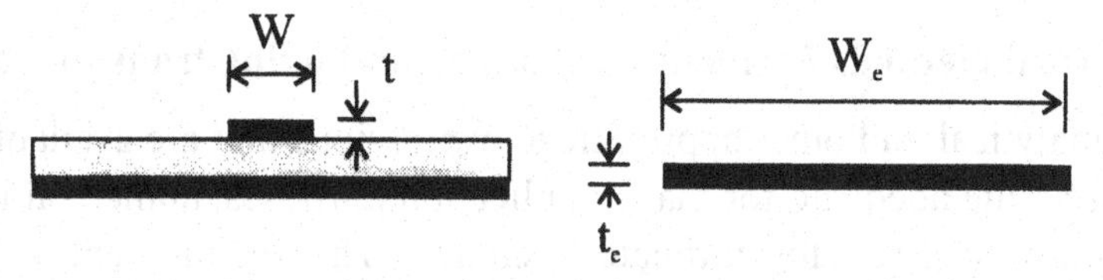

Figure 2.6.1. Microstrip line and the equivalent strip used in PEM.

Here L is the total external inductance and n is the distance normal to the conductor surface. The summation over j is just adding up the contributions from the different sides of the transmission line. This is just another way of writing the incremental inductance rule, but this time it is in the limit as the penetration depth tends to zero. The next part of the problem is to calculate the equivalent thickness t_e. This can be achieved by equating the internal impedances when the current fully penetrates both the transmission line under investigation and the equivalent strip. In this case

$$Z_i = \frac{1}{\sigma A} \tag{2.6.4}$$

where A is the effective cross-sectional area of the conductor. In the case of the microstrip in Figure 2.6.1, this is just the area Wt of the conductor. For the equivalent strip the internal impedance when there is full penetration is given by

$$Z_i = \frac{1}{\sigma W_e t_e} \tag{2.6.5}$$

Equating Equations (2.6.4) and (2.6.5) gives the equivalent thickness as

$$t_e = \frac{A}{W_e} = AG \tag{2.6.6}$$

Now that all the parameters of the equivalent strip are known, it can be used to model the internal impedance of the transmission line for different depths of penetration. Equation (1.8.3) gives the surface impedance of a film which is modified by the width to give the internal impedance of the single strip:

$$Z_i = \frac{Z_s}{W_e} \coth\left(\gamma \frac{A}{W_e}\right) = Z_s G \coth(\gamma A G) \tag{2.6.7}$$

which is the same as the internal impedance of the transmission line in question. Knowing the internal impedance and the external inductance and capacitance (the capacitance changes negligibly with field penetration) allows all the properties of the transmission line to be calculated.

This method is extremely convenient as it allows conventional analysis to be used with thin films. Its use with superconductors in both experimental and analytical work has confirmed its applicability.[8–11]

2.7 Numerical methods for analysis of superconducting transmission lines

Although analytical and other approximate expressions that are available in the literature are quite adequate for many application purposes, numerical methods are sometimes required for ultimate accuracy. The requirement is for the electromagnetic field to be calculated in the vicinity of the transmission line and from this the impedance, velocity and attenuation can be derived. A vast array of numerical methods is available for conventional conductors, and many of these are available in convenient CAD packages, where both two- and three-dimensional field solvers are available. With the increasing interest in superconductivity, a number of commercial CAD packages allow for complex impedance boundary conditions, thus allowing superconductors to be included in the problem, although they do not necessarily solve problems when there is penetration into films which are of the order of, or less than, a penetration depth thick. Conventional numerical techniques used in CAD and other numerical packages are able to cope with the problems of superconductivity by the introduction of complex conductivity.

A number of specific techniques used with superconductors have been discussed in the literature; these include finite difference,[12] partial wave synthesis or mode matching[13–15] and variational techniques.[16] Probably the most accurate technique is the spectral domain volume integral equation (SDVIE) method.[17,18] Both the mode matching technique and the SDVIE method take into account the current distribution within the volume of the superconductor, giving results which show the effect of kinetic inductance. SDVIE is computationally inefficient and other related techniques are available in order to speed up computation.[19–24] Another technique is described below which accurately predicts the current distribution on an arbitrary-shaped planar transmission line.

Figure 2.7.1 shows the cross-section of a transmission line which is a coplanar transmission line with two ground planes and a central conductor. The transmission line is split into a number of small rectangular patches and each can be thought of as behaving as a separate transmission line. The solution procedure assumes that they are coupled together, energy being distributed between the transmission lines due to their proximity. Although Figure 2.7.1 shows a coplanar line, the solution procedure described below can have any arbitrary geometry and any number of conductors in the transmission lines. The method of calculation of the properties of transmission lines which follows was first developed by Weeks[25] and subsequently adapted by Sheen[26] for superconductors. It has been used for both microstrip and coplanar lines.[27]

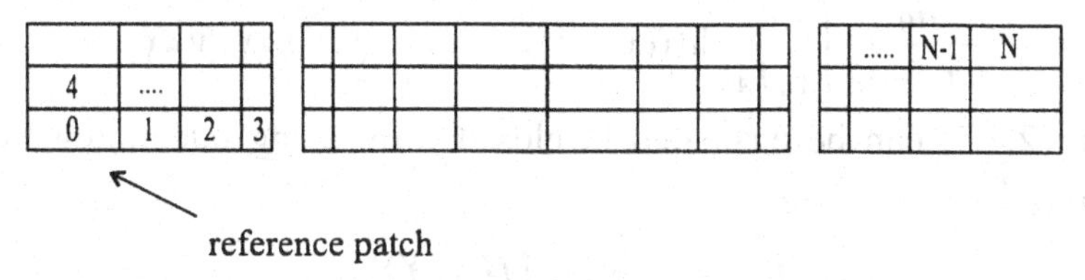

Figure 2.7.1. Example of the distribution of patches in a coplanar line.

Assuming that each line is coupled to every other line and that the total number of lines is $N+1$, then the system will obey the coupled transmission line equation

$$-\frac{\partial v_n}{\partial z} = \sum_{m=1}^{N} (r_{mn} + j\omega l_{mn}) i_m \tag{2.7.1}$$

Here v_n and i_m are the voltage (relative to the reference patch) and current vectors respectively. Each contains $N+1$ elements; they represent the voltage and currents on the individual patches. Also, r_{mn} and l_{mn} are self- and mutual resistance and inductance matrices respectively; they are per unit length parameters. Knowing the voltages on the transmission line and the inductances and resistances of the system, Equation (2.7.1) can be used to calculate the current on each individual patch by inversion of the impedance matrix. The current must be assumed to be uniform on each of the patches. An additional requirement is that the total current in the system must be zero. Defining a reference patch with subscript zero results in

$$i_0 = -\sum_{n=1}^{N} i_n \tag{2.7.2}$$

The resistance matrix is given by

$$r_{mn} = \mathrm{Re}\left(\frac{1}{\sigma A_0} + \delta_{mn}\frac{1}{\sigma A_n}\right) \tag{2.7.3}$$

Here A_n is the cross-sectional area of the nth patch and $\delta_{mn} = 1$ if $m = n$ and 0 otherwise. Also, σ is the complex conductivity of the superconductor and is assumed to be the same for each patch. However, this need not be the case; for example a normal conducting ground plane can be integrated with a superconducting signal line; in fact each patch may have a different conductivity. Similarly, the internal inductance of each patch is given by

$$_k l_{mn} = \frac{1}{\omega}\,\mathrm{Im}\left(\frac{1}{\sigma A_0} + \delta_{mn}\frac{1}{\sigma A_n}\right) \tag{2.7.4}$$

The geometric inductance consisting of the external and mutual inductance between the elements is more complex to calculate but can be found from stored energy concepts and is given by[25]

$$_g l_{mn} = -\frac{\mu_0}{4\pi A_m A_n}\int_{A_m}\int_{A_n} \ln\left((x-x')^2 + (y-y')^2\right) dx\, dy\, dx'\, dy' \tag{2.7.5}$$

Equation (2.7.5) can be evaluated in closed form using integration by parts, and is given by

$$_g l_{mn} = \frac{\mu_0}{4\pi A_m A_n}\left[f(X,Y)\right]_{limits} \tag{2.7.6}$$

where

$$f(X,Y)=\frac{X^4-6X^2Y^2+Y^4}{24}\ln(X^2+Y^2)$$
$$-\frac{XY}{3}\left(Y^2\tan^{-1}\left(\frac{X}{Y}\right)+X^2\tan^{-1}\left(\frac{Y}{X}\right)\right)+\frac{25}{24}X^2Y^2 \quad (2.7.7)$$

where $X=x-x'$ and $Y=y-y'$. The limits of the integration of Equation (2.7.5) and carried through to Equation (2.7.6) are the coordinates of the corners of the mth and nth patches. The total inductance matrix containing the internal kinetic inductance, the external self-inductance and the mutual inductance is given by

$$l_{mn}={}_k l_{mn}+{}_g l_{mn}-{}_g l_{m0}-{}_g l_{0n}+{}_g l_{00} \quad (2.7.8)$$

Now that the impedances have been defined in Equation (2.7.1), the solution for the currents proceeds by inversion of the impedance matrix

$$i=-z^{-1}\frac{\partial v}{\partial z}=j\gamma(r+j\omega l)^{-1}v \quad (2.7.9)$$

where γ is the complex propagation of the network. The resistance R and inductance L of the whole transmission line can now be calculated from

$$I_{tot}^2 R=\sum_n i_n^2 R_n \quad \text{and} \quad I_{tot}^2 L=\sum_{mn} i_m i_n L_{mn} \quad (2.7.10)$$

where $R_n=\mathrm{Re}\,(1/\sigma A_n)$ and $L_{mn}={}_g l_{mn}+{}_k l_{mn}\cdot I_{tot}$ is the total current carried on the signal or ground lines.

Figure 2.7.2 shows an example of a current distribution calculated using the above technique for a thin film coplanar line.[27] There are 405 patches in this calculation and the minimum patch size is around 1/10 of a penetration depth and is located where the current varies most rapidly, that is, at the edges. The central conductor is 200 μm wide with a gap of 73 μm, and the film is 0.35 μm thick. A penetration depth of 0.15 μm was used in the calculation. This computational technique has been combined with the Ginzburg–Landau theory[28] in order to model non-linear effects. This is clearly important for transmission lines with significant peaking of the current at the edges.

2.8 Pulse propagation on superconducting transmission lines

For transmission lines made from normal conductors which operate in a pure TEM mode, the dominant mechanism for dispersion is the change in skin depth with frequency. However, for similar superconducting transmission lines at low temperatures, and for frequencies well below the gap frequency, the penetration

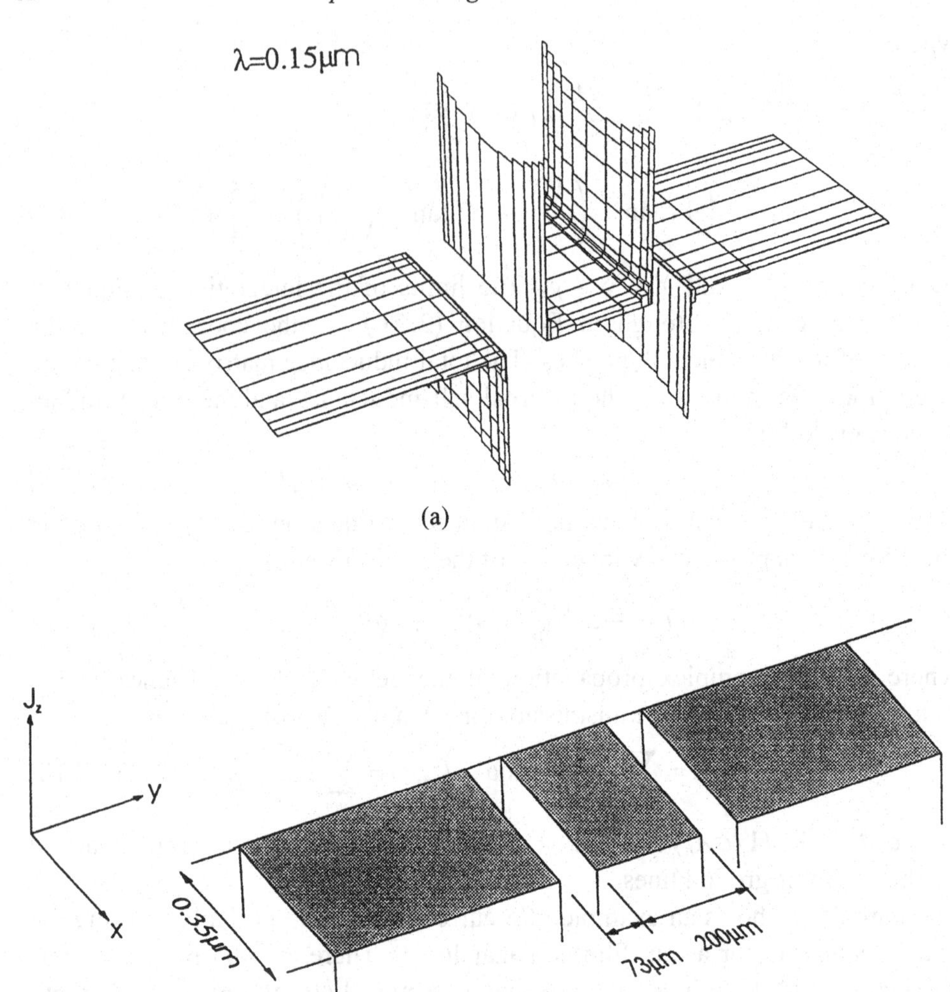

Figure 2.7.2. (*a*) Volume current distribution in the cross-section of the coplanar line shown in (*b*).

depth is frequency independent. Thus in this regime the superconducting transmission line is dispersionless, and this can be assumed for all practical microwave devices operating up to frequencies up to several hundred gigahertz, although the upper frequency limit depends upon the gap frequency of the superconductor used. If a quasi-TEM transmission line is used, such as a coplanar line or microstrip, then dispersion will occur due to the geometry itself. This occurs because part of the electromagnetic field is outside the dielectric and this travels at a different velocity from that inside the dielectric. Any change in the field distribution as a function of frequency will produce a different phase velocity. For example, if a coaxial cable is used, no dispersion occurs because the dominant mode is TEM. However, if the frequency components in a pulse travelling in a coaxial cable are high enough to excite the non-TEM modes, then dispersion will occur. However, small-diameter cables can be used to increase the non-TEM mode cut-off frequencies, as indicated in Section 3.5.

If very short pulses (< 1 ps) are propagated on superconducting transmission lines, then some frequency components within the pulse could be of the order of the gap frequency. Additional attenuation and dispersion of these higher-frequency components results, causing the pulse to change shape as it travels down the transmission line. Because the penetration depth changes for the higher-frequency components, the inductance of the line also changes, producing a slowing at higher frequencies. This can give a high-frequency oscillating tail to the pulse as it travels.

The theoretical aspects of this pulse propagation were first studied by Kautz[29] using the Mattis–Bardeen complex conductivity determined from the BCS theory. The two-fluid models are of no use in this case because the effects of the energy gap need to be considered. Kautz gives the theoretical results of pulse broadening for low-temperature superconductors and pulses from 0.2 to 1.0 ps. Significant broadening of the short pulses is observed even over distances as short as a few hundred microns, although the longer pulses obviously require a larger distance. Kautz assumes a wide microstrip so that there is no modal dispersion because the transmission line is pure TEM; the dielectric thickness used is 100 nm. The effect can be increased by narrowing the ground plane to signal line spacing, thereby increasing the effect of the kinetic inductance upon the velocity.

Experimental work has been done on both low-temperature and high-temperature superconductors and pulse broadening has been observed and modelled. Much of the work has been done on the coplanar line where the results have been complicated by intrinsic or modal dispersion, although this has been modelled in the comparisons with theory. Losses in the dielectric and radiation loss are also problems; the latter occurs especially at the higher frequencies which are, of course, the ones of more interest. Generation of short pulses can be accomplished using sharp optical pulses incident on a photoconductive switch. Using a 100 fs pulse from a colliding pulse mode-locking dye laser, Nuss has produced electrical pulses of 800 fs with 500 mV amplitude on a superconducting coplanar transmission line.[30,31] The equivalent bandwidth of the pulse is about 300 GHz. Dispersion is observed due to both the superconductor and the coplanar line, and is compared with a similar gold transmission line, in addition to the theoretical development by Kautz. Figure 2.8.1

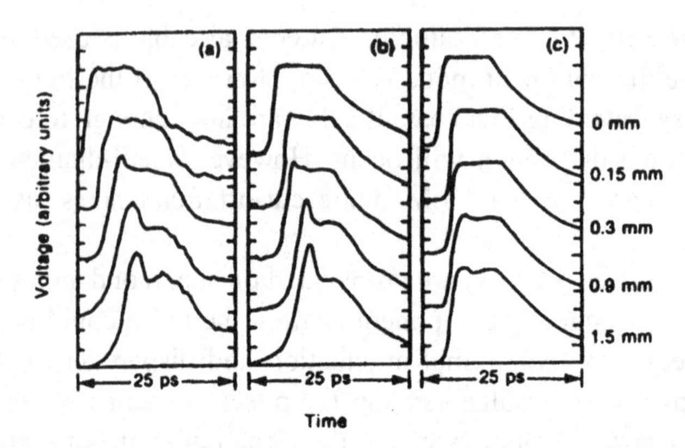

Figure 2.8.1. Sequence of pulse shapes in an indium stripline in the superconducting state ($T = 1.8$ K, $T_c = 3.4$ K, $2\Delta/\hbar \approx 230$ GHz, $\rho_n = 2.7 \times 10^{-6}$ Ω cm at 4.2 K). (*a*) Experimental results. (*b*) Computation with Mattis–Bardeen conductivity and modal dispersion taken into account. (*c*) Modal dispersion neglected. (Taken from reference [35].)

shows some similar results by the group at the University of Rochester.[32–35] Here experimental results are shown, together with the calculated distortion, in the first case with both the coplanar line and superconducting dispersion, and in the second with the coplanar line modal dispersion neglected. As can be seen, both dispersion mechanisms are important in this regime.

2.9 References

1 Pippard A. B. The surface impedance of superconductors and normal metals at high frequencies. III The relation between impedance and superconducting penetration depth, *Proc. Roy. Soc.* (London), **A191**, 399–415, 1947
2 Kraus J. D. *Electromagnetics*, McGraw-Hill, 1984
3 Swihart J. C. Field solution for a thin film superconducting strip transmission line, *J. Appl. Phys.*, **32**(3), 461–9, 1961
4 Abbas F. and Davis L. E. Propagation coefficient in a superconducting asymmetrical parallel-plate transmission line with a buffer layer, *J. Appl. Phys.*, **73**(9), 4494–9, 1993
5 Abbas F., Davis L. E. and Gallop J. C. Field solution for a thin film superconducting parallel plate transmission line, *Physica*, C**215**, 132–44, 1993
6 Cohn S. B. Problems in strip transmission lines, *IRE Proc. Microwave Theory and Techniques*, **30**, 412–24, 1942
7 Wheeler H. A. Formulas for the skin effect, *Proc. IRE*, **30**(9), 412–24, 1942
8 Lee H. Y. and Itoh T. Phenomenological loss equivalence method for planar quasi-TEM transmission lines with a thin normal conductor or superconductor, *IEEE Trans. on Microwave Theory and Techniques*, **37**(12), 1904–9, 1989

9 Kong K.-S., Kuo C. W., Kitazawa T. and Itoh T. *Electron. Lett.*, **26**(19), 1558–9, 1990

10 Baiocchi O. R., Kong K.-S., Ling H. and Itoh T. Effects of superconducting losses in pulse propagation on microstrip lines, *IEEE Microwave and Guided Wave Letters*, **1**(1), 2–4, 1991

11 Antsos D., Chew W., Riley L., Hunt B. D., Foote M. C., Bajuk L. J., Rascoe D. L. and Cooley T. W. Modelling of planar quasi-TEM superconducting transmission lines, *IEEE Trans. on Microwave Theory and Techniques*, **40**(6), 1128–31, 1992.

12 Alsop L. E., Goddman A. S., Gustavson F. G. and Miranker W. L. A numerical solution of a model for a superconductor field problem, *J. Comp. Phys.*, 216–39, 1979

13 Young B. and Itoh T. Loss reduction in superconducting transmission lines, *IEEE MTT-S Digest*, pp. 453–6, 1988

14 Kessler J., Dill R. and Russer P. Field theory investigations of high T_c superconducting coplanar waveguide transmission line resonators, *IEEE Trans. on Microwave Theory and Techniques*, **39**(9), 1566–73, 1991

15 Rauch W., Gornik E., Sölkner G., Valenzuela A. A., Fox F. and Behner H. Microwave properties of $YBa_2Cu_3O_{7-x}$ thin films studied with coplanar transmission line resonators, *J. Appl. Phys.*, **73**(4), 1866–72, 1993

16 Chang W. H. Numerical calculation of the inductances of a multi-superconductor transmission line, *IEEE Trans. on Magnetics*, **MAG-17**, 764–6, 1981

17 Lee L. H., Ali S. M. and Lyons W. G. Full wave characterisation of high-T_c superconducting transmission lines, *IEEE Trans. on Applied Superconductivity*, **2**(2), 49–57, 1992

18 Lee L. H., Ali S. M., Lyons W. G., Oates D. E. and Goettee J. D. Analysis of superconducting transmission line structures for passive microwave device applications, *IEEE Trans. on Applied Superconductivity*, **3**(1), 2782–7, 1993

19 Pond J. M., Krowne C. M. and Carter W. L. On the application of complex resistive boundary conditions to model transmission lines consisting of very thin superconductors, *IEEE Trans. on Microwave Theory and Techniques*, **37**(1), 181–9, 1989

20 Van Deventer T. E., Katehi P. B., Josefowicz J. Y. and Rensch D. B. High frequency characterisation of high temperature superconducting thin film lines, *Electromagnetics*, **11**, 255–68, 1991

21 Lee L. H., Lyons G. W., Orlando T. P., Ali S. M. and Withers R. S. Full-wave analysis of superconducting microstrip lines on anisotropic substrates using equivalent surface impedance approach, *IEEE Trans. on Microwave Theory and Techniques*, **41**(12), 2359–67, 1993

22 Klopman B. B. G., Gerritsma G. J. and Rogalla H. The propagation characteristics of wave-guiding structures with very thin superconductors; application to coplanar waveguide $Ya_2Cu_3O_{7-x}$ resonators, *IEEE Trans. on Microwave Theory and Techniques*, **41**(5), 781–91, 1993

23 Safavi-Naeini S., Faraji-Dana R. and Chow Y. L. Studies of edge current densities in regular and superconducting microstrip lines of finite thickness, *IEE Proc.*, **140**(5), 361–6, 1993

24 Lee L. H., Orlando T. P. and Lyons W. G. Current distribution in superconducting thin-film strips, *IEEE Trans. on Applied Superconductivity*, **4**(1), 41–4, 1994

25 Weeks W. T., Wu L. L., McAllister M. F. and Singh A. Resistive and inductive skin effect in rectangular conductors, *IBM J. Res. Develop.*, **23**(6), 652–60, 1979

26 Sheen D. M., Ali S. M., Oates D. E., Withers R. S. and Kong J. A. Current

distribution, resistance and strip inductance for superconducting strip transmission lines, *IEEE Trans. on Applied Superconductivity*, **1**(2), 108–15, 1991
27 Porch A., Lancaster M. J. and Humphreys R. G. The coplanar resonator technique for determining the surface impedance of $YBa_2Cu_3O_7$ thin films, *IEEE Trans. on Microwave Theory and Techniques*, **43**(2), 306–14, 1995
28 Lam C.-W., Sheen D. M., Ali S. S. and Oates D. E. Modelling the nonlinearity of superconducting strip transmission lines, *IEEE Trans. on Applied Superconductivity*, **2**(2), 58–66, 1992
29 Kautz R. L. Picosecond pulses on superconducting striplines, *J. Appl. Phys.*, **49**(1), 308–14, 1978
30 Nuss M. C. and Goossen K. W. Investigation of high temperature superconductors with terahertz bandwidth electrical pulses, *IEEE Journal of Quantum Electronics*, **25**(12), 2596–607, 1989
31 Nuss M. C., Mankiewich P. M., Howard R. E., Straughn B. L., Harvey T. E., Brandle C. D., Berkstresser G. W., Goossen K. W. and Smith P. R. Propagation of terahertz bandwidth electrical pulses on $YBa_2Cu_3O_{7-\delta}$ transmission lines on lanthanum aluminate, *Appl. Phys. Lett.*, **54**(22), 2265–7, 1989
32 Dykaar D. R., Sobolewski R., Chwalek J. M., Whitaker J. F., Hsiang T. Y., Mourou G. A., Lathrop D. K., Russek S. E. and Buhrman R. A. High frequency characterisation of thin film Y–Ba–Cu oxide superconducting transmission lines, *Appl. Phys. Lett.*, **52**(17), 1444–6, 1988
33 Dykaar D. R., Sobolewski R., Chwalek J. M., Whitaker J. F., Hsiang T. Y., Mourou G. A., Lathrop D. K., Russek S. E. and Buhrman R. A. High frequency characterisation of thin film Y–Ba–Cu oxide superconducting transmission lines, Erratum, *Appl. Phys. Lett.*, **52**(23), 2003, 1988
34 Chwalek J. M., Dykaar D. R., Whitaker J. F., Sobolewski R., Gupta S., Hsiang T. Y. and Mourou G. A. Ultra fast response of superconducting transmission lines, *IEEE Trans. on Magnetics*, **25**(2), 814–17, 1989
35 Hsiang T. Y., Whitaker J. F., Sobolewsky R., Dykaar D. R. and Mourou G. A. Propagation characteristics of picosecond electrical transits on coplanar striplines, *Appl. Phys. Lett.*, **51**(19), 1551–3, 1987

3

Superconducting cavity resonators

3.1 Introduction

A cavity resonator is any structure which is able to contain an oscillating electromagnetic field. In general, it has a number of distinct resonant frequencies which are dependent upon the geometry of the cavity. If an oscillating field is set up within a cavity it will gradually decay because of losses. These losses may be due to a number of phenomena but are mainly due to (i) the finite conductivity of the walls of the cavity, (ii) losses in any dielectric material within the cavity or (iii) radiation out of any apertures in the walls. The main reason for using superconductors in the construction of a cavity is to reduce the conduction loss and reduce this decay to a minimum. The decay of the oscillating field is inversely proportional to the quality factor of the cavity, and is discussed extensively below for the characterisation of cavities.

Cavities are of interest for a number of reasons. By producing a superconducting sample part of a cavity resonator the losses in the superconductor can be deduced. Cavities also form the main functional part of many microwave filters. By coupling a number of cavities together, a filter can be constructed. A high-Q cavity also forms the main feedback element in a microwave oscillator. In fact, cavity resonators form the most fundamental building block in the majority of microwave circuits and hence will be discussed in some detail in this chapter.

Section 3.2 first considers some fundamental definitions and relations for cavities, and how to calculate them from the electromagnetic fields in a cavity. If a cavity has superconducting walls, then any change in temperature causes the penetration depth to change, which can be viewed as an effective change in the size of the cavity, resulting in a change in resonant frequency. This change can be calculated by a simple perturbation technique and is discussed in Section 3.2.4. The following section deals in general with the problem of cavities made out of lengths of transmission lines. After this introductory material, a number of cavities are discussed in detail. In Section 3.3 the cylindrical cavity is reviewed, and a description of the calculation of the electromagnetic fields within the cavity is given, resulting in the expression for the conductor quality factor. Much detail is given for this calculation, and working through it will provide the reader with a detailed understanding of the methods of solution of cavity problems in general. In fact, much of the calculation can be used in the following two sections on dielectric and coaxial resonators. After a discussion of the helical resonator in Section 3.6, transmission line cavities are examined. This includes a cavity resonator based on stripline and microstrip. The final section of this chapter looks at the coplanar resonator in some detail. The emphasis here is placed on using the resonator for surface impedance measurements. This resonator is in fact one of the most sensitive techniques for deducing the surface resistance of a superconducting sample, and a method is also described which allows the penetration depth of the superconductor to be deduced directly. This can be done without making any assumptions about the value of the low-temperature penetration depth or using curve fitting to existing theories.

3.2 Microwave cavities and quality factors

The quality factor for a cavity resonator is defined by

$$Q_0 = \omega \frac{\textit{Energy stored in the cavity}}{\textit{Average power lost}} \tag{3.2.1}$$

The losses in a cavity arise due to a number of mechanisms. The most important are usually the losses associated with the conduction currents in the cavity walls. However, depending on the construction of the cavity, losses due to the finite loss tangent of any dielectric filling material or radiation from an open aperture can be equally, or more, important. If a cavity contains a number of loss mechanisms, then the total quality factor can be found by adding these losses together, resulting in

$$\frac{1}{Q_0} = \frac{1}{Q_c} + \frac{1}{Q_d} + \frac{1}{Q_r} \tag{3.2.2}$$

where Q_c, Q_d and Q_r are the conductor, dielectric and radiation quality factors respectively. If other loss mechanisms are present, for example conversion of energy to phonons, then this loss can be added in a similar fashion. It is quite straightforward to see how this equation can be deduced from the definition of Q given above. This discussion so far takes no account of how energy is coupled into and out of the cavity. If any connection is made to the cavity to couple energy in, then it will also remove energy from the cavity. This can be seen as another loss mechanism and can also have an associated 'external' quality factor which will be termed Q_e. It follows that

$$\frac{1}{Q_l} = \frac{1}{Q_0} + \frac{1}{Q_e} \tag{3.2.3}$$

Here Q_l is the loaded quality factor and this is what is usually measured; Q_0 is the unloaded quality factor and this is a characteristic of the cavity itself and is not dependent on the coupling to the cavity. If the cavity is used for measuring the losses of materials it is this quality factor which is of interest and the external quality factor must be extracted.

All the above quality factors will now be discussed and general expressions derived in order to calculate them if the electromagnetic field distribution is known within the cavity.

3.2.1 Conductor quality factor

The conductor Q is defined as

$$Q_c = \omega \frac{\textit{Energy stored in the cavity}}{\textit{Average power lost due to ohmic dissipation}} \tag{3.2.4}$$

With reference to Poynting's theorem for superconductors given in Section 1.7, this can be written

$$Q_c = \omega \frac{\int_{cavity} \mu |\mathbf{H}|^2 \, dv}{\int_{conductor} \sigma_1 |\mathbf{E}|^2 \, dv} \tag{3.2.5}$$

The upper integration, representing the energy stored in the cavity, is an integration throughout the whole volume of the cavity. The lower integration, representing the average power lost, is over the conductor volume only. (Although σ_1 has been used, this expression applies equally to normal conductors if σ_1 is replaced by σ.) If the cavity is formed by a number of different conductors, then each one can be treated separately, integrated over and then added together. If $E = ZH$ is used, then Equation (3.2.5) becomes

$$Q_c = \frac{\omega \mu}{\sigma_1 |Z|^2} \frac{\int_{cavity} |\mathbf{H}|^2 \, dv}{\int_{conductor} |\mathbf{H}|^2 \, dv} \tag{3.2.6}$$

The integration into the conductor, normal to its surface, can be carried out, resulting in

$$Q_c = \frac{2\omega\mu}{\lambda\sigma_1|Z|^2} \frac{\int_{cavity} |\mathbf{H}|^2 \, dv}{\int_{conductor} |\mathbf{H}|^2 \, ds} \tag{3.2.7}$$

The integration is now over the surface of the conductor. The factor of 2 arises due to the integration of the exponentially decaying magnetic field distribution inside the conductor. An approximation is involved here if the surface impedance discussed in Section 1.4 is used; that is, the curvature of the cavity walls should not be too large so that the plane wave approximation is invalid. This is usually a very good approximation because the skin depth or penetration depth is small compared with the radius of curvature of the conductors. However, for very small cavities, where the dimensions of the cavities are of the same order as the penetration depth, care must be taken in implementing Equation (3.2.7).

By substitution of the general expression for the penetration depth (Equation (1.4.15)), this equation is reduced to

$$Q_c = \frac{\omega\mu}{R_s} \frac{\int_{cavity} |\mathbf{H}|^2 \, dv}{\int_{conductor} |\mathbf{H}|^2 \, ds} = \frac{\Gamma}{R_s} \tag{3.2.8}$$

where Γ is the geometry factor of the cavity. This is an expression used extensively in the calculation of cavity Q factors. Only the magnetic field distribution within the cavity is required to deduce its conductor quality factor.

3.2.2 Dielectric quality factor

If part of the cavity contains some dielectric material, for example the substrate holding a superconducting film, then there will be a loss of energy by dissipation in this dielectric. This dissipation can be associated with a dielectric Q

$$Q_d = \omega \frac{\textit{Energy stored in the cavity}}{\textit{Average power lost in dielectric}} \tag{3.2.9}$$

This can be written in terms of the field integrals

$$Q_d = \frac{1}{\tan(\delta)} \frac{\int_{cavity} \mu|\mathbf{H}|^2 \, dv}{\varepsilon \int_{dielectric} |\mathbf{E}|^2 \, dv} \tag{3.2.10}$$

where the integration over $|\mathbf{H}|^2$ is throughout the whole cavity and the integration over $|\mathbf{E}|^2$ is just within the volume of the dielectric within the cavity. The loss tangent of the dielectric is given by

$$\tan(\delta) = \frac{\sigma_d}{\omega\mu} \tag{3.2.11}$$

and σ_d is the conductivity of the dielectric. If the dielectric completely fills the cavity, then the two integrations in Equation (3.2.10) cancel as they both represent the energy stored in the cavity. The dielectric Q is then simply

$$Q_d = \frac{1}{\tan(\delta)} \tag{3.2.12}$$

Examples of some loss tangents for dielectrics used in conjunction with superconductors are given in Appendix 2.

3.2.3 Radiation quality factor

If a cavity is not fully enclosed, then it will radiate; it is important to investigate this radiation and consider whether it has any effect upon the cavity function. A radiation quality factor can be defined as

$$Q_r = \omega \frac{\textit{Energy stored in the cavity}}{\textit{Average power radiated}} \tag{3.2.13}$$

The average power radiated is given by the integration of the Poynting vector over any spherical surface surrounding the cavity:

$$Q_r = \omega\mu \frac{\int_{cavity} |\mathbf{H}|^2 \, dv}{\oint_{sphere} \frac{1}{2} \operatorname{Re}(\mathbf{E} \times \mathbf{H}) \, ds} \tag{3.2.14}$$

If the integration takes place in free space far away from the cavity, then the expression for the quality factor can be reduced to

$$Q_r = 2\omega\sqrt{\mu\varepsilon} \frac{\int_{cavity} |\mathbf{H}|^2 \, dv}{\oint_{sphere} |\mathbf{H}|^2 \, ds} \tag{3.2.15}$$

3.2.4 Cavity perturbations

It is sometimes convenient to know what effect the addition or removal of a small volume of cavity material has on the resonant frequency of a cavity. This could be the introduction of a sample, or the change in the effective size of a sample due to a change in superconducting penetration depth. A general expression for the frequency shift, $\Delta\omega = \omega_0 - \omega$, with the removal of a small volume Δv of any cavity is[1]

$$\frac{\Delta\omega}{\omega} = \frac{\int_{\Delta v} \mu|\mathbf{H}|^2 - \varepsilon|\mathbf{E}|^2\, dv}{\int_{v} \mu|\mathbf{H}|^2 - \varepsilon|\mathbf{E}|^2\, dv} \tag{3.2.16}$$

The numerator here is an integration over the added volume, and is proportional to the electric and magnetic energies removed by the perturbation. The denominator is over the whole volume of the cavity and is proportional to the total energy stored. For a superconducting cavity where the perturbation is a change in penetration depth, the change in magnetic energy results in a frequency shift given by

$$\frac{\Delta\omega}{\omega} = -\frac{\Delta X_s}{2\omega\mu} \frac{\int_{walls} |\mathbf{H}|^2\, ds}{\int_{cavity} |\mathbf{H}|^2\, dv} \tag{3.2.17}$$

where the integration in the numerator is over the cavity walls and the integration in the denominator is throughout the whole cavity volume. Here ΔX_s is the change in the surface reactance of the cavity walls due to the change in penetration depth. This, together with Equation (3.2.8), results in

$$R_s + j\Delta X_s = \frac{\Gamma}{Q_c} - 2j\Gamma\frac{\Delta\omega}{\omega} \tag{3.2.18}$$

where the geometry factor is given by

$$\Gamma = \omega\mu \frac{\int_{cavity} |\mathbf{H}|^2\, dv}{\int_{walls} |\mathbf{H}|^2\, ds} \tag{3.2.19}$$

With the calculation of this geometry factor, the frequency shift and Q_c are directly related to the change in surface reactance and surface resistance respectively. Equation (3.2.18) assumes that penetration depth is small compared with the dimensions of the cavity. Equation (3.2.18) is useful in relating the measured quantities of the shift in frequency and the Q of the cavity to the fundamental quantities of surface resistance and change in surface reactance.

3.2.5 Quality factors of cavities based on TEM waveguides

A cavity based on TEM waveguides can be formed by having a short section of TEM transmission line such as coaxial cable or stripline. The cavity is formed by having either a short circuit or an open circuit at the ends of the transmission line. This creates a standing wave in the section of the transmission line by the reflections from the ends. For two open- or short-circuit ends, for example, the field variations correspond to an integer number of half-wavelengths. The equations given above can be used to calculate the Q of such a cavity, but can be immediately simplified because the field in the z direction along the transmission line is always given by

$$\mathbf{H} = \mathbf{H}_t \cos(\beta z) \cos(\omega t) \tag{3.2.20}$$

$$\mathbf{E} = \mathbf{E}_t \sin(\beta z) \cos(\omega t) \tag{3.2.21}$$

where $\mathbf{H}_t$ and $\mathbf{E}_t$ represent the transverse fields across the waveguide. The integration along the waveguide in Equation (3.2.8) can now be carried out, resulting in

$$Q_c = \frac{\omega\mu}{R_s} \frac{\int_{\textit{across cavity}} |\mathbf{H}_t|^2 \, ds}{\int_{\textit{around conductor}} |\mathbf{H}_t|^2 \, dl} \tag{3.2.22}$$

The integrations are now reduced to a surface integration of the transverse field and a line integration of $\mathbf{H}$ along the surface of the conductor in the cross-section of the guide. Similar arguments apply for the dielectric quality factor. Equation (3.2.22) is very general and can be used for any TEM guide. Another approach to the problem is from a circuit point of view and is detailed below.

The general propagation constant for a TEM transmission line is[2]

$$\gamma = \surd((R + j\omega L)(G + j\omega C)) \tag{3.2.23}$$

where R is the line series resistance per unit length, G is the line shunt conductance per unit length, and L and C are the inductance and capacitance per unit length respectively. For low-loss lines, this can be approximated by

$$\gamma = j\omega\surd(LC)\left(1 - j\frac{R}{2\omega L} - j\frac{G}{2\omega C}\right) \tag{3.2.24}$$

The real part of this expression is just the loss per unit length down the waveguide, and is given by

$$\alpha = \frac{R}{2}\sqrt{\left(\frac{C}{L}\right)} + \frac{G}{2}\sqrt{\left(\frac{L}{C}\right)} = \frac{R}{2Z_0} + \frac{GZ_0}{2} \tag{3.2.25}$$

The impedance of a TEM line is given by[2]

$$Z_0 = \sqrt{\left(\frac{R + j\omega L}{G + j\omega C}\right)} \tag{3.2.26}$$

Substitution of Equation (3.2.26) (when expanded for low-loss transmission lines) into Equation (3.2.25), with only the lowest-order terms taken, results in

$$\alpha = \frac{R}{2Z_0} + \frac{GZ_0}{2} \tag{3.2.27}$$

For a series resonant circuit the conductor quality factor is given by

$$Q_c = \frac{\omega L}{R} \tag{3.2.28}$$

For the moment, ignoring the dielectric loss characterised by the conductance G, and combining the first term in Equation (3.2.27) with Equation (3.2.28) and the imaginary part of Equation (3.2.24) (which is the propagation constant (ω/c)), results in

$$Q_c = \frac{1}{\alpha_c}\frac{\omega}{2c} = \frac{\pi}{\alpha_c \lambda} \tag{3.2.29}$$

Here λ is the free space wavelength in the guide. An identical expression can be gained by using the second term in Equation (3.2.27) with the Q of a parallel resonant circuit to obtain the dielectric quality factor:

$$Q_d = \frac{\pi}{\alpha_d \lambda} \tag{3.2.30}$$

The above expressions simplify the calculation of the quality factors of TEM transmission lines and are used later in this chapter.

3.3 The cylindrical cavity resonator

The cylindrical cavity resonator has traditionally been used in microwave systems because of its high Q and convenient shape. The cavities are used in many microwave systems such as stabilised oscillators, filters and wave meters and for the measurement of the properties of materials, including superconductors. This section looks at the cylindrical cavity in some detail, deriving the expressions for the fields within the cavity and the resonant frequencies. The fields are then used to calculate the Q. The reason for the detail is that the cylindrical cavity represents an excellent example of a cavity problem, it is not too complicated and can be presented in only a few pages; additionally, it gives an insight into how these types of problem can be solved. The ideas and formulations can be applied to many other cavities such as the dielectric resonator and the coaxial cavity discussed in the following sections. For these cavities detailed derivations are not included, only the results being presented.

In order to find the electromagnetic fields within the cylindrical cavity resonator shown in Figure 3.3.1, the solution of the wave equation for a circular waveguide will first be found. Two ends will then be put on to a section of circular waveguide to form the cavity.

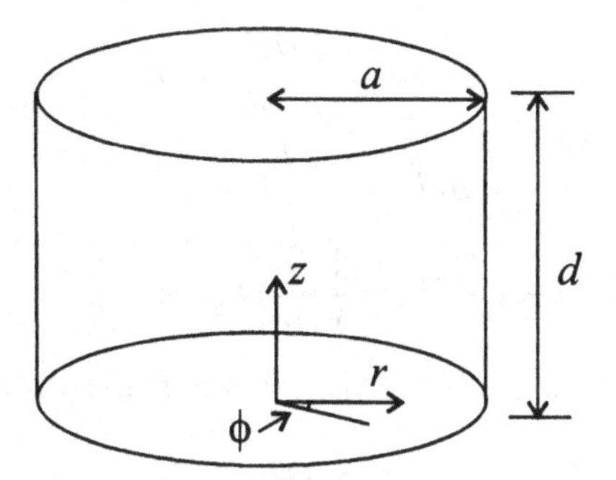

Figure 3.3.1. Cylindrical cavity.

3.3.1 Circular waveguide

Consider the following general expression for the electric and magnetic fields in cylindrical coordinates. The equations assume a wave travelling in the positive z direction which is cosinusoidal:

$$\mathbf{E} = (E_r(r,\phi)\mathbf{a}_r + E_\phi(r,\phi)\mathbf{a}_\phi + E_z(r,\phi)\mathbf{a}_z)e^{-j\beta_z z}e^{j\omega t}$$

$$\mathbf{H} = (H_r(r,\phi)\mathbf{a}_r + H_\phi(r,\phi)\mathbf{a}_\phi + H_z(r,\phi)\mathbf{a}_z)e^{-j\beta_z z}e^{j\omega t} \tag{3.3.1}$$

This expression allows any variation of field amplitude across the circular waveguide, but only a cosinusoidal variation along the guide represented by $e^{-j\beta_z z}$; β_z is the propagation constant in the direction of the guide, and the guide is assumed to be lossless. The objective is to find $E(r, \phi)$, $H(r, \phi)$ and β_z for the boundary conditions and dimensions of the waveguide. Firstly, Maxwell's equations can be modified by substituting Equation (3.3.1) into each in turn. The time variation is dropped in the following derivation but remains implicit in the equations.

In cylindrical coordinates Gauss's law, with the substitution of Equation (3.3.1), becomes

$$\frac{1}{r}\frac{\partial}{\partial r}rE_r + \frac{1}{r}\frac{\partial E_\phi}{\partial \phi} = j\beta_z E_z \tag{3.3.2}$$

Similarly, for the magnetic field:

$$\frac{1}{r}\frac{\partial}{\partial r}rH_r + \frac{1}{r}\frac{\partial H_\phi}{\partial \phi} = j\beta_z H_z \tag{3.3.3}$$

Using the same substitutions in Faraday's law, the following three equations can be obtained. Each equation is derived by comparing the coefficients of $\mathbf{a}_r$, $\mathbf{a}_\phi$ and $\mathbf{a}_z$ after the substitution:

$$-j\omega\mu H_r = \frac{1}{r}\frac{\partial E_z}{\partial \phi} + j\beta_z E_\phi \tag{3.3.4}$$

$$j\omega\mu H_\phi = \frac{\partial E_z}{\partial r} + j\beta_z E_r \tag{3.3.5}$$

$$-j\omega\mu H_z = \frac{1}{r}\frac{\partial}{\partial r} rE_\phi - \frac{1}{r}\frac{\partial E_r}{\partial \phi} \tag{3.3.6}$$

Finally, using Ampère's law in a similar manner, the following three equations are obtained:

$$j\omega\varepsilon E_r = \frac{1}{r}\frac{\partial H_z}{\partial \phi} + j\beta_z H_\phi \tag{3.3.7}$$

$$-j\omega\varepsilon E_\phi = \frac{\partial H_z}{\partial r} + j\beta_z H_r \tag{3.3.8}$$

$$j\omega\varepsilon E_z = \frac{1}{r}\frac{\partial}{\partial r} rH_\phi - \frac{1}{r}\frac{\partial H_r}{\partial \phi} \tag{3.3.9}$$

Now take Equations (3.3.5) and (3.3.7) and eliminate H_ϕ to give

$$E_r = \frac{-j\omega\mu}{\beta^2 - \beta_z^2}\left\{\frac{1}{r}\frac{\partial H_z}{\partial \phi} + \frac{\beta_z}{\omega\mu}\frac{\partial E_z}{\partial r}\right\} \tag{3.3.10}$$

Here the free space propagation constant is $\beta = \omega\sqrt{(\mu\varepsilon)}$. Now take Equations (3.3.4) and (3.3.8) and eliminate H_r to give

$$E_\phi = \frac{j\omega\mu}{\beta^2 - \beta_z^2}\left\{\frac{\partial H_z}{\partial r} - \frac{\beta_z}{\omega\mu}\frac{1}{r}\frac{\partial E_z}{\partial \phi}\right\} \tag{3.3.11}$$

Again use Equations (3.3.5) and (3.3.7) but this time eliminate E_r to give

$$H_\phi = \frac{-j\omega\varepsilon}{\beta^2 - \beta_z^2}\left\{\frac{\partial E_z}{\partial r} + \frac{\beta_z}{\omega\varepsilon}\frac{1}{r}\frac{\partial H_z}{\partial \phi}\right\} \tag{3.3.12}$$

Finally, use Equations (3.3.4) and (3.3.8) again, this time eliminating E_ϕ to give

$$H_r = \frac{j\omega\varepsilon}{\beta^2 - \beta_z^2}\left\{\frac{1}{r}\frac{\partial E_z}{\partial \phi} - \frac{\beta_z}{\omega\varepsilon}\frac{\partial H_z}{\partial r}\right\} \tag{3.3.13}$$

Equations (3.3.10)–(3.3.13) enable the calculation of E_r, E_ϕ, H_r and H_ϕ provided E_z and H_z are known. The solution now proceeds to try to find E_z and H_z from the wave equation and the appropriate boundary conditions. The wave equation in cylindrical coordinates for the z component of the electric field is

$$\frac{1}{r}\frac{\partial}{\partial r}\left(r\frac{\partial E_z}{\partial r}\right) + \frac{1}{r^2}\frac{\partial^2 E_z}{\partial \phi^2} = -(\beta^2 - \beta_z^2)E_z \tag{3.3.14}$$

The solutions obtained will be for transverse magnetic (TM) fields. Using a similar equation to Equation (3.3.14) but for the magnetic field leads to transverse electric (TE) solutions. The solution for each proceeds in exactly the same way, but the following concentrates on TM waves. The solution of Equation (3.3.14) can be assumed to be

$$E_z = f_1(r) f_2(\phi) e^{-j\beta_z z} \tag{3.3.15}$$

The solution proceeds using the standard separation of variables technique. Taking the appropriate derivatives of Equation (3.3.15) and substituting them into Equation (3.3.14) gives

$$\frac{r}{f_1(r)} \left\{ r \frac{\partial^2 f_1(r)}{\partial r^2} + \frac{\partial f_1(r)}{\partial r} \right\} + (\beta^2 - \beta_z^2) r^2 = -\frac{1}{f_2(\phi)} \frac{\partial^2 f_2(\phi)}{\partial \phi^2} \tag{3.3.16}$$

The equation has been separated so that the left-hand side only contains functions of r and the right-hand side only contains functions of ϕ. By putting both sides equal to a constant n^2, two separate differential equations can be obtained. For the right-hand side,

$$-\frac{1}{f_2(\phi)} \frac{\partial^2 f_2(\phi)}{\partial \phi^2} = n^2 \tag{3.3.17}$$

which has the solution

$$f_2(\phi) = K_1 \cos(n\phi) + K_2 \sin(n\phi) \tag{3.3.18}$$

where K_1 and K_2 are constants of integration. The left-hand side of Equation (3.3.16) is

$$\frac{r}{f_1(r)} \left\{ r \frac{\partial^2 f_1(r)}{\partial r^2} + \frac{\partial f_1(r)}{\partial r} \right\} + (\beta^2 - \beta_z^2) r^2 = n^2 \tag{3.3.19}$$

Now make the substitution

$$(\beta^2 - \beta_z^2) r^2 = \beta_t^2 r^2 = w^2 \tag{3.3.20}$$

into Equation (3.3.19), giving

$$w^2 \frac{\partial^2 f_1(z)}{\partial w^2} + w \frac{\partial f_1(z)}{\partial w} + (w^2 - n^2) f_1(z) = 0 \tag{3.3.21}$$

This is Bessel's equation, and after back substitution of Equation (3.3.20) it has a solution[3]

$$f_1(r) = K_3 J_n(\beta_t r) + K_4 Y_n(\beta_t r) \tag{3.3.22}$$

Here J_n and Y_n are Bessel functions of the first and second kind and K_3 and K_4 are constants. Two of the constants can immediately be removed if we observe that Y_n goes to infinity as r tends to zero; then K_4 must be equal to zero for the field in the waveguide centre to be finite. Also, for simplicity, only the $\cos(n\phi)$ variation is taken so that $K_2 = 0$. Under these circumstances, the z-directed electric field is found using Equations (3.3.15), (3.3.18) and (3.3.22):

$$E_z = E_0 J_n(\beta_t r) \cos(n\phi) e^{-j\beta_z z} \tag{3.3.23}$$

E_0 is the combination of constants K_1 and K_3. The boundary conditions are that on the walls of the waveguide the tangential components of the electric field should go to zero, assuming perfect conducting walls. For this we require $J_n(\beta_t r) = 0$ on the walls, which occurs when $\beta_t a = v_{nm}$. Then the field variation is

$$E_z = E_0 J_n\left(\frac{v_{nm} r}{a}\right) \cos(n\phi) e^{-j\beta_z z} \tag{3.3.24}$$

Here v_{nm} is the mth zero of the Bessel function of order n. These values are tabulated in many texts,[3] but a few of the low-order ones are shown in Table 3.3.1. The same boundary condition also defines the z-directed propagation constant from Equation (3.3.20):

$$\beta_t^2 = \beta^2 - \beta_z^2 = \frac{v_{nm}^2}{a^2} \tag{3.3.25}$$

Therefore, given the radius of the waveguide a and the free space propagation constant β, the propagation constant in the z direction can be calculated. This of course also enables the phase velocity ω/β_z to be calculated.

The other components of the fields can now be found using Equations (3.3.10)–(3.3.13), with $H_z = 0$, giving the transverse magnetic fields as

$$E_r = Z_{TM} H_\phi = -E_0 \frac{j\beta_z}{\beta_t} J'_n\left(\frac{v_{nm} r}{a}\right) \cos(n\phi) e^{-j\beta_z z} \tag{3.3.26}$$

$$E_\phi = -Z_{TM} H_r = E_0 \frac{j\beta_z}{\beta_t^2} \frac{n}{r} J_n\left(\frac{v_{nm} r}{a}\right) \sin(n\phi) e^{-j\beta_z z} \tag{3.3.27}$$

where Z_{TM} is the impedance of the transverse magnetic mode given by

$$\frac{E_r}{H_\phi} = -\frac{E_\phi}{H_r} = Z_{TM} = \frac{\beta_z}{\omega\varepsilon} = \frac{\beta_z}{\beta}\eta \tag{3.3.28}$$

where η is the impedance of free space. Equations (3.3.25)–(3.3.28) yield the solution to the problem of the TM modes in a circular waveguide. TE modes are calculated in a similar manner and the fields are given by

$$H_z = H_0 J_n\left(\frac{v'_{nm} r}{a}\right) \cos(n\phi) e^{-j\beta_z z} \tag{3.3.29}$$

Table 3.3.1. *Zeros of the Bessel function v_{nm} and the zeros of the derivative of the Bessel function v'_{nm}*

v_{nm}	$n = 0$	$n = 1$	v'_{nm}	$n = 0$	$n = 1$
$m = 1$	2.40482	3.83171	$m = 1$	3.83170	1.84118
$m = 2$	5.52007	7.01559	$m = 2$	7.01558	5.33144

$$H_\phi = \frac{E_r}{Z_{TE}} = H_0 \frac{j\beta_z}{\beta_t^2} \frac{n}{r} J_n \left(\frac{v'_{nm} r}{a}\right) \sin(n\phi) e^{-j\beta_z z} \tag{3.3.30}$$

$$H_r = -\frac{E_\phi}{Z_{TE}} = -H_0 \frac{j\beta_z}{\beta_t} J'_n \left(\frac{v'_{nm} r}{a}\right) \cos(n\phi) e^{-j\beta_z z} \tag{3.3.31}$$

Here v'_{nm} is the mth zero of the derivative of the Bessel function J'_n, and some values are given in Table 3.3.1. The TE mode impedance is given by

$$\frac{E_r}{H_\phi} = -\frac{E_\phi}{H_r} = Z_{TE} = \frac{\omega\mu}{\beta_z} = \frac{\beta}{\beta_z}\eta \tag{3.3.32}$$

The transverse propagation constant for the TE mode is given by Equation (3.3.25), with v_{nm} replaced by v'_{nm}.

3.3.2 Cylindrical cavity

When two walls are placed on the ends of a piece of a circular waveguide a resonant cavity is formed. On the surface of these end walls, assuming they are perfectly conducting, a perfect reflection occurs. The longitudinal fields within the cavity can be found by the sum of forward- and backward-propagating waves. Thus the fields in Equations (3.3.24), (3.3.26) and (3.3.27) must be modified by replacement of the term $e^{-j\beta_z z}$ by the following standing wave:

$$A^+ e^{-j\beta_z z} + A^- e^{j\beta_z z} \tag{3.3.33}$$

The constants A^+ and A^- represent the amplitudes of the forward- or backward-propagating waves in the cavity and their magnitudes are equal in this case. In addition, the values of E_ϕ and E_r must vanish on the surfaces at the ends of the cavity. For this to occur $\beta_z z$ must equal $l\pi z/d$, where l is an integer. Thus, taking into account the directions of the backward-propagating waves, and using Equations (3.3.24), (3.3.26), (3.3.27) and (3.3.33), the TM fields in a cylindrical cavity are

$$E_z = E_0 J_n \left(\frac{v_{nm} r}{a}\right) \cos(n\phi) \cos\left(\frac{l\pi z}{d}\right) \tag{3.3.34}$$

$$E_r = -E_0 \frac{j\beta_z}{\beta_t} J'_n \left(\frac{v_{nm} r}{a}\right) \cos(n\phi) \sin\left(\frac{l\pi z}{d}\right) \tag{3.3.35}$$

$$H_\phi = -\frac{E_0}{Z_{TM}} \frac{j\beta_z}{\beta_t} J'_n \left(\frac{v_{nm} r}{a}\right) \cos(n\phi) \cos\left(\frac{l\pi z}{d}\right) \tag{3.3.36}$$

$$E_\phi = E_0 \frac{j\beta_z}{\beta_t^2} \frac{n}{r} J_n \left(\frac{v_{nm} r}{a}\right) \sin(n\phi) \sin\left(\frac{l\pi z}{d}\right) \tag{3.3.37}$$

$$H_r = -\frac{E_0}{Z_{TM}}\frac{j\beta_z}{\beta_t^2}\frac{n}{r}J_n\left(\frac{v_{nm}r}{a}\right)\sin(n\phi)\cos\left(\frac{l\pi z}{d}\right) \tag{3.3.38}$$

The constants A have been absorbed into the constant E_0. Similar equations apply for the TE modes and are given below:

$$H_z = H_0 J_n\left(\frac{v'_{nm}r}{a}\right)\cos(n\phi)\sin\left(\frac{l\pi z}{d}\right) \tag{3.3.39}$$

$$E_r = H_0 Z_{TE}\frac{j\beta_z}{\beta_t^2}\frac{n}{r}J_n\left(\frac{v'_{nm}r}{a}\right)\sin(n\phi)\sin\left(\frac{l\pi z}{d}\right) \tag{3.3.40}$$

$$H_\phi = H_0\frac{j\beta_z}{\beta_t^2}\frac{n}{r}J_n\left(\frac{v'_{nm}r}{a}\right)\sin(n\phi)\cos\left(\frac{l\pi z}{d}\right) \tag{3.3.41}$$

$$E_\phi = H_0 Z_{TE}\frac{j\beta_z}{\beta_t}J'_n\left(\frac{v'_{nm}r}{a}\right)\cos(n\phi)\sin\left(\frac{l\pi z}{d}\right) \tag{3.3.42}$$

$$H_r = -H_0\frac{j\beta_z}{\beta_t}J'_n\left(\frac{v'_{nm}r}{a}\right)\cos(n\phi)\cos\left(\frac{l\pi z}{d}\right) \tag{3.3.43}$$

The resonant frequency of the cavity for TM modes can now be calculated simply from Equation (3.3.25):

$$f = \frac{c}{2\pi}\sqrt{\left(\left(\frac{l\pi}{d}\right)^2 + \left(\frac{v_{nm}}{a}\right)^2\right)} \tag{3.3.44}$$

Analysis of the TE modes shows that a similar expression applies for their resonant frequency, except that v_{nm} is replaced by v'_{nm}, where the v'_{nm} are the zeros of the derivative of the Bessel function J'_{nm}. It can be seen that only certain frequencies are allowed for the values of n, m and l. To illustrate this, a mode chart can be drawn. Equations (3.3.44) can be rewritten as

$$(2af_{nml})^2 = c^2\left\{\left(\frac{v_{nm}}{\pi}\right)^2 + \left(\frac{l}{2}\right)^2\left(\frac{2a}{d}\right)^2\right\} \tag{3.3.45}$$

and $(2af_{nml})^2$ is plotted against $(2a/d)^2$. This is shown in Figure 3.3.2 for a number of low-order modes. For any size of cavity the resonant frequency of any of the modes can be read off this chart.

The Q of the cylindrical cavity can now be calculated using Equation (3.2.8), that is,

$$Q_c = \frac{\omega\mu}{R_s}\frac{\int_0^d\int_0^a\int_0^{2\pi}|H_r|^2 + |H_\phi|^2 + |H_z|^2 r\,d\phi\,dr\,dz}{\int_{cavity\ walls}|\mathbf{H}|^2\,ds} \tag{3.3.46}$$

By substitution of the field equations given above, the conductor unloaded

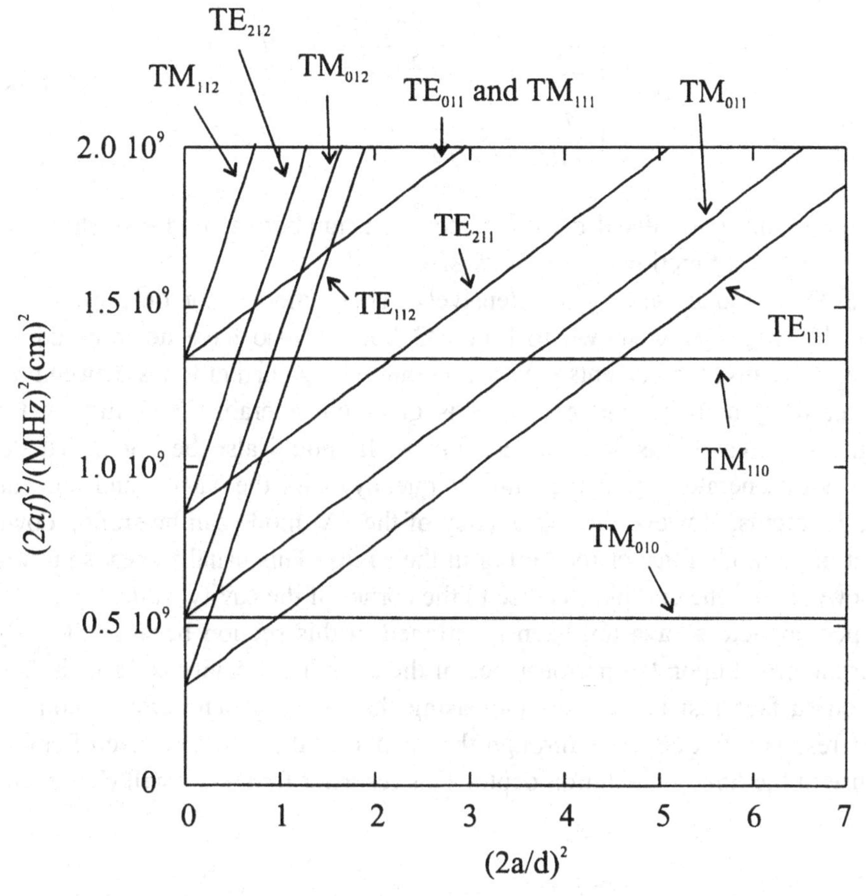

Figure 3.3.2. Mode chart for the cylindrical cavity.

quality factor of a cylindrical cavity can be calculated. The integrations are simple but the algebra is somewhat involved so only the results are presented here. For TE modes this result is

$$Q_{nml} = \frac{\eta}{2R_s} \frac{\left[1 - \left(\dfrac{n}{v'_{nm}}\right)^2\right]\left[v'^2_{nm} + \left(\dfrac{l\pi a}{d}\right)^2\right]^{3/2}}{\left[v'^2_{nm} + \dfrac{2a}{d}\left(\dfrac{l\pi a}{d}\right)^2 + \left(1 - \dfrac{2a}{d}\right)\left(\dfrac{nl\pi a}{v'_{nm}d}\right)^2\right]} \tag{3.3.47}$$

where η is the intrinsic impedance of the material contained in the cavity. For the TM modes

$$Q_{nml} = \begin{cases} \dfrac{\eta}{2R_s} \dfrac{[v_{nm}^2 + (l\pi a/d)^2]^{1/2}}{(1 + 2a/d)} & l \neq 0 \\ \dfrac{\eta}{2R_s} \dfrac{v_{nm}}{(1 + a/d)} & l = 0 \end{cases} \qquad (3.3.48)$$

A graph of the normalised quality factor of a number of modes is shown in Figure 3.3.3 as a function of the cavity size $2a/d$.

The TE_{01l} modes are used extensively. These modes not only possess a reasonably high Q, as shown in Figure 3.3.3, but also have no longitudinal currents. Because the currents are only azimuthal, no current flows between the cylinder body and the end plates. This gives considerable flexibility in the design of the cavity, as will be seen below. It should also be noted that the TE_{0ml} is degenerate (i.e. at the same frequency) with the TM_{1ml} and this can cause problems. However, the frequency of the TM mode can be shifted down by putting a mode filter or mode trap in the cavity. This usually means putting a groove around the end plates close to the corner of the cavity walls.

Superconductors have not been mentioned in this section because the only important effect upon the performance of the cylindrical cavity is through their reduced surface resistance, thus increasing the cavity Q. The effect upon the cavity resonant frequency is through the change in the effective size of cavity because of the finite penetration depth. This resonant frequency will change the

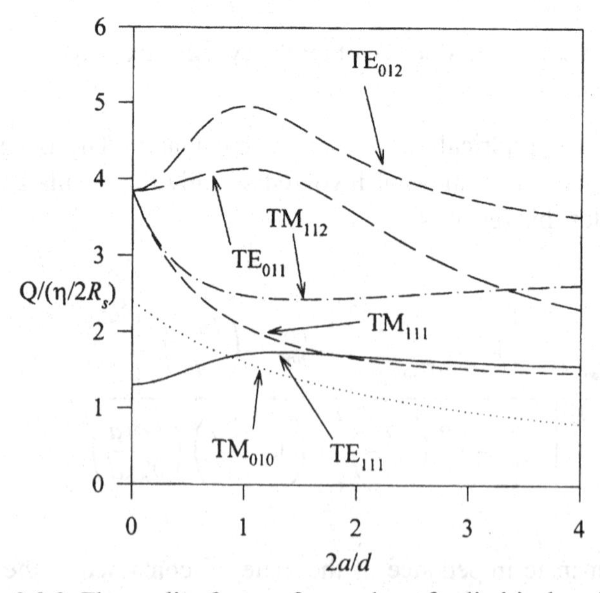

Figure 3.3.3. The quality factor of a number of cylindrical cavity modes.

function of temperature as the penetration depth alters. In the design of cavities, this can usually be ignored, but it has been used to measure the properties of superconducting films, as discussed below.

3.3.3 Measurements of HTS samples using superconducting cylindrical cavities

Cylindrical cavities have become one of the most important devices for measuring the surface impedance of superconducting samples. They have been used at various resonant frequencies for the measurement of thin and thick film, bulk polycrystalline and single crystals. Measurements have been made as a function of temperature, microwave power and d.c. magnetic fields at various laboratories around the world. The following describes some of these techniques; another brief survey is also given by Portis.[4,5]

The most obvious method of measuring the surface impedance, or in fact trying to produce the highest Q for application purposes, is to make the entire cavity from superconducting material. Thick film and bulk polycrystalline material have to be used in the construction of a number of such cavities because of the practical limitations on the deposition of thin films on large curved substrates. Such cavities have been produced by a number of groups, probably one of the first by Kennedy in 1988.[6] The cavity operated at 8 GHz and measured the surface resistance of bulk polycrystalline $YBa_2Cu_3O_{7-x}$ as about 4 mΩ at 20 K. Another cavity[7] operating at 13 GHz gave unloaded Qs of 34 000 at 20 K and 10 000 at 77 K, corresponding to surface resistance values of 23 mΩ and 70 mΩ respectively. An all-YBCO polycrystalline thick film cavity with a Q of 4×10^5 at 10 GHz and 77 K has been used in a down converter.[8] This impressive Q is also accompanied by good power handling capabilities.

Owing to the difficulty of the construction of whole cavities made of HTS, the most common way of determining the surface resistance of a planar sample using a cylindrical cavity is to replace one of the end plates of the cavity with the sample. This is now a standard technique in many laboratories. Bohn[9] describes a number of cavities using the TE_{012} and TE_{011} modes made out of copper and niobium. The cavities provide the ability to measure at frequencies from 2.65 GHz to 29 GHz. Measurements are performed on $Bi_2Sr_2CaCu_2O_x$ and $Bi_4Sr_3Ca_3Cu_6O_x$ samples and, in the particular publication referenced, the 2212 thick films gave a surface resistance of about 20 mΩ at 20 K and 11.6 GHz. Müller[10,11] and Klein[12] have used end plate replacement for the measurement of HTS surface impedance extensively. Figure 3.3.4 shows two of

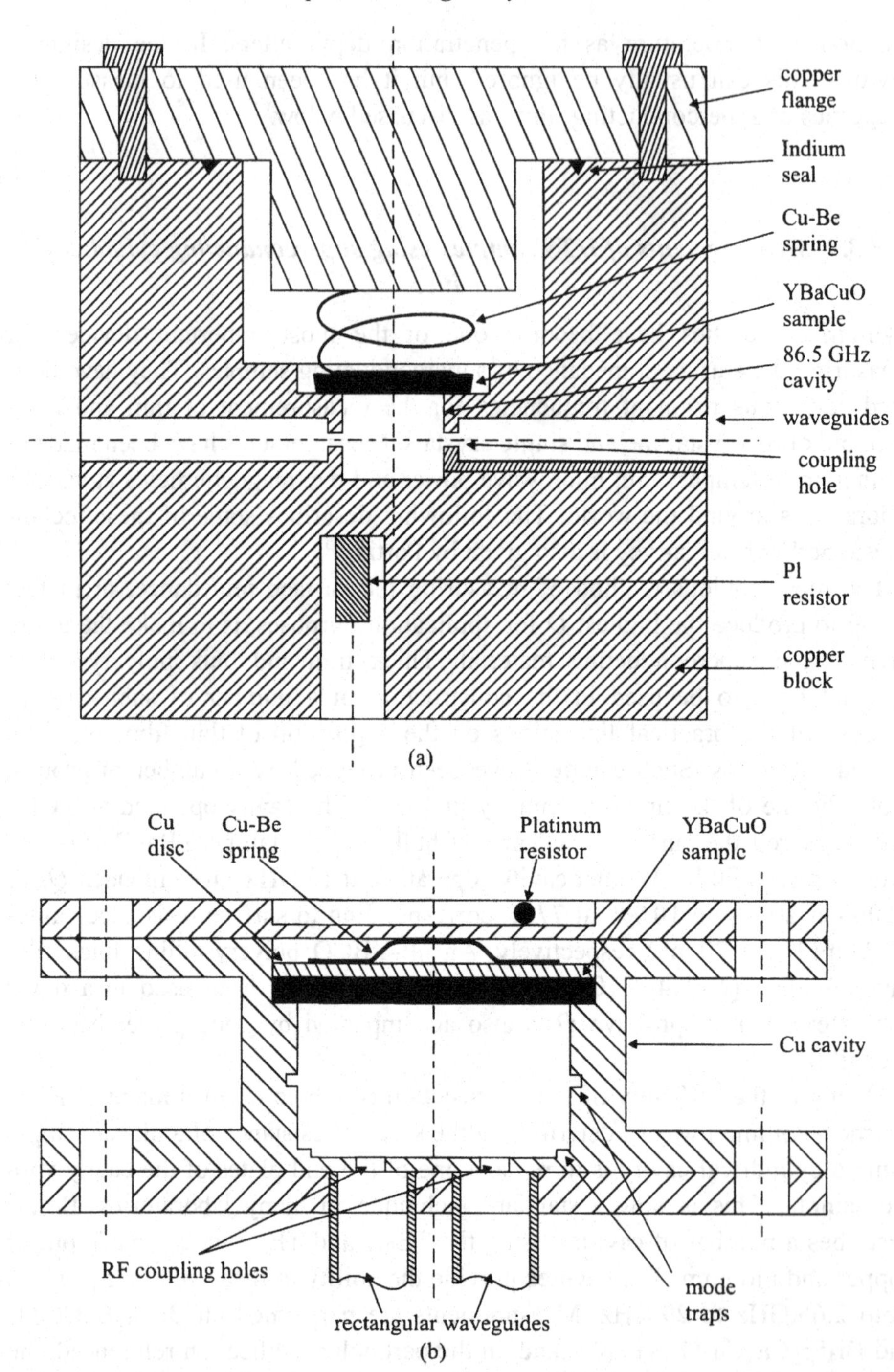

Figure 3.3.4. Cylindrical copper cavities constructed at Wuppertal University for testing planar samples: (*a*) the 86.5 GHz cavity operating in the TE_{013} mode and (*b*) the 21.4 GHz cavity operating in the TE_{011} mode. (Taken from reference [10].)

their cavities. The high-frequency cavity operates in the TE_{013} mode at 86.5 GHz, and the lower-frequency one in the TE_{011} mode at 21.4 GHz.

By measuring the Q of an all-copper (or superconducting) cavity and then by measuring the Q of the cavity with the end plate replaced by HTS, the surface resistance can be calculated using the expressions given in Section 3.2. In addition, the change in frequency can be measured as a function of temperature and can be attributed to a change in cavity size; if thermal expansion of the cavity is taken into account, then the change in penetration depth can be calculated.

Both cavities in Figure 3.3.4 are fed by waveguides and energy is transferred into and out of the cavity by small coupling holes. The size of the holes is determined experimentally to give weak coupling so that the total insertion loss of the cavity is high and the measured loaded Q only needs a small correction factor (Equation (3.2.3)) in order to derive the unloaded Q. Another method of coupling energy into and out of the cavities is by a small loop antenna or probe; these are normally used when the feed to the cavity is via coaxial cable.

It is important to keep the thermal mass and heat input to the cavity structure at a minimum, thus removing the need for high cooling powers. This can be achieved by minimising the size and wall thickness of any feeding waveguides or coaxial cable. Stainless steel cables can be used to minimise the thermal input to the cavity.

A mode filter or trap is shown in Figure 3.3.4. This trap separates the TE_{01l} from the TM_{11l} mode, removing the degeneracy. The sample is mounted so that it overlaps the cavity ends and is pressed into place by a beryllium–copper spring. Because the TE_{01l} modes are used, currents do not travel across this contact. One of the limitations of this type of cavity is that the cavity and the sample are at the same temperature, which ultimately limits the sensitivity, requiring that the losses due to the HTS sample be not much more than an order of magnitude larger than the losses in the cavity body. Oxygen-free, high-conductivity (OFHC) copper is the preferred choice of cavity material, but higher sensitivity can be obtained by using a superconducting cavity. The use of a superconducting cavity precludes measurements in d.c. magnetic fields. A niobium cavity with the end plate held away from the main cavity has been constructed at Wuppertal University. Such a cavity is extremely sensitive, allowing precise low-temperature measurements. The major problem with the design of such a cavity is to prevent field leakage from the gap between the end plate and the cavity body. It is also possible to design cylindrical cavities of variable size so that measurements can be made as a function of frequency,[13] and to load the cavity with dielectric to alter its resonant frequency over a small range.[14]

It must be remembered that when making measurements of surface resistance, it is the effective surface impedance that is being measured. If the HTS film is not many penetration depths thick, which is usually the case, then the field in the cavity penetrates the film, the dielectric substrate and any other backing material. It is quite straightforward to extract these effects using the theory in Section 1.8 which is also discussed by Drabeck,[15,16] Hartemann[17] and Klein.[18]

At low frequencies, cylindrical cavities become large and it may not be convenient to use very large samples for a full end plate replacement. It is, however, possible to replace the end plate of a cavity partially. This is usually either for a disc-shaped sample placed on the end plate or for the sample placed below a circular opening in the end plate. In this case, either smaller samples and/or larger (lower frequency) cavities can be used. Clearly, the expression for Q needs to be considered for this case and is given by Martens.[19] He discusses such a cavity for the measurement of the surface resistance of thick films using a cavity with a piston to vary the cavity length, resulting in measurement as a function of frequency over the whole X-band range. Also, Li[20] has used such a cavity to look at the surface resistance of polycrystalline material. Here the sample was mounted on a plate, on the end plate of the cavity, which had independent temperature control from the cavity. This allowed the cavity to be kept cold whilst the sample was heated, giving increased sensitivity.

The final technique for the measurement of surface impedance using a cylindrical cavity is the cavity perturbation technique. Here a small sample of material is placed inside a cavity and is heated independently from the cavity, allowing maximum sensitivity. To achieve this the sample is usually placed on a sapphire rod and heated with a small resistance heater whilst the cavity is kept at a low temperature. The technique is particularly useful for single crystals, where only small sizes are available, and has become one of the major analysis techniques for looking at the microwave properties of high-quality single crystals. The technique was also one of the first used to measure the microwave properties of HTS materials. In 1987 Hagen[21] placed a small sample of YBCO into a 1.3 GHz resonant cavity and observed its effect. Improvements in the technique have lead to well-developed experimental apparatus at a number of laboratories.[22–28] The cavities have also been used extensively to measure the microwave surface impedance as a function of an applied d.c. magnetic field.

Figure 3.3.5 shows a cavity set-up for the measurement of single crystals by the cavity perturbation technique. At the centre of the diagram is the copper cavity with a sapphire rod and the sample mounted on it. This rod protrudes into the cavity through a small hole in the base plate. On the sapphire rod

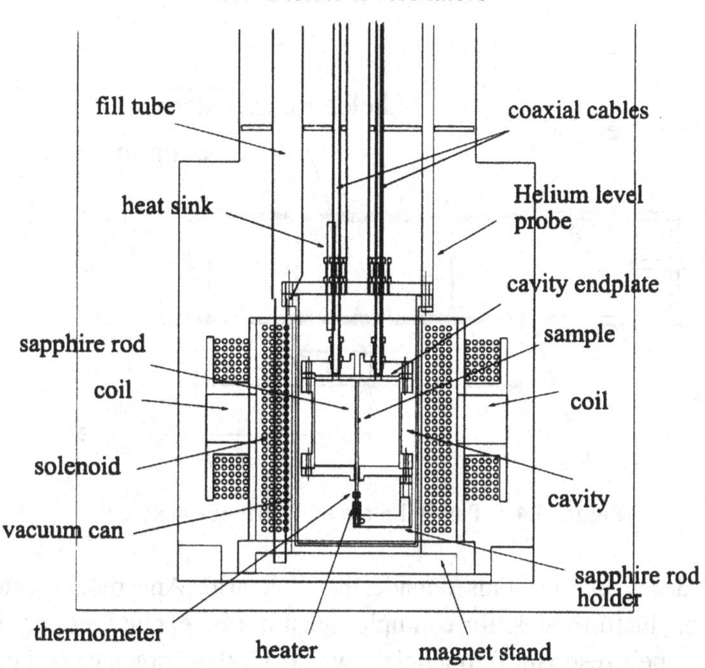

Figure 3.3.5. A cylindrical cavity and associated cryogenic apparatus for the measurement of the properties of single-crystal HTS samples using the cavity perturbation technique.[29]

outside the cavity are mounted a thermometer and a heater. Coupling to the cavity is by two stainless steel coaxial cables with loop antennas just inside the cavity. The cavity and heater are enclosed in a vacuum can outside of which are two superconducting magnets, allowing a d.c. field to be placed on the sample in two different orthogonal orientations. The magnets are immersed in helium inside a conventional cryostat. Measurements made using this apparatus are discussed in references [23 and 29].

3.4 Dielectric resonators

Another type of resonator which has become popular with the emergence of high-temperature superconductors is the dielectric resonator. This resonator has been used for surface impedance measurements, for stabilising oscillators and for use in filters of various types. The dielectric resonator has one of the highest Qs of all the resonators discussed in this chapter. Figures 3.4.1 and 3.4.2 show the basic types of dielectric resonator. In both cases the modes of interest are mainly contained in the dielectric material, with an approximately

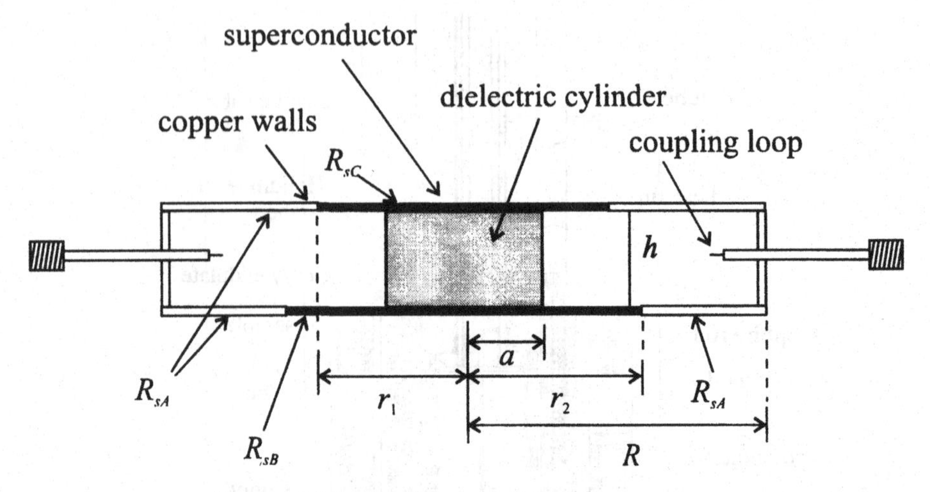

Figure 3.4.1. Parallel plate dielectric resonator.

exponential decay of the fields outside the dielectric. Any modes external to the dielectric, shield modes, for example, must not be excited, or care must be taken to keep their resonant frequencies away from the resonance of interest.

The parallel plate resonator shown in Figure 3.4.1 consists of a low-loss dielectric cylinder with superconducting plates placed on the top and bottom. Usually, the structure is held together with copper–beryllium springs and the coupling is through the side walls via coaxial cable with loop antenna termination.

The alternative dielectric resonator is shown in Figure 3.4.2. In this case the dielectric cylinder is placed on the superconducting film, with the normal metal

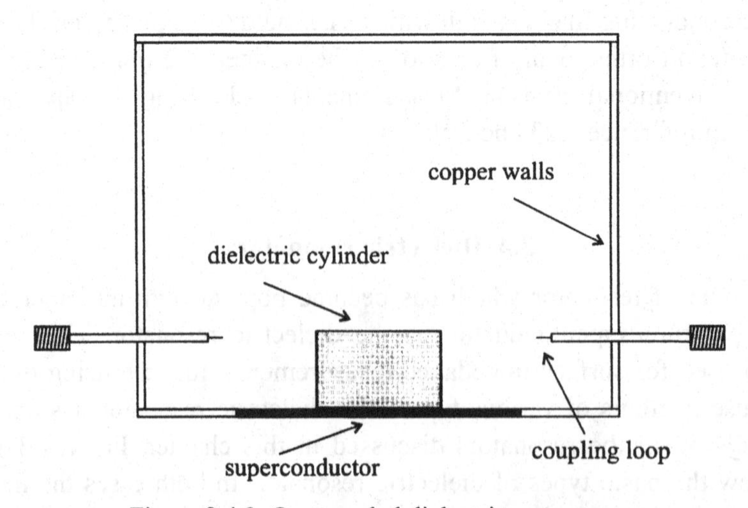

Figure 3.4.2. Open-ended dielectric resonator.

walls of the cavity a sufficiently large distance away so as not to affect the fields appreciably. The dielectric cylinder can be glued to the superconducting film, and a removable polymer has been found to be suitable and not affect the quality factor of the cavity.[30] With both types of cavity, care must be taken to evaluate the effect of the copper walls. The open-ended cavity has the advantage that only one superconducting film is required, and this can be important if the cavity is used for the measurement of surface impedance; otherwise a calibrated film is required. This open-ended resonator has the higher Q of the two resonators, since there is no loss from the second film. For the highest Q the dielectric cylinder can be removed from the superconducting film into the centre of the cavity and the walls of the cavity removed far from the dielectric. In this case the Q is only dependent on the loss tangent of the dielectric and can be very high if sapphire is used as the dielectric at low temperature. However, such a cavity is large in size and problems occur when mounting the dielectric. Even the cavity of Figure 3.4.2 can have problems with mechanical stability. However, the parallel plate cavity is much less susceptible to vibrational noise because of its more rigid construction. It is also much smaller, making it more appropriate for application purposes. Also, by moving the upper superconducting plate away from the dielectric cylinder it is possible to tune the cavity; up to 10% change in resonant frequency has been demonstrated.[30]

Sapphire ($\varepsilon_r \approx 9.4$, $\tan\delta < 2 \times 10^{-9}$ at 4×2 K[31]) is the principal dielectric used in the cavities because of its very low loss at low temperatures. However, smaller cavities can be produced if a higher dielectric constant material is used such as lanthanum aluminate ($\varepsilon_r \approx 24$). Using such cavities the dielectric loss tangent of lanthanum aluminate has been found to be as low as 3×10^{-6} at 18 GHz and 77 K.[30] The additional advantage in using a higher dielectric loss material is that the field is more confined and the effect of the cavity walls becomes less of a problem.

It should be noted that dielectric resonators have been used routinely in the microwave electronics industry for filters and stabilising oscillators. Materials such as $BaTi_4O_9$, $Ba_2Ti_9O_{20}$, or (Zr, Sn)TiO_4, with dielectric constants around 40 and loss tangents around 10^{-4}, are used.[32] These dielectric cylinders are not used in cavities of the types discussed above, but are just placed close to a microstrip transmission line in a microwave circuit. The proximity couples the resonator into the circuit.

Dielectric resonator cavities have also been used for many years, principally for the measurement of dielectric properties of the dielectric cylinder,[33,34] and only more recently have they found applications using high-temperature superconductors. There has also been considerable work using low-temperature superconductors,[35–37] and it is these devices which still provide the highest Q. Theoretical aspects, particularly for the parallel plate cavity, are thus well developed and expressions exist in the literature for the field distribution and quality factors for various configurations. Mode charts have been given by Kobayashi[38] and further extensive discussion is given by Kajfez.[39] A number of publications regarding HTS dielectric resonator cavities also discuss the calculation for both the parallel plate dielectric resonator[40,41] and the open dielectric resonator.[30] The modes of interest are mainly the TE_{01n}. Here there are closed loops of transverse electric field centred on the cylinder axis, and the loops of the magnetic field lie in the plane containing the axis. This results in only circumferential currents in the superconductor and surrounding packaging. There are low Q modes which should be avoided, and these TM_{lm0} modes lose energy by radiation transverse to the axis direction.[42]

The fields for the TE_{011} mode for the parallel plate geometry are given by[41]

$$E_\phi(r,\phi,z) = jA\frac{\omega\mu h}{\pi}M(r)\sin(\pi z/h) \tag{3.4.1}$$

$$H_r(r,\phi,z) = AM(r)\cos(\pi z/h) \tag{3.4.2}$$

$$H_z(r,\phi,z) = AN(r)\sin(\pi z/h) \tag{3.4.3}$$

where

$$M(r) = \begin{cases} \dfrac{J_1(ur/a)}{J_1(u)} & 0 \leqslant r \leqslant a \\[2ex] \dfrac{K_1(Rv/a)I_1(vr/a) - I_1(Rv/a)K_1(vr/a)}{K_1(Rv/a)I_1(v) - I_1(Rv/a)K_1(v)} & a \leqslant r \leqslant R \end{cases} \tag{3.4.4}$$

$$N(r) = \begin{cases} -\dfrac{uh}{\pi a}\dfrac{J_0(ur/a)}{J_1(u)} & 0 \leqslant r \leqslant a \\[2ex] -\dfrac{vh}{\pi a}\dfrac{K_1(Rv/a)I_0(vr/a) + I_1(Rv/a)K_0(vr/a)}{K_1(Rv/a)I_1(v) - I_1(Rv/a)K_1(v)} & a \leqslant r \leqslant R \end{cases} \tag{3.4.5}$$

Here a is the radius of the dielectric, R is the radius of the whole cavity, h is the height of the cavity and A is a constant; J, I and K are the Bessel functions using the usual notation. The parameters u and v are determined by the solution of the simultaneous equations

$$\frac{J_1(u)}{uJ_0(u)} = -\frac{1}{v}\frac{K_1(Rv/a)I_1(v) - I_1(Rv/a)K_1(v)}{K_1(Rv/a)I_0(v) + I_1(Rv/a)K_0(v)} \tag{3.4.6}$$

and

$$u = \sqrt{\left((\varepsilon_r - 1)\left(\frac{\pi a}{h}\right)^2 - \varepsilon_r v^2\right)} \tag{3.4.7}$$

The resonant frequency for the TE_{011} mode is given by

$$f = \frac{c}{2a}\sqrt{\left(\left(\frac{a}{h}\right)^2 + \left(\frac{v}{\pi}\right)^2\right)} \tag{3.4.8}$$

The losses in the cavity, and hence the Q, can be calculated directly by Equation (3.2.8) or by the incremental inductance rule[43] described in Section 2.5. Expressions are given in references [40, 41 and 44] and below. These authors also consider the maximum current on the surface of the superconducting plate. This is particularly important if the resonators are used for the power dependent surface impedance measurements of a superconductor. The Q and frequency shift for the TE_{011} mode using the nomenclature shown in Figure 3.4.1 are[41]

$$\frac{1}{Q} = \frac{(1 - C_1)R_{sC} + (C_1 + C_2)R_{sA} + (1 - C_2)R_{sB}}{G} + \frac{\varepsilon_r \tan\delta}{\varepsilon_r + W} \tag{3.4.9}$$

where

$$G = 480\pi^2 \left(\frac{hf}{c}\right)^3 \frac{\varepsilon_r + W}{1 + W} \tag{3.4.10}$$

$$W = \frac{J_1^2(u)}{K_1^2(v)} \frac{K_0(v)K_2(v) - K_1^2(v)}{J_1^2(u) - J_0(u)J_2(u)} \tag{3.4.11}$$

$$C_n = \left(\frac{r_n}{a}\right)^2 \frac{K_0(r_n v/a)K_2(r_n v/a) - K_1^2(r_n v/a)}{K_0(v)K_2(v) - K_1^2(v)} \frac{W}{1 + W} \qquad a \leqslant r_n \leqslant R \tag{3.4.12}$$

Here $n = 1$ or $n = 2$ in order to define C_1 or C_2 in Equation (3.4.9), R_{sA}, R_{sB} and R_{sC} are the surface resistance of the copper surround, the lower superconducting plate and the upper superconducting plate, respectively, as shown in Figure 3.4.1. Fletcher[41] also gives an expression for the change in resonant frequency and takes the effects of the end walls of the cavity into account.

These cavities have been used for the measurements of surface impedance values of superconductors[45] but for other applications in oscillators and filters, for example, the unloaded quality factor is the important performance criterion. As stated above, these cavities have very high Q values; values of unloaded Q of 2×10^6 at 90 K and 1.4×10^7 at 4.2 K and 5.52 GHz have been demonstrated using $Tl_2Ba_2CaCu_2O_8$ films and a sapphire resonator.[40,46] For YBCO films and a lanthanum aluminate dielectric, Qs of 4.5×10^7 and 1.3×10^5 have been obtained[30] at 10 K and 77 K respectively; both measurements were at 11.6 GHz. These values are low in comparison with what can be achieved using low-temperature superconductors; values greater than 10^9 are possible[35] at 2 K and 10 GHz, and at 4.2 K values substantially higher then 10^8 are achievable.

3.5 Coaxial cavity

The coaxial cavity resonator shown in Figure 3.5.1 consists of an outer tube containing an inner wire. It can be a length of coaxial cable which has either a short or open circuit at each end, producing standing waves in the cavity body, the length of the cavity usually corresponding to an integer number of half-wavelengths. A coaxial cable normally supports TEM waves, and so does the coaxial cavity. However, as the radius of the cavity increases, TE and TM waves begin to appear. The coaxial cavity has been used very successfully in determining the surface resistance of HTS wires. Normally a few modes are used in order to determine the frequency dependence of the surface resistance. The cavity can be used in the frequency range of a few hundred megahertz to many tens of gigahertz. The lower limit is due to the size increasing and becoming impracticable to handle; the upper limit is due to the size becoming smaller, but the use of higher harmonics extends its frequency range. For lower frequencies, the helical cavity can be used, and this is discussed in Section 3.6. The coaxial cavity is reviewed briefly below, and the expressions for the quality factor for the TEM modes are derived. A brief review of a number of coaxial cavities used for HTS measurements is also given.

The starting point for the calculation of the quality factor of the coaxial resonator shown in Figure 3.5.1 are the equations for the transverse electromagnetic fields within a coaxial cable (the TE and TM fields are discussed below). These equations are derived by considering standing waves caused by reflections at the end of a section of coaxial cable[47] and are

$$\mathbf{E} = \frac{V}{r} \frac{1}{\ln(b/a)} \mathbf{a}_r \cos(\beta z) \cos(\omega t) \tag{3.5.1}$$

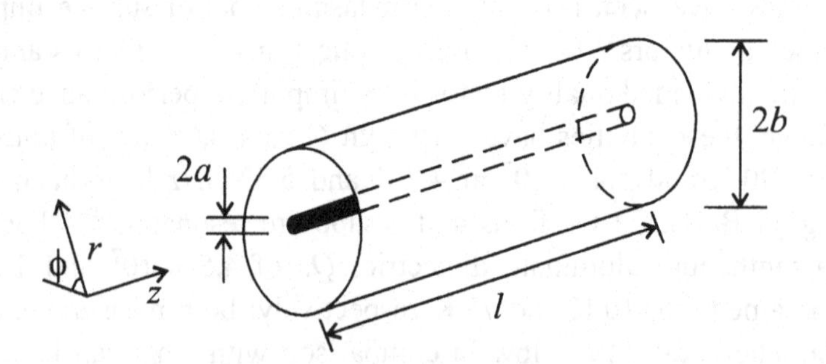

Figure 3.5.1. Coaxial resonator.

$$\mathbf{H} = \frac{I}{2\pi r}\mathbf{a}_\phi \sin(\beta z)\cos(\omega t) \tag{3.5.2}$$

Here, a is the radius of the inner conductor and b is the inner radius of the outer conductor, V is the voltage between the inner and outer conductors and I is the current in the cable in the z direction; conventional cylindrical coordinates are used.

Following the definitions described in Section 3.2, the Q of the resonator can be calculated. First, the energy stored in the resonator is given by

$$\frac{1}{2}\mu\int_{cavity\ volume}|\mathbf{H}|^2\,dv = \frac{1}{2}\mu\int_0^l\int_0^{2\pi}\int_a^b\left(\frac{I}{2\pi r}\right)^2 \sin^2(\beta z) r\,dr\,d\phi\,dz \tag{3.5.3}$$

If the cavity length l is assumed to be composed of an integer number n of half-wavelengths, then the integrations result in

$$\frac{1}{2}\mu\int_{cavity\ volume}|\mathbf{H}|^2\,dv = \frac{\mu I^2 nl\ln(b/a)}{8\pi} \tag{3.5.4}$$

The power loss due to dissipation on the centre conductor, the outer conductor and any end walls are given by

$$\begin{aligned}\frac{1}{2}\int_{wall\ surfaces}R_s|\mathbf{H}|^2\,ds &= \frac{1}{2}R_{sa}\int_0^l\int_0^{2\pi}\left(\frac{I}{2\pi a}\right)^2\sin^2(\beta z)a\,d\phi\,dz \\ &+ \frac{1}{2}R_{sb}\int_0^l\int_0^{2\pi}\left(\frac{I}{2\pi b}\right)^2\sin^2(\beta z)b\,d\phi\,dz \\ &+ R_{se}\int_0^{2\pi}\int_a^b\left(\frac{I}{2\pi r}\right)^2 r\,dr\,d\phi\end{aligned} \tag{3.5.5}$$

The first and second terms on the right-hand side represents the power loss on the inner and outer conductor walls respectively. The last term represents the power loss on the end walls. The surface resistances R_{sa}, R_{sb} and R_{sc} are the surface resistances for the inner conductor, the outer walls and the ends of the cavity respectively. It should be noted that this is only relevant if the ends of the cavity are closed; the ends of the cavity could also be open, in which case the radiation from the open ends may have to be considered. The fields differ slightly depending on whether the cavity is open or short circuited; in the open-circuit case; the magnetic field goes to zero at the ends; in the short-circuit case, it has a maximum. The integrations in Equation (3.5.5) are trivial and result in

$$\frac{1}{2}\int_{wall\ surfaces}R_s|\mathbf{H}|^2\,ds = \frac{I^2 nl}{8\pi}\left\{\frac{R_{sa}}{a}+\frac{R_{sb}}{b}+\frac{4R_{se}\ln(b/a)}{nl}\right\} \tag{3.5.6}$$

Hence using Equations (3.2.8), (3.5.4), and (3.5.6), the quality factor of the cavity is

$$Q_c = \frac{\omega\mu \ln(b/a)}{\dfrac{R_{sa}}{a} + \dfrac{R_{sb}}{b} + \dfrac{4R_{se}\ln(b/a)}{nl}} \tag{3.5.7}$$

If the cavity is an open circuit, the last term in the denominator does not need to be included. Figure 3.5.2 shows typical values of the unloaded conductor quality factor using Equation (3.5.7) as a function of b/a for a number of different values of inner conductor surface resistance. It can be seen that the Q increases rapidly as the cavity outer radius becomes larger than the inner wire radius, but the rate of increase of Q gradually becomes slower. Some design considerations for optimum Q coaxial cavities are discussed by Davis.[48]

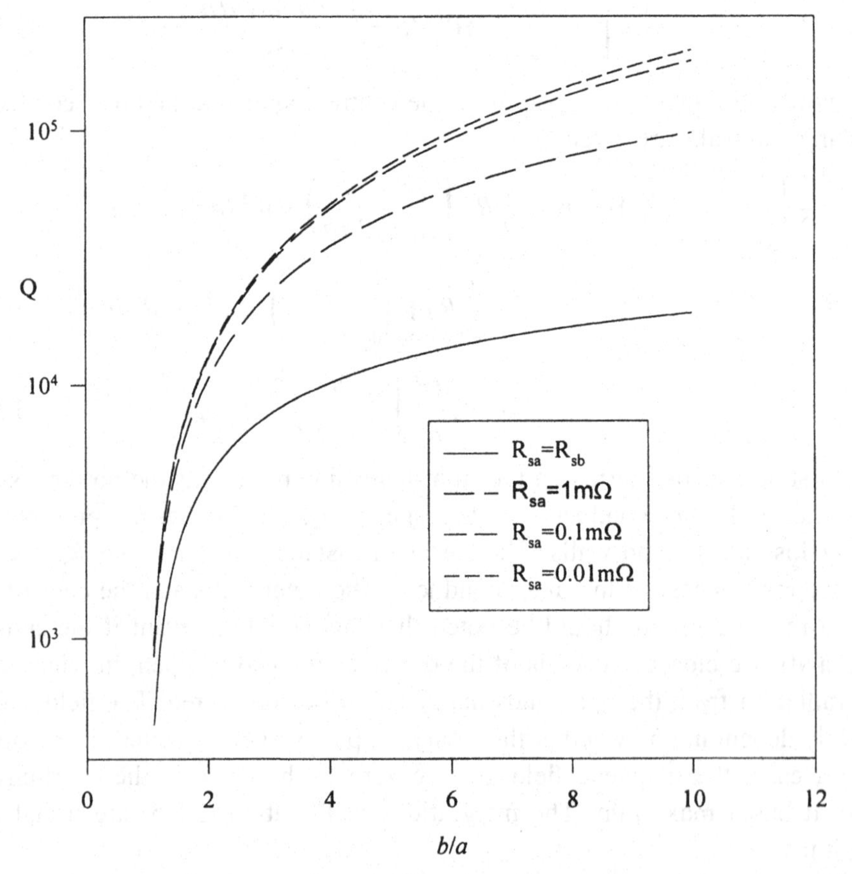

Figure 3.5.2. The quality factor of a coaxial cavity with $a = 1$ mm and a copper shield at 77 K ($R_s = 8.7$ mΩ) as a function of cavity size b/a for a number of different surface resistance values of the inner conductor. The resonant frequency is 10 GHz.

It can be seen from Figure 3.5.2 and from Equation (3.5.7) that, as the radius of the outer conductor increases, then its contribution to the Q decreases and the Q becomes more dependent upon the surface resistance of the inner conductor. However, if the surface resistance of the inner conductor is substantially lower than the outer walls, then the Q is dominated by the losses in the outer walls. Most of the HTS coaxial cavities demonstrated have outer conducting walls made from normal metal; it is thus advantageous for maximum Q to have large radii cavities. A problem which may occur with such large radii cavities is that other TE and TM modes will eventually start to appear. Provided they are not close to the modes of interest, these may not be a problem; in fact it is possible to use the modes in HTS surface impedance determination.[49,50]

The TE and TM fields in a coaxial waveguide are found in exactly the same way as in the case of the circular waveguide given in Section 3.3, except that the boundary conditions are different. For the coaxial cavity, the value of K_4 cannot be put to zero in Equation (3.3.22) because $r = 0$ is excluded from the region of the field. Referring to Section 3.3, the z component of the electric field for the TM mode can then be written:

$$E_z = E_0(J_n(\beta_t r) + K_4 Y_n(\beta_t r)) \cos(n\phi) e^{-j\beta_z z} \tag{3.5.8}$$

For the TE mode, the z-directed component of the magnetic field is

$$H_z = H_0(J_n(\beta_t r) + K_4' Y_n(\beta_t r)) \cos(n\phi) e^{-j\beta_z z} \tag{3.5.9}$$

The other components of the fields can be found by the method outlined in Section 3.3. The calculation then proceeds in exactly the same way as for the cylindrical cavity, but using Equations (3.5.8) and (3.5.9) to find the fields, the Qs and the resonant frequency of the modes. The results of such a calculation are given by Marcuvitz.[51]

To a good approximation, the cut-off wavelength for the TM modes in a coaxial cable are given by[52]

$$\lambda_{c,mn} = \frac{2(b-a)}{m} \tag{3.5.10}$$

and, for TE modes,

$$\begin{aligned} \lambda_{c,nl} &= \frac{\pi(b+a)}{n} \qquad n = 1, 2, 3, \ldots \\ \lambda_{c,nm} &= \frac{2(b-a)}{m-1} \qquad m = 2, 3, 4, \ldots \end{aligned} \tag{3.5.11}$$

A practical coaxial resonator is shown in Figure 3.5.3.[53] The resonator is designed to allow measurements of a large number of harmonics without interference with TE or TM modes; hence the cavity is rather long and thin. The cavity consists of a copper tube, with the dielectric insulator PTFE

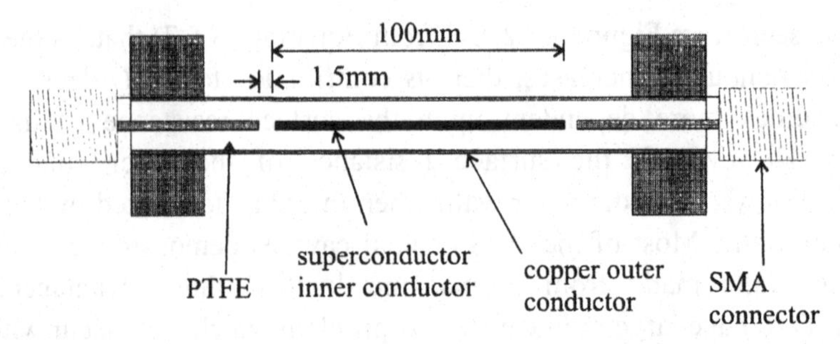

Figure 3.5.3. Construction details of a coaxial cavity for TEM mode operation.

supporting an inner conductor of YBCO superconducting wire. The length of the wire is 100 mm and it has a fundamental resonant frequency of 1.15 GHz. The slight increase in resonant frequency from that calculated from the length of the wire and the dielectric constant of the PTFE is due to the fields extending slightly beyond the end of the wire. The radius of the wire is about 0.5 mm and the diameter of the outer conductor is 4.35 mm. Coupling to the cavity is by a capacitive gap from the 50 Ω feed cables at either end of the resonator. The gap between the superconducting wire and the feed wire is adjusted experimentally so that the resonator is weakly coupled, and so that any measurements of loaded Q require only a small correction to calculate the unloaded Q.

Figure 3.5.4 shows the transmitted response of the cavity; 17 harmonics can be observed in the frequency range 1–20 GHz. The modes appear to be free

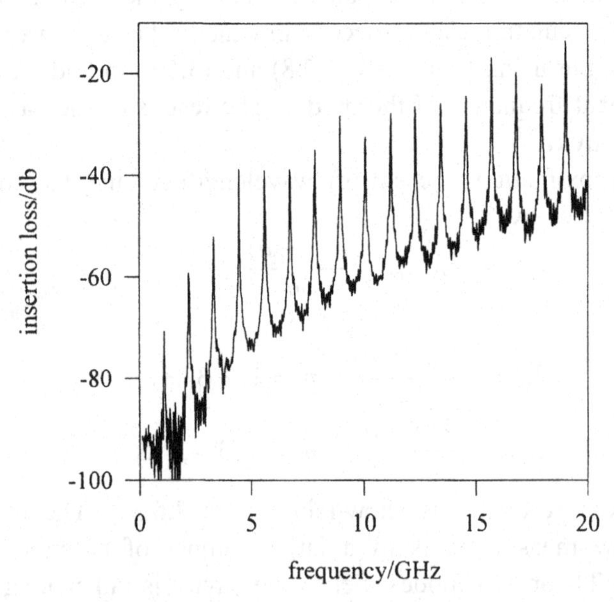

Figure 3.5.4. Insertion loss of the coaxial cavity shown in Figure 3.5.3.

from dispersion and spurious responses. The overall increase in the insertion loss is due to the change in reactance of the capacitive coupling gap as the frequency increases.

From each of the resonances shown in Figure 3.5.4 a value of Q can be calculated; for the fundamental, this is 1500. The effect of the loss tangent of the PTFE, which supports the central wire, is small at these values of Q, although it is possible to take it into account, as is done in reference [53], where the results of using the cavity are discussed.

A number of other coaxial cavities have been developed for the measurement of superconducting wire. The first was designed by Pippard in 1950.[54] Gallop[50,49] used the TM_{010}, TM_{011}, TM_{112}, TM_{111} and TE_{111} modes to investigate both a YBCO and a BSCCO bulk polycrystalline rod with a brass outer cavity. The first three modes produce longitudinal currents in the rods, just like TEM modes; the latter three modes produce currents which circulate in the superconducting rods, allowing an idea to be gained about any directional dependence of the surface resistance. A typical value obtained for the surface resistance for a YBCO is 20 mΩ at 10 GHz and 77 K. Bohn[55,56] also discusses a coaxial cavity; the fundamental operating frequency this time is substantially lower at 150 MHz, since this cavity is 0.8 m long, with about 0.5 m-long samples. The sample is placed in a quartz tube and lowered into the cavity, and the whole cavity is filled with liquid cryogen and its temperature varied by adjusting the vapour pressure. TEM modes are used. Other superconducting coaxial cavities are discussed in references [57–60].

3.6 Helical cavity resonators

At low frequencies, the coaxial resonator described in the preceding section increases substantially in length. This does not necessarily mean that it occupies a large volume, but the length makes it unusable in many applications. The coaxial cable can be meandered or coiled in order to allow it to occupy a more convenient volume. However, another solution is to place a helix inside a single outer shield; the resonant frequency should then be dependent on the total length of coiled wire. This is the helical resonator and it has been used substantially in low-frequency filter[61,62] and oscillator applications, as well as in the measurement of superconducting[63,64] and dielectric[65,66] properties. The device provides a high Q in a reasonably small volume. The helical resonator is a slow-wave structure, more of which will be discussed in Section 5.8.

Figure 3.6.1 shows a helical resonator; traditionally the helix is shorted to the outer shield at one end to form a quarter-wave resonator. In this case it is possible to tune the cavity by mechanically moving the position of the short

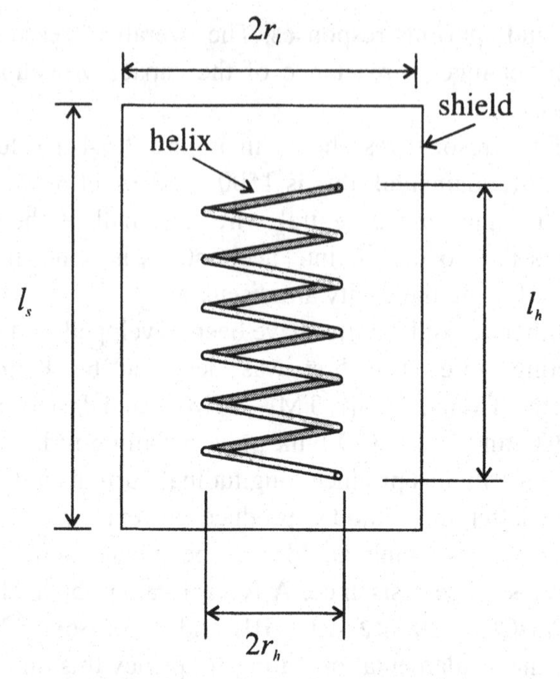

Figure 3.6.1. The helical cavity resonator.

along the helix.[67,68] However, this connection is difficult to make when high-temperature superconductors are used, so half-wavelength resonators are usually made, as shown in Figure 3.6.1.

The helical resonator consists of an outer cylindrical shield inside which is placed a helix, usually supported by a low-loss dielectric former. Coupling to the resonator can be by a number of techniques, and a loop terminated coaxial cable is often used.

The resonant frequencies of practical use of the helical resonator lie between around 10 MHz and several gigahertz. At the higher-frequency end, the coaxial resonator is generally more convenient; below the lower frequency a simple *LC* circuit usually suffices. If a high Q is not the main concern, then a lumped element *LC* circuit can be used up to higher frequencies. Some design equations are given below for helical resonators. It is interesting that in a study in 1959 by Macalpine,[69] on copper helical resonators, he arrived at the following conclusion: 'The unloaded Q is equal to 50 times the shield diameter in inches times the square root of the resonant frequency in megacycles. Shield length is about 30 percent greater than its diameter.' With the surface resistance of superconductors being much lower than copper, substantial increases in Q can be obtained. For example, for a thick film superconductor with a crossover frequency of 20 GHz, Macalpine's statement can be changed to: 'The unloaded Q is equal to 1.4×10^8 times the shield diameter in inches times the square of the resonant frequency in megacycles.' These statements represent order of magnitude estimates, but show the difference between HTS and copper. Another interesting comparison which Macalpine makes is to demonstrate the space savings of the helical resonator over a coaxial resonator; for his quarter-wave resonator he says '... a helical resonator at 10 megacycles with an unloaded Q of 1000 is about six inches in diameter by about eight inches in length. In contrast, a TEM mode coaxial-line resonator would be 25 feet in length by three inches in diameter.'

Other technologies which are in competition with the helical resonator are crystal resonators and surface acoustic wave (SAW) devices; both are high-performance, well-developed technologies, and are adequate for the majority of applications. SAW filters cover much of the frequency range described above for helical resonators, and only in specialist applications should a helical resonator filter be chosen over a SAW device. At the higher frequencies, planar resonators of microstrip or stripline, discussed in Section 3.7, should also be considered. A meandered microstrip, made from HTS, can have resonant frequencies in the subgigahertz region and still occupy a small area and have a reasonable Q. One application where these other techniques are not applicable is the measurement of the surface resistance of superconducting wire in the lower-frequency range. Helical resonators made from niobium have been used to good effect in high-field accelerator cavities,[70] where Qs of 3×10^8 have been obtained at 135 MHz.

The complex physical structure of the helical resonator makes it very difficult to analyse, and no exact solutions to the problem exist. The structure can be analysed numerically, and a number of general three-dimensional electromagnetic field analysis programs exist and can cope with this problem. However, using such programs, it is difficult to design to a specific resonant frequency without running the program many times with different parameters. Below are a set of design equations based simply on a lumped element model. This model gives reasonable agreement with experiment and further analysis can be performed numerically if required. There have been a number of attempts to analyse the helical resonator, two of the more successful being the sheath and tape models. In the sheath model,[71,72] the helix is modelled by an inner cylinder conducting only in the direction of the helix. In the tape model,[73,74] the wire of the helix is represented by a flat tape on the inner cylinder. Both these models have the shortcoming that they do not describe the field close to the helix. The problem has also been formulated in terms of helical coordinates[75] but, again, only approximate results can be obtained. The analysis given below is based on modelling the helical resonator with lumped elements; this follows Macalpine[69] but appears here in a modified form.[63,76]

The inductance per unit length, L, of a helix in the presence of a cylindrical shield is given by[77]

$$L = L_0\left[1 - \frac{2}{3}\left(\frac{r_h}{r_s}\right)^2 \frac{l_h}{l_s} K_1\right] \tag{3.6.1}$$

Here, r_h and r_s are the radius of the helix and shield, l_h and l_s are the length of the helix and shield, and K_1 is a factor to take the non-uniformity of the field within the helix into account.[78] Here it is taken to equal 1. The inductance L_0 is the inductance of a helix without a shield and is given by

$$L_0 = \mu_0 N^2 \pi \left(\frac{r_h}{l_h}\right)^2 \tag{3.6.2}$$

where N is the number of turns on the helix. The capacitance between the helix and the shield can be approximated by considering the capacitance between two cylinders; this is given by

$$C = \frac{2\pi\varepsilon_0}{\ln(r_s/r_h)} K_2 \tag{3.6.3}$$

K_2 is another geometrical factor which can account for the fringing field at the end of the helix and is taken as 1 in the calculations given below. The inter-turn capacitance is not taken into account in this formulation, and this will be more important in tightly wound helices. The loss resistance per unit length due to cylindrical shield can be shown to be approximately

$$R_{shield} = R_{ss}\frac{\pi N^2}{6 l_h}\left(\frac{r_h}{(r_s^2 l_s/2)^{1/3}}\right)^4 \tag{3.6.4}$$

and the loss resistance of the helix is given by

$$R_{helix} = R_{sh} N \frac{r_h}{a l_h} K_3 \tag{3.6.5}$$

K_3 is another constant, the proximity factor,[78] accounting for the enhancement of the surface magnetic field of the helix owing to screening currents induced in the adjacent helix turns; also, a is the radius of the wire. In both Equations (3.6.4) and (3.6.5), R_{ss} and R_{sh} are the surface resistance of the material that the shield and helix and wire are constructed from, the subscript referring either to the shield or the helix. It is now quite easy to combine these equations in the usual way to calculate the resonant frequency and unloaded Q of a half-wavelength helical resonator:

$$f_m = \frac{m}{2 l_h}\frac{1}{\sqrt{(LC)}} \tag{3.6.6}$$

and

$$Q_0 = \frac{2\pi f_m L}{R_{helix} + R_{shield}} \tag{3.6.7}$$

where m is the mode number (an integer). It should be noted that the field confinement around the central helix makes this component have a much larger effect on the Q than the shield. Figure 3.6.2 shows how the Q can vary as a function of both the helix radius and the radius of the wire forming the helix. As can be seen, both quantities have a large effect on the quality factors obtained.

A number of HTS helical resonators have been built and tested, and these have been made from either thick film or bulk polycrystalline materials. If the helices were made out of thin film material, the resultant resonator would obviously have a much higher Q, but, clearly, it is a very difficult task to coat helical dielectric with thin film HTS. Table 3.6.1 shows the performance of a number of resonators, including some copper ones for comparison.

It is interesting to note the number of different types of HTS helix, and some of these are shown in Figure 3.6.3. The bulk wire is made by viscous extrusion

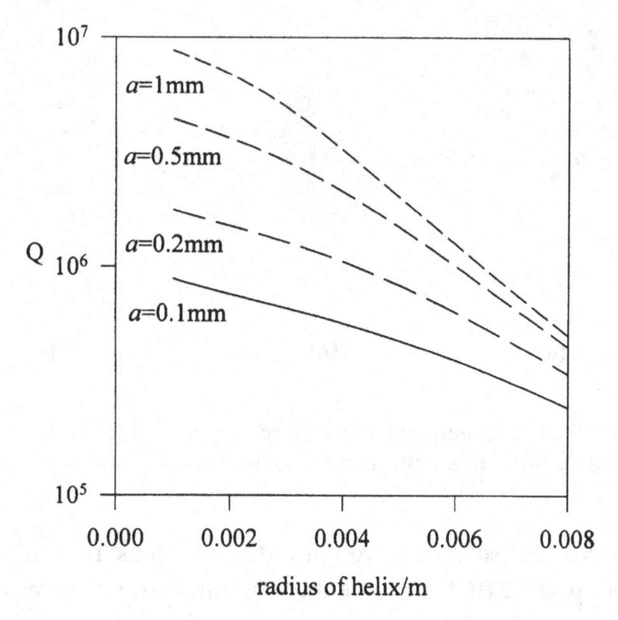

Figure 3.6.2. The unloaded quality factor of a helical cavity at 300 MHz, with the helix made from a superconductor with a crossover frequency of 10 GHz ($R_{sh} = 7.8\ \mu\Omega$) and the outer conductor made of copper at 77 K ($R_{sh} = 1.5\ \text{m}\Omega$) as a function of the radius of the helix. Each line represents different wire radii. Also $r = 16$ mm, $l_h = 20$ mm, $l_s = 32$ mm and $N = 15$. All the constants K_1, K_2 and K_3 are assumed to be unity.

Table 3.6.1. *A number of HTS helical resonators*

Type of helix	Resonant frequency/MHz	Unloaded Q at 77 K	N	a/mm	l_h/mm	r_h/mm
YBCO wire[64]	355	16000	–	–	–	–
Copper wire[64]	355	2000	–	–	–	–
YBCO wire[76]	420	12000	25	0.35	25	5.5
YBCO thick film wire[76]	415	2200	5.5	0.9	32	7
Copper wire[76]	470	3500	20	0.5	25	5
YBCO patterned thick film[76]	280	700	7	2 × 0.04	32	7.5

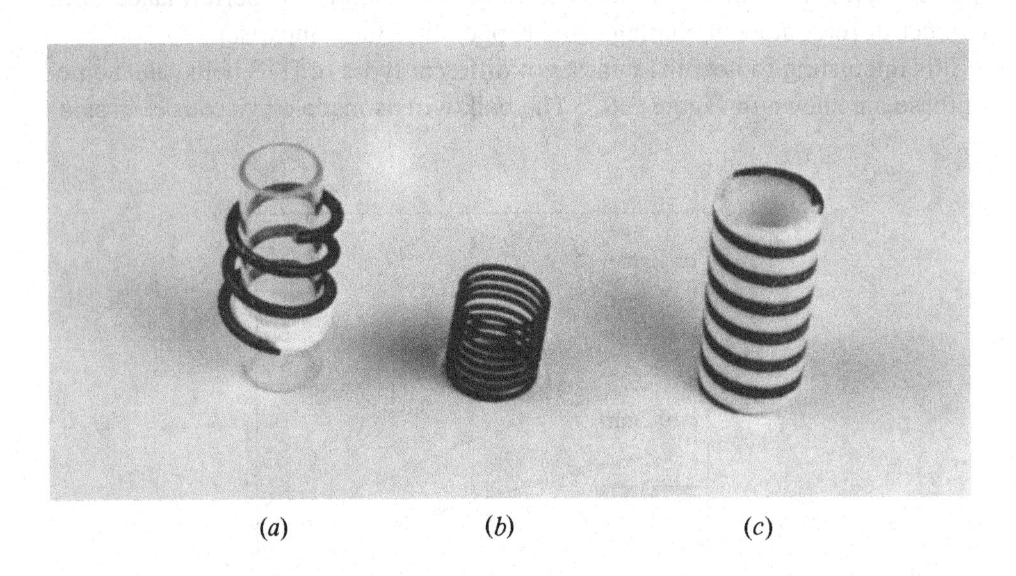

(*a*) (*b*) (*c*)

Figure 3.6.3. (*a*) Thick film helix coated on zirconia wire on a quartz mount. (*b*) Bulk wire helix. (*c*) Thick film helix patterned on a zirconia cylinder.

of YBCO powder and solvents through a die; the thick film material is made either by painting the YBCO mixture on a zirconia wire or by producing a film helix on the surface of a zirconia cylinder. The latter helix is essentially a tape, but because part of the zirconia cylinder is uncovered the microwave field enters it, limiting the Q of the cavity due to the high losses in the zirconia. It can be seen from Table 3.6.1 that high Qs can be obtained even with non-optimised materials; substantial improvements can be made by using better optimised thick film and bulk materials. The Q values are substantially lower than those of Figure 3.6.3 because the surface resistance of the material is higher and the frequency dependence is not necessarily frequency-squared, as assumed in the calculation of R_s for the helix in this diagram.

3.7 Cavities constructed from microstrip and stripline

3.7.1 Wide microstrip

The wide microstrip resonator is shown in Figure 3.7.1; it consists of a length of wide microstrip transmission line, discussed extensively in Section 2.2. The wide microstrip is so-called because the width of the line (w) is much larger than the spacing between the upper and lower conductors (h). The microstrip shown is an open circuit at both ends and hence reflections can occur at these points. The result is standing waves set up in the section of transmission line, with voltage peaks as the ends and a current maximum in the centre. Like other transmission line cavities with open-circuit ends, the fundamental mode is when the resonator is a half-wavelength long. Other modes occur at integer values of half-wavelengths ($n\lambda/2$). Wide microstrip is discussed in this section because of the ease of analysis, and estimations can be made of the quality factors of this type of cavity. This resonator is very similar to the general microstrip resonator in the next section. Although the usual resonance is longitudinal, it is also possible to obtain resonances across the width of the resonator; the first such mode occurs when the width is a half-wavelength. This can either be used in the design of a multi-resonance device or, if not wanted the modes moved to a higher frequency by narrowing the resonator.

If energy is coupled into the resonator, then a standing electromagnetic field

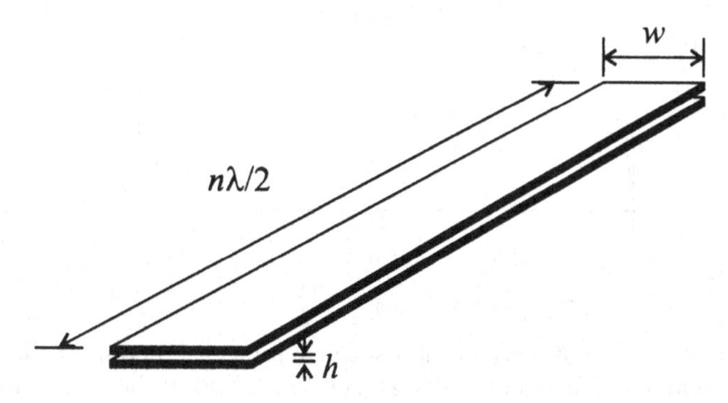

Figure 3.7.1. Wide microstrip cavity resonator.

is set up. This field is spatially invariant but oscillates with time. The quality factor is then given by Equation (3.2.29), that is,

$$Q_c = \frac{\omega}{2\alpha_c c} \tag{3.7.1}$$

Substitutions can be made for α and from Equation (2.3.6), and assuming that the film is a thin film superconductor results in the expression

$$Q_c = \frac{\omega\mu h}{2R_s} \frac{\left(1 + \frac{2\lambda}{h}\coth(t/\lambda)\right)}{\left(\coth(t/\lambda) + \frac{t/\lambda}{\sinh^2(t/\lambda)}\right)} \tag{3.7.2}$$

Because the dielectric completely fills the cavity, the dielectric Q is given by the inverse of the loss tangent and because the microstrip is wide and thin, the radiation losses are negligible. Figure 3.7.2 shows how the Q of a microstrip cavity varies as a function of the film thickness for a number of different conductor spacings.

It can be seen from Figure 3.7.2 that the quality factor is highly dependent on both the film thickness and the spacing between films. As the film thickness drops below a penetration depth, the quality factor can be substantially

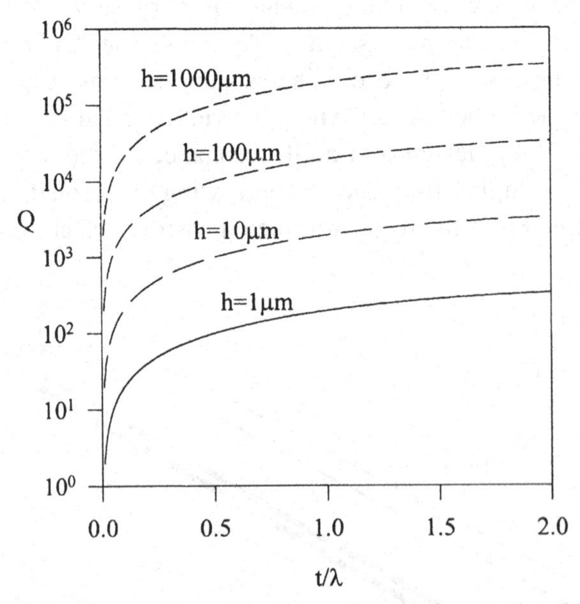

Figure 3.7.2. Quality factor of a wide microstrip cavity as a function of the thickness of the film for a number of different spacings $h\,(h \gg \lambda)$. The surface resistance used is 100 $\mu\Omega$ and the frequency 10 GHz.

reduced; in addition, as the spacing between the films decreases, the quality factor again drops substantially. This reduction is due to leakage through the film. However, for reasonably thick films with larger film spacing, the Q values are quite high and appropriate for high-performance applications.

3.7.2 Microstrip and stripline resonators

Conventional microstrip resonators differ from wide microstrip resonators only in the fact that the width of the microstrip line can no longer be taken as large compared with substrate thickness. Thus the analytical calculations of the preceding section are no longer valid, although they can still be used as a guide to the performance of the resonators. Practical resonators more often fall into this category rather than being wide microstrip. A stripline resonator is the same as a microstrip, but it uses a stripline transmission line, discussed in Section 2.4. Here there are two ground planes rather than the single one found in the microstrip.

The great advantage of using these planar resonators over the three-dimensional cavity resonators discussed previously in this chapter is their smaller size and their ability to integrate into conventional microwave circuitry. The penalty paid is the reduction in Q and a poorer power handling capability over the larger three-dimensional resonators such as the cylindrical cavity or dielectric resonator. The stripline is more complex than the microstrip, requiring two substrates, and contacts are more difficult to apply. However, it has the advantage that the propagation mode is pure TEM and is therefore dispersion-free in the case of a superconductor. There is also virtually no radiation as it is almost fully enclosed. It also has a lower propagation velocity, and hence the resonators are smaller. This lower velocity arises because the field is totally enclosed within the dielectric, whereas in the case of the microstrip some of the field is external to the substrate. In comparison with the coplanar resonator discussed in the next section the current is more evenly distributed and therefore the power handling is better; however, two-sided deposition is required, as opposed to one for the coplanar resonator.

The idea of the microstrip resonator is not new; early resonators, as well as a large proportion of the microstrip resonators made using high-temperature superconductors, were used in order to find the surface impedance of the superconductor. Such a technique was introduced by Young[79] in 1960. This was later taken up by Mason and Gould[80] and Henkels and Kircher,[81] all three studying low-temperature superconductors. In 1971 DiNardo looked at both linear and ring microstrip resonators as well as a coaxial resonator from the application viewpoint of attaining high Q devices.[82] Unloaded quality factors in excess of 500 000 at 14 GHz and 1.8 K were reported for lead microstrip resonators; this reduced to just over 200 000 at 4.2 K.

Soon after the discovery of HTS, a number of groups looked at the microstrip resonator again,[83–87] and since then there has been considerable effort around the world using the device; a few examples of the work are given in references [88–105]. The majority of the work looks at the properties of the superconductors, but some resonators have been used for oscillator stabilisation, discussed in Section 8.2, and also for filters, discussed in Chapter 5.

An example of a stripline resonator used extensively for the characterisation of superconductors is the one designed at MIT.[106–108] This is shown in Figure 3.7.3. The structure consists of three separate superconducting films, clamped together in a copper package. The central strip is a meander line with a 35 Ω characteristic impedance; a gap at the end of the line enables feed lines to the

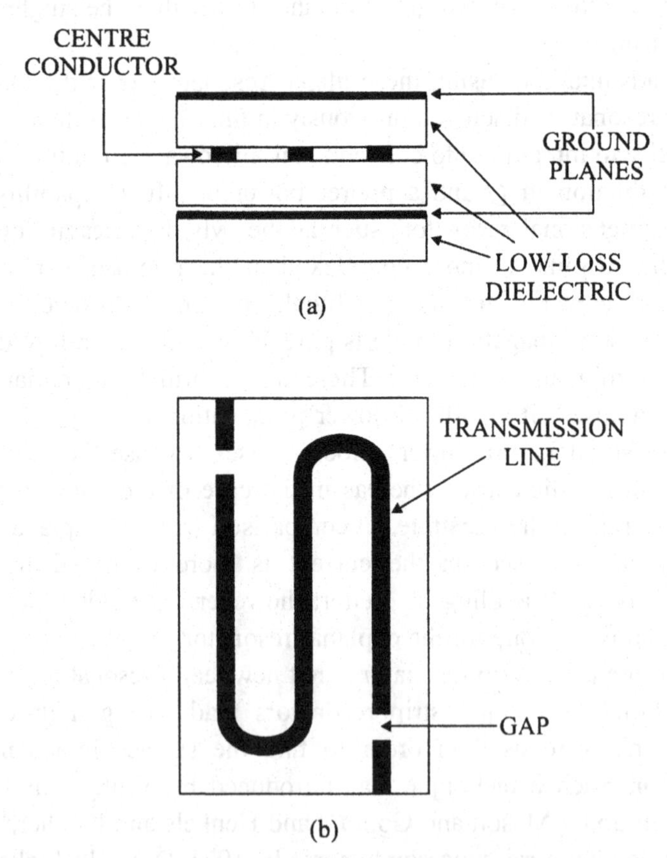

Figure 3.7.3. A schematic of a stripline resonator in the form of a meander line: (*a*) cross-section, (*b*) top view showing the superconductor track. (Taken from reference [108].)

device to be patterned on to the substrate. The fundamental resonant frequency of the device is 1.5 GHz, although many of the harmonics of this can be used. The whole resonator fits on a 1-cm-square $LaAlO_3$ substrate. Although the measured quality factor (and hence the surface impedance) is a weighted average of the contributions from three films, because of the concentration of current on the central strip it contributes 80%. In order to determine the surface impedance from such a resonator, the electromagnetic fields need to be calculated numerically, since analytic calculations with the required accuracy are not available. Figure 3.7.4 shows the current distribution on this stripline resonator using the method of coupled transmission lines described in Section 2.7. The determination of surface impedance from planar resonators in relation to the coplanar resonator is discussed more extensively in the next section.

The quality factors of resonators in the literature vary greatly because of the different quality of films and geometries of the microstrip or stripline. The different frequencies of operation also complicate matters, although this can be removed by assuming that the surface resistance is proportional to f^2. Thus the Q is inversely proportional to frequency. This is evident for the wide microstrip in Equation (3.7.2). A survey of the literature shows that the unloaded Q values for YBCO resonators at 77 K and normalised to 10 GHz lie in the range of several thousands to several tens of thousands. This rises as the temperature is decreased. For niobium resonators at 4.2 K, these figures rise by a factor of

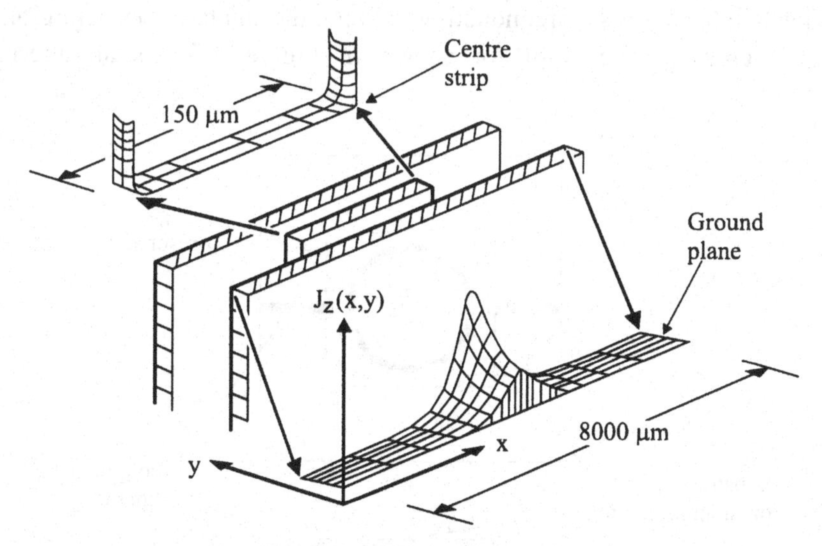

Figure 3.7.4. Calculated current distribution (J_z) for the centre conductor and ground plane for the resonator shown in Figure 3.7.3. The scale is different for the central strip and the ground plane. The parameters used in the calculation are $\lambda = 0.16$ μm with a film thickness of 0.30 μm. (Taken from reference [108].)

about 10. A few examples of high-Q resonators are given throughout this section.

The dielectric loss is not a limiting factor in the performance of microstrip resonators since the loss tangents of the substrates are fairly low. For very-high-Q resonators an inverted microstrip structure with a sapphire dielectric layer may help improve the performance. For the ultimate in performance in terms of Q the radiation from the structure (or the absorption of energy due to the proximity of the box walls) must be considered and optimised. The radiation can be reduced by reducing the substrate thickness.[109,110] Box resonances must be kept well away from the resonance of interest, so as not to have any modal coupling. The box must also be made as large as possible in order to reduce wall losses. These two criteria may not be compatible and careful design is needed. An optimum Q exists which is a compromise between having thin substrates (reducing the radiation loss but increasing the conductor loss) and having thick substrates (which increase the radiation loss, but reduce the loss due to the finite surface resistance of the HTS sample). The use of stripline removes the radiation loss almost entirely. An example of a microstrip resonator, where the shield or box is important, is given in reference [111]. Here, for the particular geometry of a microstrip using a YBCO thin film, a Q of about 26 000 is obtained at 77 K and 5.7 GHz, which increases to 60 000 at 10 K as the temperature decreases. A sudden increase in Q occurs at 4.2 K, increasing the Q to about 1.7×10^5. The sudden jump is due to a niobium shield which becomes superconducting, making the shield losses negligible.

The microstrip ring resonator shown in Figure 3.7.5 is an alternative

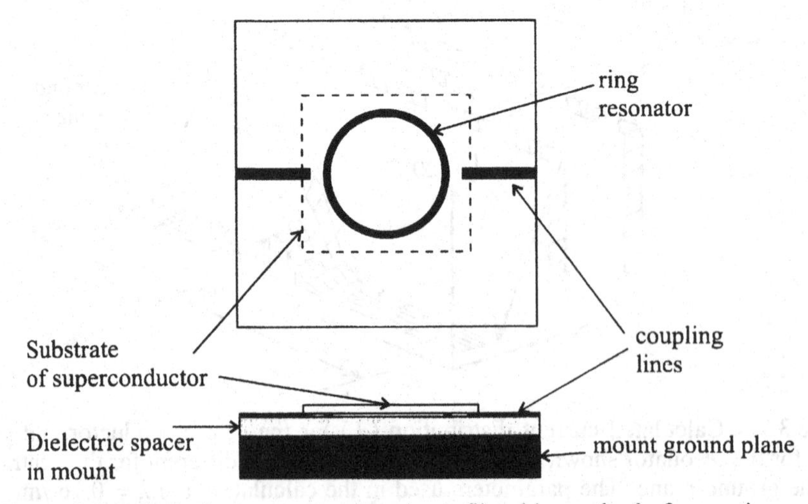

Figure 3.7.5. Split ring resonator showing a flip-chip method of mounting.

structure for a microstrip resonator and a number of groups have developed this for application purposes.[112–116] The main problem with the resonator is the appearance of a degenerate fundamental mode; such modes can be separated by perturbations of the field around the circumference of the ring. In addition to the degenerate fundamental mode which appears with a current distribution such that there is a single full wavelength around the ring, another resonance at approximately the same frequency can appear due to two half-wavelength resonances caused by reflections from the two coupling lines. Splitting of the ring or using coupling of the lines at different points around the ring helps to alleviate this latter problem. It is possible to vary the frequency of resonance of such a ring resonator by placing a variactor diode in series with a break and applying a bias current. Changes in frequency of 30% have been demonstrated using a normal conducting ring.[117] It is in fact possible to use the multiple modes in a ring resonator to build dual mode filters, as discussed in Section 5.3.

Figure 3.7.5 shows a mounting arrangement for a ring resonator for the measurement of the properties of the superconductor.[118] The ring, patterned from a thin film, is flip-chip mounted on to the test structure, where the coupling lines and ground plane lie. The test structure ground plane can either be a normal conductor or preferably a superconductor. This test rig allows quick and easy mounting of devices for testing. Niobium films using this structure with a superconducting ground plane give an unloaded Q of about 140 000 at 3.5 GHz and 4.2 K.

3.8 Coplanar resonators

The coplanar resonator is shown in Figure 3.8.1. It consists of two ground planes and a central strip of superconductor deposited on the surface of a substrate. The central strip is an open circuit at each end, forming the resonant section of the coplanar transmission line. It is also possible to make the central strip a meander line or a spiral, so reducing the resonant frequency. For the case of the resonator shown in Figure 3.8.1, the ground planes have silver evaporated on to each end in order to connect them to the package. The package also connects both ground planes. The resonator shown resonates at about 8 GHz and fits on a 1-cm-square MgO substrate.

One advantage of the coplanar resonator over other types of superconducting planar resonator is that it is easy to produce, as only one side of the substrate requires deposition of the superconductor. When used as a characterisation tool for superconductors, it also has the advantage of very high sensitivity as the resonator consists wholly of the superconductor of interest. Because of the

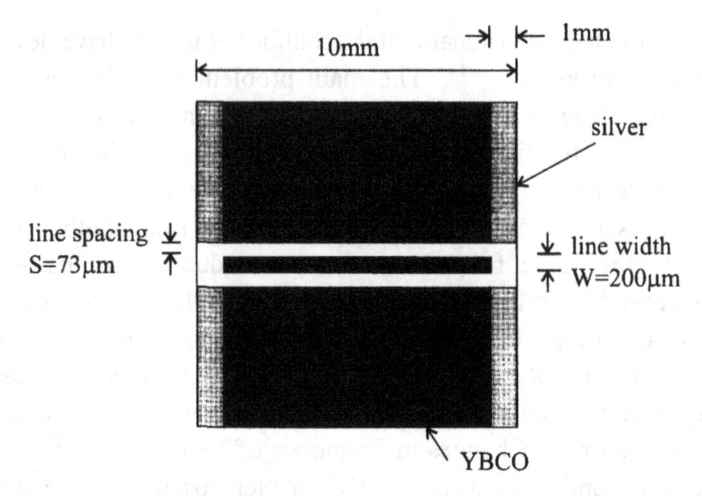

Figure 3.8.1. The coplanar resonator.

large peaking of the current at the edges of the central strip, it is also possible to study the effects of patterning[119] and nonlinearity.[120] In fact, it is found that the material is generally insensitive to patterning. Although the coplanar resonator cannot generate Qs as high as the larger three-dimensional cavities, with the correct choice of geometry it is possible to produce high-quality factors. For example, the resonator shown in Figure 3.8.1, with a good epitaxial film, has produced Qs of 45 000 at 15 K and 6500 at 77 K, both at 8 GHz. Such Qs are of the same order as other types of planar superconducting resonator. The small size of the coplanar resonator makes it particularly attractive for integration with other planar components to form the basis filters and oscillators. This is, of course, true for other planar resonators, such as the microstrip resonator discussed in Section 3.7.

One of the main applications of the coplanar resonator with relation to superconductors has been the measurement of surface impedance. This section now concentrates on this application, since both filters and oscillators are discussed elsewhere. The first reports of using the resonator for surface impedance measurements of high-temperature superconductors appeared in 1989,[121,122] since then the technique has been developed and is used by many groups around the world. The discussion which follows concentrates on a technique which can determine the surface resistance and surface reactance,[123–125] and hence the penetration depth of the superconductor. In this technique the absolute value of the penetration depth is measured. This is in contrast to other techniques, which use data from the change in the resonant frequency of a cavity to determine the change in penetration depth, and use curve fitting techniques in order to determine λ_0, the penetration depth at absolute zero. In this, the technique is particularly powerful as there are few methods for determining the absolute penetration depth of superconductors.

For the case of an infinitely thick substrate, infinitely wide ground planes and thin conductors, the propagation characteristics can be deduced analytically by conformal mapping techniques, resulting in an inductance and capacitance per unit length of[126]

$$C = 4\varepsilon_0 \frac{(\varepsilon_r + 1)}{2} \frac{K(k)}{K(k')} \quad \text{and} \quad L = \frac{\mu_0}{4} \frac{K(k')}{K(k)} \tag{3.8.1}$$

Here K is the complete elliptic integral of the first kind and

$$k = \frac{W}{W + 2S} \quad \text{and} \quad k'^2 = 1 - k^2 \tag{3.8.2}$$

where W is the width of the central strip and S is the spacing between the strip and the ground plane. These equations can be used to estimate the impedance ($\sqrt{(L/C)}$) and velocity ($1/\sqrt{(LC)} = c_0/\sqrt{((\varepsilon_r + 1)/2)}$) of a coplanar line. With these approximations, the incremental inductance rule described in Section 2.5 can be used to calculate the losses. Also, the PEM method (described in Section 2.6) can give an estimate of the effect of using thin films. However, for surface impedance measurements these estimates may not be adequate and a numerical solution taking the volume current distribution into account is required. Both partial wave synthesis[127] and the method of coupled lines[123] have been used to solve this problem for the coplanar resonator. These methods are discussed in Section 2.7; the discussion below uses the latter.

In Section 3.2.4 it was shown that for a resonator the surface impedance is connected to the resonator bandwidth and the frequency shift by

$$R_s + j\Delta X_s = \frac{\Gamma}{f}(f_B - 2j\Delta f) \tag{3.8.3}$$

Here f_B is the resonance bandwidth and Δf is the frequency shift due to a change in penetration depth. This expression is only valid when the penetration depth is small compared with the dimensions of the resonator, and in the case under discussion it needs to be modified. Two resonator geometry factors are now required such that

$$R_s + j\Delta X_s = \Gamma_1(\lambda) f_B - 2j\Gamma_2(\lambda)\Delta f \tag{3.8.4}$$

Here, f in the denominator of Equation (3.8.3) has been absorbed into Γ_1 and Γ_2. Now Γ_1 and Γ_2 are interpreted in terms of the inductance and resistance of the coplanar transmission line as[123]

$$\Gamma_1(\lambda) = 2\pi\mu_0 \frac{g_1(\lambda)}{g_2(\lambda)} \quad \text{and} \quad \Gamma_2(\lambda) = 2\pi\mu_0 \frac{g_1(\lambda)}{\partial g_1(\lambda)/\partial \lambda} \tag{3.8.5}$$

where

$$L(\lambda) \approx \mu_0 g_1(\lambda) = \frac{X_s}{\omega\lambda} g_1(\lambda) \quad \text{and} \quad R(\lambda) \approx \frac{1}{2}\mu_0^2 \sigma_1 \omega^2 \lambda^3 g_2(\lambda) = R_s g_2(\lambda) \tag{3.8.6}$$

The approximation $\sigma_1 \ll \sigma_2$ has been used here to simplify the equations. Thus, using the numerical method outlined in Section 2.7, both $R(\lambda)$ and $L(\lambda)$ can be determined, allowing Equation (3.8.4) to be used to determine R_s and ΔX_s from the bandwidth of the resonance and frequency shift, respectively, of the resonator.

However, in order to calculate the film surface impedance at a particular temperature the value of λ at that temperature is required. In principle this could be done by measuring the exact frequency of operation and calculating λ from the geometry of the resonator. However, this is impossible because the size measurements of the resonator would have to be made on a scale of less then a penetration depth and the modelling would have to include end effects at the open circuits of the resonator. The penetration depth needs to be deduced from measurements of the shift in frequency only. This can be achieved by measuring the properties of two different coplanar resonators, one with a ground plane spacing different from the other.

The frequency shift as a function of temperature is given by

$$\frac{\Delta f(T)}{f(T)} = \frac{1}{2}\frac{\Delta\varepsilon_r}{\varepsilon_r} + \frac{\Delta l}{l} + \frac{1}{2g_1}\frac{\partial g_1}{\partial\lambda}\Delta\lambda \tag{3.8.7}$$

The first term on the right-hand side represents a change in dielectric constant with respect to temperature, the second a change in the length of the resonator and the third the change in the penetration depth. It has been shown for the geometry and materials of the coplanar resonator under discussion that the first two terms are negligible.[123] The technique of extracting the penetration depth can be explained with reference to Figure 3.8.2.

Two calibration curves can be plotted, one for the resonator with a ground plane spacing of 73 μm as shown in Figure 3.8.2 (a) and another for the resonator with a ground plane spacing of 12 μm, as shown Figure 3.8.2 (b). The curves show the change in resonant frequency as a function of penetration depth for a number of different penetration depths at absolute zero (λ_0). Measurements can be made of a change in resonant frequency from a resonant frequency at low temperature. These will be different for each resonator because of the different geometry, but each can be plotted on the abscissa of each of the graphs shown in Figure 3.8.2. For each of these frequency changes, there will only be one value of penetration depth which corresponds to the same λ_0. An example is shown in Figure 3.8.2, where the change in frequency is 0.017 and 0.006 for the wide and narrow gap resonators respectively. If a line

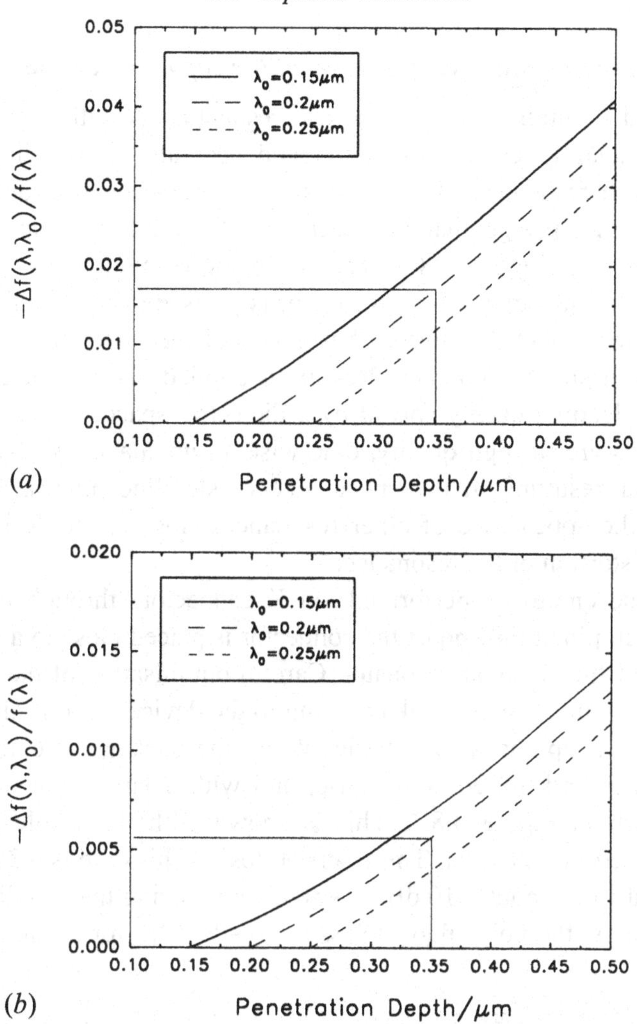

Figure 3.8.2. Frequency shift as a function of penetration depth for a number of different λ_0: (*a*) ground plane spacing 73 μm, (*b*) ground plane spacing 12 μm.

is drawn on the graphs at the appropriate points, then the only λ_0 and λ to which these two values correspond are 0.2 μm and 0.35 μm respectively. This gives the penetration depth at any temperature and is repeated for every temperature where measurements are made.

The surface resistance then needs to be deduced. The bandwidth due to conductor losses can be calculated as a function of λ from Equation (3.8.6) for a given number of values of σ_1. Hence σ_1 can be deduced from the measured value of f_B, and the deduced value of λ at any temperature. Then R_s is deduced using $R_s = \mu_0^2\omega^2\sigma_1\lambda^3/2$.

3.8.1 Packaging and measurements of the coplanar resonator

The package and mounting of the coplanar resonator, as with other planar resonators, is crucial to correct operation and accurate surface impedance measurements. The resonance of interest must be free from spurious modes and interference. A package that has been found suitable for the coplanar resonator is shown in Figure 3.8.3. The resonator is mounted so that the evaporated silver on the edges of the resonator is pressed against ledges in the packaging, indium is placed between the brass and the silver to maintain a good evenly spread pressure contact. Pressure is applied via beryllium copper springs, with the loading evenly spread by a dielectric spacer. It is important that these contacts are of high quality, otherwise an imbalance of the ground planes may occur resulting in the appearance of slot line modes. Slot line modes result in the appearance of other resonances close to the desired one; this may cause distortion of the resonance.

Coupling of the energy is performed via K-connectors through the brass housing. The small pin at the end of the connector is placed close to and above the central strip of the coplanar resonator. Careful optimisation of the position of this pin results in the desired weak coupling to the device. Care is also taken to ensure that the coupling is symmetric. When the package is optimised a resonance which is undistorted, symmetric, and with a large dynamic range appears, as shown in Figure 3.8.4. This is shown with the whole package mounted in a closed cycle cooler. The insertion loss in this case is −35 dB but it can be adjusted from about −10 dB to −50 dB by moving the coupling pins. At low temperatures, the noise floor is less than −90 dBm for an input power

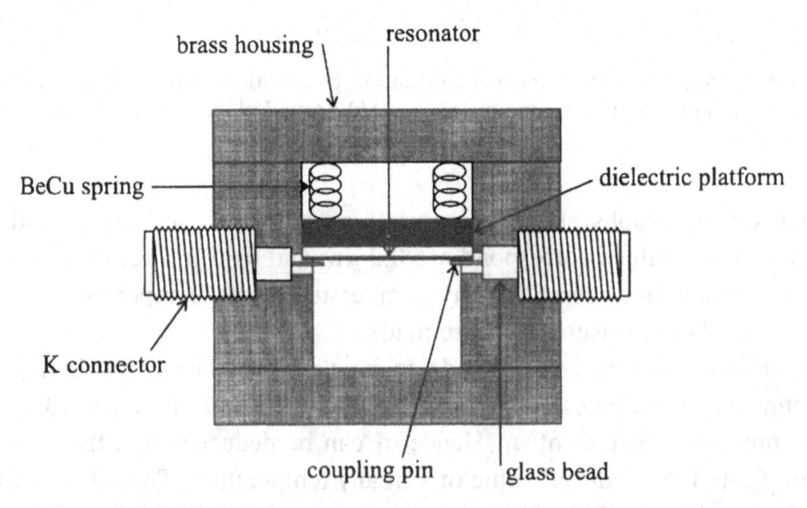

Figure 3.8.3. The coplanar resonator mounted in the brass package.

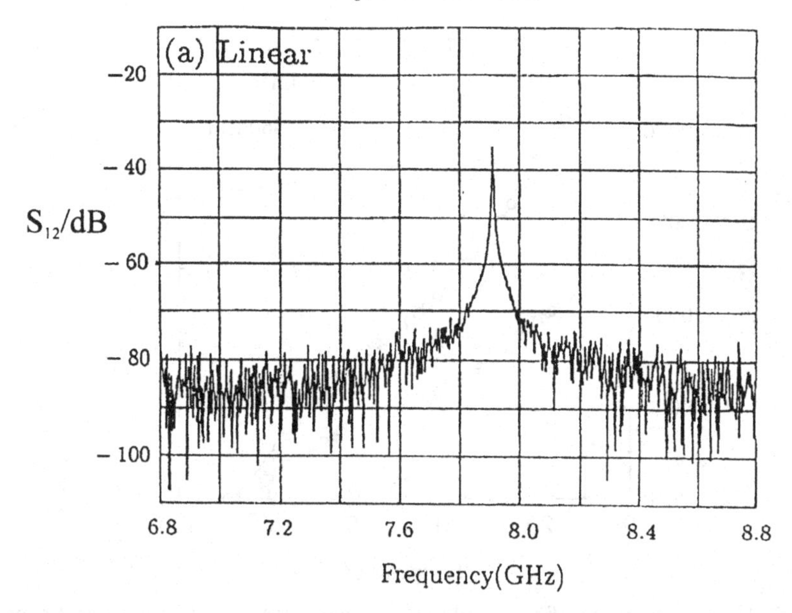

Figure 3.8.4. Measurement of S_{21} of the packaged coplanar resonator at 77 K.

of −10 dBm and insertion loss −20 dBm. This results in a dynamic range in excess of 60 dB. Measurements are taken as given in detail in Chapter 4.

Measurements of the unloaded Q are shown in Figure 3.8.5. These are translated to surface resistance and penetration depth measurements by the above technique and are shown in Figures 3.8.6 and 3.8.7. The film on which the measurements are made is a YBCO film 0.35 μm thick and produced by coevaporation. Further details of the production are given in reference [123]. A much fuller set of measurements, including the microwave power dependence and d.c. magnetic field dependence, is given in Appendix 1. It can be seen that the unloaded Q varies from around 45 000 at 12 K down to 6500 at 77 K, high enough for high-performance applications when used in filters and oscillators.

Figure 3.8.6 shows the measurements of the surface resistance. The values obtained at 8 GHz range from 23 $\mu\Omega$ at 15 K to 110 $\mu\Omega$ at 77 K, with a systematic error of ±30% due mainly to the measurement of the value of the gap between the ground plane and signal line in the narrow resonator. The error in the results due to the finite loss tangent of the MgO has been corrected for in Figure 3.8.6. This is accomplished by measuring the unloaded quality factor for a number of different modes at different frequencies. By assuming that the surface resistance is quadratic in frequency, and that the loss tangent is independent of frequency, the loss tangent value can be extracted at each temperature. The value obtained is about 5×10^{-6} at 15 K, although this varies by about 50% from substrate to substrate.

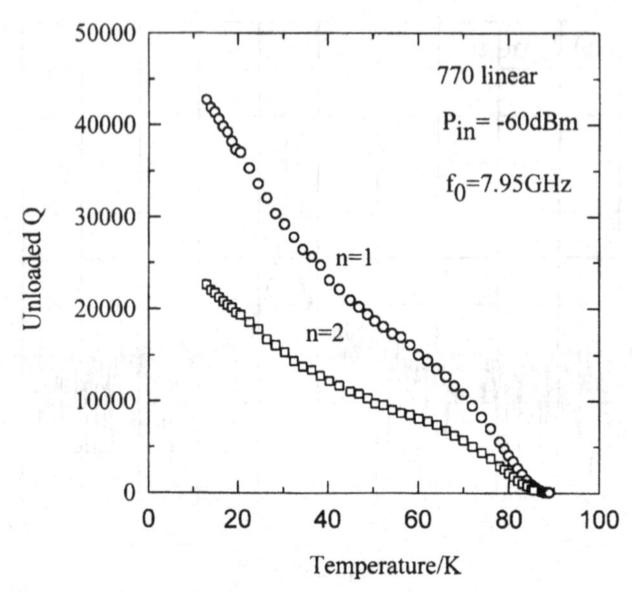

Figure 3.8.5. The unloaded Q of the coplanar resonator; the fundamental and second harmonics are shown at frequencies of about 8 and 16 GHz. The input power to the resonator (P_{in}) is -60 dBm.

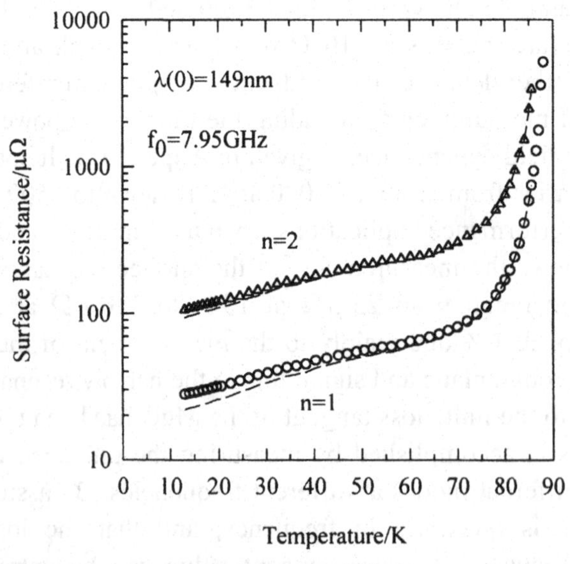

Figure 3.8.6. The surface resistance of the YBCO film measured using the coplanar resonator. The solid line includes the effects of a small temperature dependent loss tangent of the MgO substrate which is estimated to be around 5×10^{-6} at 15 K.

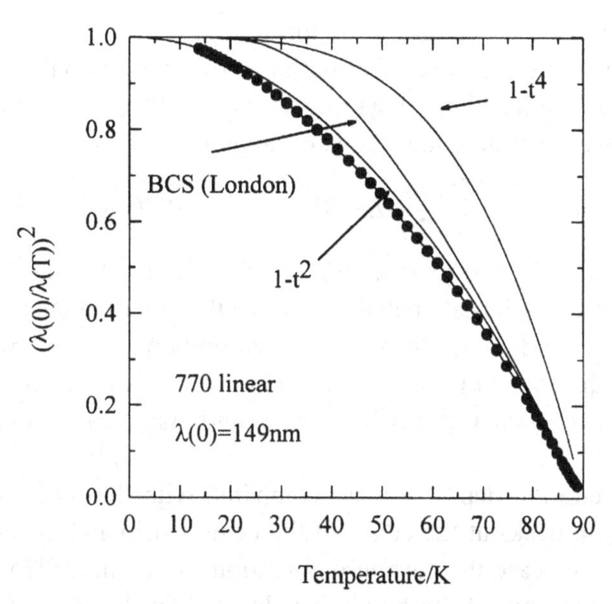

Figure 3.8.7. Measurements of the penetration depth of YBCO film compared with that of the BCS theory in the London limit and a $1 - t^2$ variation ($t = T/T_c$).

The results of the calculation of the penetration depth are shown in Figure 3.8.7. In this case the value of λ_0 is about 150 nm, consistent with other methods of measuring the penetration depth. In general λ_0 varies between 150 and 220 nm for different films, the ones with the highest loss having the largest penetration depth. The error in these results is around $\pm 10\%$.

3.8.2 Radiation losses in a coplanar resonator

As with any other resonator with external fields there will be losses by radiation from the structure. These can be estimated analytically for the coplanar resonator based on the discussion given in Section 3.3.3. The coplanar line can be considered as a number of parallel wires, each carrying a current which is cosinusoidal in magnitude along the length. The wires together make up the current distribution along the resonator. In this case, the magnitude of the electric field at a large distance from the resonator is given by[128]

$$|E(r, \theta, \phi)| = \frac{\eta}{2\pi r} \frac{\cos(\pi \cos(\theta)/2)}{\sin(\theta)} e^{-j\beta r} \int J(x) \exp(j\beta x \sin(\theta) \sin(\phi))\, dx \tag{3.8.8}$$

Here the resonator lies in the x, z plane with the z axis along its length and there is assumed to be no current variation normal to the plane in which the resonator sits. $J(x)$ is the current distribution across the resonator. In order to find the total power radiated, the Poynting vector needs to be integrated on the surface of a sphere containing the resonator, giving

$$P_{tot} = \int_0^{\pi}\int_0^{2\pi} \frac{1}{2\eta}|E(r,\theta,\varphi)|^2 r^2 \sin\theta \, d\theta \, d\phi \tag{3.8.9}$$

Using Equations (3.8.8) and (3.8.9) and knowing the current distribution, the total power radiated can be calculated and hence the radiation quality factor for the resonator can be found. The current distribution can be found by the numerical techniques given in Section 2.7. However, this is not convenient for the estimation of the radiation quality factor, and analytical results are more appropriate.

When the penetration depth is small compared with the film thickness the current distribution peaks at the edges of the central strip and the edges of the ground plane. In this case the current distribution can be modelled as just two pairs of wires. Each wire of the first pair is located on the edge of the central strip and each wire of the other pair is located on the edge of the ground plane next to the central strip. This approximation is valid at low temperatures when the penetration depth is small. In this case the radiation quality factor can be deduced to be

$$Q_r \approx \frac{2.49}{n^4}\frac{Z_0}{\eta}\frac{l^4}{S^2(S+W)^2} \tag{3.8.10}$$

The approximation that βS and βW must be much less than 1 has been made, which is valid for the coplanar resonator under discussion. Here Z_0 is the characteristic impedance of the line in the absence of the dielectric. For the coplanar resonator in Figure 3.8.1, the radiation quality factor in the above approximation is 8×10^6. This is much larger than the measured quality factors, even at low temperatures, and the radiation can therefore be neglected.

Another approximation for the current distribution valid close to T_c is that it is constant across the width of the central strip and decays away exponentially into the ground plane with decay length d, that is $J(x) = J_0 \exp(-x/d)$. Under this approximation, assuming that d is much less than the width of the ground planes, the radiation Q is reduced by a factor ν compared with the parallel wire case given in Equation (3.8.10). The factor ν is given by

$$\nu \approx \left(1 + \left(\frac{12d^2 + 6d(2S+W) + W^2}{6S(S+W)}\right)\right)^2 \tag{3.8.11}$$

This only reduces the radiation quality factor to 10^5, even when λ is as large as 10 μm. Larger values of penetration depth occur much closer to T_c, when the Q

of the resonator is substantially lower than this. This makes the radiation quality factor much greater than the conductor quality factor for all temperatures for the resonator in question.

The addition of a box surrounding the coplanar resonator can only improve matters provided that it is not too close to the resonator and that the box modes resonate at frequencies far away from the resonant frequencies of interest. Hence the calculations on the radiation loss given above, provided care is taken with the box design, provide the worst-case scenario.

It should be noted that there is an optimum line spacing, S, in order to maximise the Q of the coplanar resonator. If the line spacing is too wide, then the Q is reduced due to radiation losses or losses in the box walls; if the spacing is too narrow then the increase in conductor losses dominates, also reducing the Q. It is found that this optimum spacing is 150 μm for the coplanar resonator shown in Figure 3.8.1.[123] The Q is, however, only 5% higher than the Q of the resonator shown which had a 50 Ω line impedance and ground plane spacing of 73 μm.

3.9 References

1 Harrington R. F. *Time-harmonic Electromagnetic Fields*, McGraw-Hill, 1961
2 Kraus J. D. *Electromagnetic*, McGraw-Hill, 1984
3 Abramowitz M. and Stegun I. A. *Handbook of Mathematical Functions*, Dover Publications, New York, 1972
4 Portis A. M., Cooke D. W. and Gray E. R. RF properties of high temperature superconductors: cavity methods, *J. Supercond.*, **3**, 297, 1990
5 Portis A. M. *Electrodynamics of High Temperature Superconductors*, World Scientific, 1992
6 Kennedy W. L. and Sridhar S. Low temperature microwave surface resistance of $Y_1Ba_2Cu_3O_y$ and $La_{1.85}Sr_{0.15}CuO_4$, *Solid State Communications*, **68**(1), 71–75, 1988
7 Lancaster M. J., Wu Z., Maclean T. S. M. and Alford N. McN. High temperature superconducting cavity for measurement of surface resistance, *Cryogenics*, **30**, 1048–50, 1990
8 Phillips W. A., Jedamzik D., Lamacraft K., Zammattio S., Greed R. B., Hedges S. J., Whitehead P. R., Nicholson B. F., Button T. W., Smith P. A. and Alford N. McN. An integrated 11 GHz cryogenic down converter, *IEEE Trans. on Applied Superconductivity*, **5**(2), 2283–9, 1995
9 Bohn C. L., Delaven J. R., Balachandran U. and Lanagan M. T. Radio frequency surface resistance of large area Bi–Sr–Ca–Cu–O thick films on Ag plates, *Appl. Phys. Lett.*, **53**(3), 304–6, 1989
10 Müller G. Microwave Properties of high-T_c superconductors, *4th Workshop on RF Superconductivity, KEK, Tsukuba-Shi*, 1989
11 Müller G., Brauer D. J., Eujen R., Hein M., Klein N., Piel H., Ponto L., Klein U. and Peiniger M. Surface impedance measurements on high T_c superconductors, *IEEE Trans. Magnetics*, **25**(2), 2402–5, 1989

12 Klein N., Müller G., Piel H., Roas B., Schultz L., Klein U. and Peiniger M. Millimetre wave surface resistance of epitaxially grown $YBa_2Cu_3O_{7-\delta}$ thin films, *Appl. Phys. Lett.*, **54**(8), 757–9, 1989
13 Martens J. S., Beyer J. B. and Ginley D. S. Microwave surface resistance of $YBa_2Cu_3O_{6.9}$ superconducting films, *Appl. Phys. Lett.*, **55**, 1822, 1988
14 Zahopoulos C., Kennedy W. L. and Sridhar S. Performance of a fully superconducting microwave cavity made of the high T_c superconductor $Y_1Ba_2Cu_3O_y$, *Appl. Phys. Lett.*, **52**, 2168, 1988
15 Drabeck L., Holczer K., Grüner G., Chang J. J., Scalapino D. J., Inam A., Wu X. D., Nazar L. and Venkatesan T. Surface resistance of laser deposited $YBa_2Cu_3O_7$, *Phys. Rev.*, B**42**(16), 10020–9, 1990
16 Drabeck L., Holczer K., Grüner G. and Scalapino D. J. Ohmic and radiation losses in superconducting films, *J. Appl. Phys.*, **68**(2), 892–9, 1990
17 Hartemann P. Effective and intrinsic surface impedances of high T_c superconducting thin films, *IEEE Trans. on Appl. Superconductivity*, **2**(4), 229–35, 1992
18 Klein N., Chaloupka H., Müller G., Orbach S., Piel H., Roas B., Schultz L., Klein U. and Peiniger M. The effective microwave surface impedance of high T_c thin films, *J. Appl. Phys.*, **67**(1), 6940–5, 1990
19 Martens J. S., Beyer J. B. and Ginley D. S. Microwave surface resistance of $YBa_2Cu_3O_{6.9}$ superconducting films, *Appl. Phys. Lett.*, **52**(21), 1822–4, 1988
20 Li Q., Rigby K. W. and Rzchowski M. S. Temperature-, magnetic-field-, and power-dependant microwave resistance of $YBa_2Cu_3O_{7-\delta}$, *Phys. Rev.*, B**39**(10), 6607–11, 1989
21 Hagen M., Hein M., Klein N., Michalke A., Müller G., Piel H., Röth R. W., Mueller F. M., Sheinberg H. and Smith J. L. Observation of RF superconductivity in $Y_1Ba_2Cu_3O_{9-\delta}$ at 3 GHz, *J. Magnetism and Magnetic Materials*, **68**, L1–L5, 1987
22 Rubin D. L., Green K., Gruschus J., Kirchgessner J., Moffat D., Padamsee H., Sears J., Shu Q. S., Schneemeyer L. F. and Waszczak J. V. Observation of a narrow superconducting transition at 6 GHz in crystals of $YBa_2Cu_3O_7$, *Phys. Rev.*, B**38**(10), 6538–42, 1988
23 Exon N., Gough, C. E., Porch, A. and Lancaster, M. J. Measurements of the surface impedance of high temperature superconducting single crystals, *IEEE Trans. Applied Superconductivity*, **3**(1), 1442–5, 1993
24 Gallop J. C., Quincey P. G. and Radcliffe W. J. Surface impedance properties of *in situ* HTS thin films measured in a superconducting cavity, *Supercond. Sci. Technol.*, **4**, 574–6, 1991
25 Chang L. D., Moskowitz M. J., Hammond R. B., Eddy M. M., Olson W. L., Casavant D. D., Smith E. J. and Robinson M. Microwave surface resistance in Tl-based superconducting thin films, *Appl. Phys. Lett.*, **55**(13), 1357–9, 1989
26 Sridhar S. and Kennedy W. L. Novel technique to measure the microwave response of high T_c superconductors between 4.2 and 200 K, *Rev. Sci. Instrum.*, **59**, 531, 1988
27 Rubin D. L., Green K., Gruschus J., Kirchgessner J., Moffat D., Padamsee H., Sears J., Shu Q. S., Schneemeyer L. F. and Waszczak J. V. Observation of a narrow superconducting transition at 6 GHz in crystals of $YBa_2Cu_3O_7$, *Phys. Rev.*, B**38**, 6538, 1988
28 Aswasrhi A. M., Carini J. P., Alavi B. and Gruner G. *Solid State Commun.*, **67**, 373, 1988

29 Exon N. Microwave Properties of Single Crystal High Temperature Superconductors PhD thesis University of Birmingham UK, 1994
30 Tellmann N., Klein N., Dähne U., Scholen A., Schulz H. and Chaloupka H. High Q $LaAlO_3$ dielectric resonator shielded by YBCO-films, *IEEE Trans. on Applied Superconductivity*, **4**(3), 143–8, 1994
31 Braginsky V. B. and Panov V. I. Superconducting resonators on sapphire, *IEEE Trans. on Magnetics*, **MAG-15**(1), 30–2, 1979
32 Trinogga L. A., Kaizhou G. and Hunter I. C. *Practical Microstrip Circuit Design*, Ellis Horwood, England, 1991
33 Kobayashi Y. and Katoh M. Microwave measurements of dielectric properties of low loss materials by the dielectric rod resonator method, *IEEE Trans. on Microwave Theory and Techniques*, **MTT-33**(7), 586–92, 1985
34 Courtney W. Analysis and evaluation of a method of measuring the complex permitivity and permeability of microwave insulators, *IEEE Trans. on Microwave Theory and Techniques*, **18**, 476–85, 1970
35 Blair D. G. and Jones S. K. A high Q sapphire loaded superconducting cavity resonator, *J. Phys. D: Appl. Phys.*, **20**, 1559–6, 1987
36 Blair D. G. and Sanson A. M. High Q tuneable sapphire loaded cavity resonator for cryogenic operation, *Cryogenics*, **29**, 1045–9, 1989
37 Mann A. G., Giles A. G., Blair D. G. and Buckingham M. J. Ultra-stable cryogenic sapphire dielectric microwave resonators: mode frequency-temperature compensation by residual paramagnetic impurities, *J. Phys.*, D**25**, 1105–9, 1992
38 Kobayashi Y. and Tanaka S. Resonant modes of a dielectric rod resonator short-circuited at both ends by parallel conducting plates, *IEEE Trans. on Microwave Theory and Techniques*, **MTT-28**, 1077–85, 1980
39 Kajfez D. and Guillon X. Dielectric Resonators, Oxford, MS: Vector Fields, 1990
40 Shen Z.-Y., Wilker C., Pang P., Holstein W. L., Face D. and Kountz D. J. High T_c superconductor–sapphire microwave resonator with extremely high Q values up to 90 K, *IEEE Trans. on Microwave Theory and Techniques*, **40**(12), 2424–32, 1992
41 Fletcher R. and Cook J. Measurement of surface impedance versus temperature using a generalised sapphire resonator technique, *Rev. Sci. Instrum.*, **65**(8), 2658–66, 1994
42 Edgcombe C. J. Measurement of surface resistance of high T_c superconductor by use of a dielectric resonator, *Electron. Lett.*, **27**(10), 815–17, 1991
43 Kajfez D. Incremental frequency rule for computing the Q-factor of shielded TE_{0np} dielectric resonator, *IEEE Trans. on Microwave Theory and Techniques*, **MTT-32**, 941–3, 1984
44 Kobayashi Y., Aoki T. and Kabe Y. Influence of conductor shields on the Q factors of a TE_0 dielectric resonator, *IEEE Trans. on Microwave Theory and Techniques*, **33**, 1361–6, 1985
45 Kobayashi Y., Imai T. and Kayano H. Microwave measurement of temperature and current dependencies of surface impedance of high-T_c superconductor, *IEEE Trans. on Microwave Theory and Techniques*, **39**(9), 1530–8, 1991
46 Wilker C., Shen Z.-Y., Nguyen V. X. and Brenner M. S. A sapphire resonator for microwave characterisation of superconducting thin films, *IEEE Trans. Applied Superconductivity*, **3**, 1457–60, 1993
47 Kraus J. D. *Electromagnetics*, McGraw-Hill, 1985
48 Davis L. E. and Smith P. A. Q of a coaxial cavity with a superconducting inner conductor, *IEE Proc.*, A**138**(6), 313–19, 1991

49 Radcliffe W. J., Gallop J. C., Langham C. D., Alford N. McN and Button T. W. Multi-mode microwave measurements on a coaxial cavity with high-temperature superconductor centre conductor, *Supercond. Sci. Technol.*, **3**, 151–4, 1990
50 Gallop J. C., Radcliffe W. J., Button T. W. and Alford N. McN. Microwave surface impedance in a coaxial cavity as a material characterisation technique, *IEEE Trans. on Magnetics*, **27**(2), 1310–12, 1991
51 Marcuvitz N. *Waveguide Handbook*, McGraw-Hill, New York, 1951
52 Elliott R. S. *An Introduction to Guided Waves and Microwave Circuits*, Prentice-Hall, New Jersey, 1993
53 Woodall P., Lancaster M. J., Maclean T. S. M., Gough C. E. and Alford N. McN. Measurements of the surface resistance of $YBa_2Cu_3O_{7-\delta}$ by the use of a coaxial resonator, *IEEE Trans. on Magnetics*, **27**(2), 1264–7, 1991
54 Pippard A. B. *Proc. R. Soc. London*, **A203**, 98, 1950
55 Bohn C. L., Delayen J. R., Dos Santos D. I., Lanagan M. T. and Shepard K. W. RF properties of high-T_c superconductors, *IEEE Trans. on Magnetics*, **MAG25**, 2406–9, 1989
56 Delayen J. R., Goretta K. C., Poeppel R. B. and Shepard K. W. RF properties of an oxide–superconductor half wave resonant line, *Appl. Phys. Lett.*, **52**(11), 930–2, 1988
57 Peterson G. E., Stawicki R. P. and Paek U. C. *J. Am. Ceram. Soc.*, **72**, 704–6, 1989
58 Ganne J. P., Kormann R., Labeyrie M., Lainee F. and Lloret B. Frequency dependence of microwave surface resistance of YBCO superconducting ceramics, *Physica*, **C162–4**, 1541–2, 1989
59 Klein N. *Appl. Phys. Lett.*, **54**, 757–9, 1989
60 Labeyrie M. *et al. Journees d'Etudes*, at ISMRa, Caen, France 13–14 Sept. 1988
61 Spencer N. P. A review of the design of helical resonator filters, *J. Inst. Elect. Radio. Eng.*, **57**(5), 213–220, 1987
62 Rawat B. and Miller R. E. Improved design of a helical resonator filter for 450–500 MHz band land mobile communications, *IEEE Trans. on Vehicular Technology*, **VT-33**(1), 32–6, 1984
63 Porch A., Lancaster M. J., Maclean T. S. M., Gough C. E. and Alford N. McN. Microwave resonators incorporating ceramic YBa2CuO3 Helices, *IEEE Trans. on Magnetics*, **27**(2), 2948–51, 1991
64 Peterson G. E., Stawicki R. P. and Alford N. McN. Helical resonators containing high T_c ceramic superconductors, *Appl. Phys. Lett.*, **55**(17), 1798–1800, 1989
65 Meyer W. Dielectric measurements on polymeric materials by using superconducting microwave resonators, *IEEE Trans. Microwave Theory and Tech.*, **MTT25**(12), 1092–9, 1977
66 Deri R. J. Dielectric measurements with helical resonators, *Rev. Sci. Instrum.*, **57**(1), 82–6, 1985
67 Frenois C. Broadband tunable cavities with helical microstrip lines, *J. Phys. E: Sci Instrum.*, **17**, 35–9, 1984
68 Fossheim K. and Holt R. M. Broadband tuning of helical resonant cavities, *J. Phys. E: Sci. Instrum.*, **11**, 892–3, 1978
69 Macalpine W. W. and Schildknecht R. O. Coaxial resonators with helical inner conductors, *Proc. IRE*, **47**, 2099–105, 1959
70 Cauvin B., Coret M., Fouan J. P., Girard J., Girma J. L., Leconte Ph., Lussignol Y., Moreau R., Passérieux J. P., Ramstein G. and Wartski L. Fabrication, tests and RF control of the 50 superconducting resonators of the Saclay heavy ion linac, *3rd Workshop on RF Superconductivity, Argonne*, pp. 379–87, 1987

71 Miley D. J. and Beyer J. B. Field analysis of helical resonators with constant bandwidth filter application, *IEEE Trans. Parts, Mats., Pack*, **PMP5**(3), 127–32, 1969
72 Sichak W. Coaxial line with helical inner conductor, *Proc. IRE*, **42**, 1315–19, 1954
73 Sensiper S. Electromagnetic wave propagation on helical structures, *Proc. IRE*, **43**, 149–61, 1955
74 Stark L. Lower modes of a concentric line having a helical inner conductor, *J. Appl. Phys.*, **25**(9), 1155–62, 1954
75 Sollfrey W. Wave propagation on helical wires, *J. Appl. Phys.*, **22**(7), 905–10, 1951
76 Lancaster M. J., Maclean T. S. M., Wu Z., Porch A., Woodall P. and Alford N. McN. Superconducting microwave resonators, *IEE Proc.*, H**139**(2), 149–56, 1992
77 Howe G. W. O. The effect of screening cans of the effective inductance and resistance of coils, *Wireless Engineer*, **11**, 115–17, 1934
78 Terman F. E. *Radio Engineers Handbook*, McGraw-Hill, New York, p. 143, 1943
79 Young D. R., Swihart J. C., Tansal S. and Meyers N. H. *Sol. State Electron.*, **1**, 378, 1960
80 Mason P. V. and Gould R. W. *J. Appl. Phys.*, **40**, 2039, 1969
81 Henkels W. H. and Kircher C. J. Penetration depth measurements on type II superconducting films, *IEEE Trans. on Magnetics*, **MAG-13**, 63–6, 1977
82 DiNardo A. J., Smith J. G. and Arams F. R. Superconducting microstrip high Q resonators, *J. Appl. Phys.*, **42**(1), 186–9, 1971
83 DiIorio M. S., Anderson A. C. and Tsaur B.-Y. rf surface resistance of Y–Ba–Cu–O thin films, *Phys. Rev.*, B**38**(10), 7019–22, 1988
84 Anlage S. M., Sze H., Snortland H. J., Tahara S., Langley B., Eom C.-B., Beasley M. R. and Taber R. Measurements of the magnetic penetration depth in $YBa_2Cu_3O_{7-\delta}$ thin films by the microstrip resonator technique, *Appl. Phys. Lett.*, **54**(26), 2710–12, 1989
85 Taber R. C. *Rev. Sci. Instrum.*, **61**, 2200–6, 1990
86 McAvoy B. R., Wagner G. R., Adam J. D., Talvacchio J. and Driscoll M. Superconducting stripline resonator performance, *IEEE Trans. on Magnetics*, **25**(2), 1104–6, 1989
87 Martens J. S., Hohenwarter G. K. G., McGinnis D. P., Beyer J. B. and Ginley D. S. A transmission line resonator to measure the microwave surface resistance of $YBa_2Cu_3O_{7-x}$, *IEEE Trans. on Magnetics*, **25**(2), 984–6, 1989
88 Pinto R., Goyal N., Pai S. P., Apte P. R., Gupta L. C. and Vijayaraghavan R. Improved performance of Ag-doped $Y_1Ba_2Cu_3O_{7-\delta}$ thin film microstrip resonators, *J. Appl. Phys.*, **73**(10), 5105–8, 1993
89 Pinto R., Apte P. R., Gupta L. C., Vijayaraghavan R., Easwar K. and Sarkar B. K. Microstrip resonator using sputtered $Y_1Ba_2Cu_3O_{7-\delta}$ film on $\langle 100 \rangle$ MgO, *Superconductor Science and Technology*, **4**, 557–8, 1991
90 Gallop J. C. and Radcliffe W. J. Characterisation of microwave surface impedance of high temperature superconductors, *Superconductor Science and Technology*, **4**, 568–73, 1991
91 Kuhn M., Klinger M., Baranyak A. and Hinken J. H. HTSC inverted and conventional geometry microstrip resonator for UHF frequencies, *IEEE Trans. on Magnetics*, **MAG-27**, 2809–12, 1991
92 Reible S. A. and Wilker C. W. Parallel plate resonator for accurate RF surface loss measurements, *IEEE Trans on Magnetics*, **27**(2), 2813–16, 1991
93 Willemsen B. A., Derov J. S., Silva J. H. and Sridhar S. Non-linear response of suspended high temperature superconducting thin film microwave resonators, *IEEE Trans. on Applied Superconductivity*, **5**(2), 1753–5, 1995

94 Mueller C. H., Miranda F. A., Toncich S. and Bhasin K. B. YBCO X-band microstrip linear resonators on (1102) and (1100)-oriented sapphire substrates, *IEEE Trans. on Applied Superconductivity*, **5**(2), 2559–62, 1995
95 Clark J. H., Donaldson G. B., Gallop J. C. and Bowman R. M. Temperature dependence of the surface impedance of pulsed-laser-deposited YBCO films, *IEEE Trans. on Applied Superconductivity*, **5**(2), 2555–8, 1995
96 Reznik A. N., Zharov A. A. and Chernobrovtseva M. D. Non-linear thermal effects in the HTSC microwave stripline resonator, *IEEE Trans. on Applied Superconductivity*, **5**(2), 2579–82, 1995
97 Miura S., Yoshitake T., Tsuge H. and Inui T. Properties of shielded microstrip line resonators made from double-sided $Y_1Ba_2Cu_3O_x$ films, *J. Appl. Phys.*, **76**(7), 4440–2, 1994
98 Hammond R. B., Negrete G. V., Schmidt M. S., Moskowitz M. J., Eddy M. M., Strother D. D. and Skoglund L. Superconducting Tl–Ca–Ba–Cu–O thin film microstrip resonator and its power handling performance at 77 K, *IEEE MTT-S Int. Microwave Symp. Digest*, **2**, 867–70, 1990
99 Newman H. S., Chrisey D. B., Horwitz J. S., Weave B. D. and Reeves M. E. Microwave devices using $YBa_2Cu_3O_{7-\delta}$ films made by pulsed laser deposition, *IEEE Trans. on Magnetics*, **MAG-27**, 2540–3, 1991
100 Rensch D. B., Josefowicz J. Y., Macdonald P., Niel C. W., Hoefer W. and Krajenbrink F. Fabrication and characterisation of high T_c superconducting X-band resonators and bandpass filters, *IEEE Trans. on Magnetics*, **MAG-27**, 2553–6, 1991
101 Wilker C., Shen Z.-Y., Pang P., Face D. W., Holstien W. L., Matthews A. L. and Laubacher D. B. 5 GHz high temperature superconductor resonators with high Q and low power dependence up to 90 K, *IEEE Trans. on Microwave Theory and Techniques*, **MTT-39**, 1462–7, 1991
102 Young K. H., Negrete G. V., Hammond R. B., Inam A., Ramesch R., Hart D. L. and Yonezawa Y. *Appl. Phys. Lett.*, **58**, 1789, 1991
103 Gergis L. S., Kobrin P. H., Cheung J. T., Sovero E. A., Lastufka C. L., Deakin S. and Lopez J. *Physica*, **C175**; 603, 1991
104 Goettee J. D., Skocpol W. J. and Oates D. E. Two GHz microstrip thin film resonators of niobium and $YBa_2Cu_3O_7$, *IEEE Trans. on Applied Superconductivity*, **5**(2), 2547–50, 1995
105 Lynch J. T., Withers R. S., Anderson A. C., Wright P. V. and Reible S. A. Multigigahertz-bandwidth linear-frequency-modulated filters using a superconductive stripline, *Appl. Phys. Lett.*, **43**(3), 319–21, 1983
106 Oates D. E., Anderson A. C. and Shih B. S. Superconducting stripline resonators and high T_c materials, *IEEE MTT-S Digest*, 627–30, 1991
107 Oates D. E. and Anderson A. C. Surface impedance measurements of $YBa_2Cu_3O_{7-x}$ thin films in stripline resonators, *IEEE Trans. on Magnetics*, **27**(2), 867–71, 1991
108 Oates D. E., Anderson A. C., Sheen D. M. and Ali S. M. Stripline resonator measurements of Z_s versus H_{rf} in $YBa_2Cu_3O_{7-x}$ thin films, *IEEE Trans. on Microwave Theory and Techniques*, **39**(9), 1522–9, 1991
109 Abbas F. and Davis L. E. Radiation Q-factor of high T_c superconducting parallel plate resonators, *Electron. Lett.*, **29**(1), 105–7, 1993
110 Gopinath A. Maximum Q-factor of microstrip resonators, *IEEE Trans. on Microwave Theory and Techniques*, **MTT-29**(2), 128–31, 1981
111 Yoshitake T., Tsuge H. and Inui T. Effects of microstructure on microwave

properties in Y–Ba–Cu–O microstrip resonators, *IEEE Trans. on Applied Superconductivity*, **5**(2), 2571–4, 1995

112 To H. Y., Valco G. J. and Bhasin K. B. 10 GHz $YBa_2Cu_3O_{7-\delta}$ superconducting ring resonators on $NdGaO_3$ substrates, *Superconductor Science and Technology*, **5**, 421–6, 1992

113 Takemoto J. H., Jackson C. M., Hu R., Burch J. F., Daly K. P. and Simon R. W. Microstrip resonators and filters using high T_c superconducting thin films on $LaAlO_3$, *IEEE Trans. on Magnetics*, **MAG-27**, 2549–52, 1991

114 Takemoto J. H., Oshita F. K., Fetterman H. R., Kobrin P. and Sovero E. Microstrip ring resonator technique for measuring microwave attenuation in high-T_c superconducting thin films, *IEEE Trans. on Microwave Theory and Techniques*, **37**(10), 1650–2, 1989

115 Chorey C. M., Kong K. S., Bhasin K. B., Warner J. D. and Itoh T. YBCO superconducting ring resonators at millimetre frequencies, *IEEE Trans. on Microwave Theory and Techniques*, **MTT-39**, 1480–7, 1991

116 Lee S. Y., Kang K. Y. and Ahn D. Fabrication of superconducting dual mode resonator for satellite communications, *IEEE Trans. on Applied Superconductivity*, **5**(2), 2563–6, 1995

117 Al-Charchafchi S. H. and Dawson C. P. Varactor tuned microstrip ring resonators, *IEE Proc.*, H**136**(2), 165–8, 1989

118 Andreone A., DiChiara A., Peluso G., Santoro M., Attanasio C., Maritato L. and Vaglio R. Surface impedance measurements of superconducting (NbTi)N films by a ring microstrip resonator technique, *J. Appl. Phys.*, **73**(9), 4500–06, 1993

119 Avenhaus B., Porch A., Lancaster M. J., Hensen S., Lenkens M., Orbach-Werbig S., Müller G., Dähne U., Tellmann N., Klein N., Dubourdieu C., Thomas O., Karl H., Stritzker B., Edwards J. A. and Humphreys R. Microwave properties of YBCO thin films, *IEEE Trans. on Applied Superconductivity*, **5**(2), 1737–40, 1995

120 Porch A., Lancaster M. J., Humphreys R. G. and Chew N. G. Non-linear microwave surface impedance of Patterned $YBa_2Cu_3O_7$ thin films, *J. Alloys and Compounds*, **195**, 563–5, 1993

121 Valenzuela A. A., Daalmans B. and Roas B. High Q coplanar transmission line resonator of $YBa_2Cu_3O_{7-x}$ on $LaAlO_3$, *Electron. Lett.*, **25**(21), 1435–6, 1980

122 Valenzuela A. A. and Russer P. High Q coplanar transmission line resonator of $YBa_2Cu_3O_{7-x}$ on MgO, *Appl. Phys. Lett.*, **55**(10), 1029–31, 1989

123 Porch A., Lancaster M. J. and Humphreys R. Coplanar resonator technique for the determination of the surface impedance of patterned thin films, *IEEE Trans. on Microwave Theory and Techniques*, **43**(2), 306–14, 1995

124 Porch, A. and Lancaster M. J. Surface impedance measurements of $YBa_2Cu_3O_7$ thin films using coplanar resonators, *IEEE Trans. Appl. Superconductivity*, **3**(1), 1719–22, 1993

125 Porch A., Lancaster M. J. and Humphreys R. Microwave surface impedance of patterned YBCO thin films, *Physica*, B**194–6**, 1605–6, 1994

126 Collin R. E. *Foundations for Microwave Engineering*, McGraw-Hill, 1992

127 Gieres G., Kessler J., Kraus J., Roas B., Russer P., Soelkner G. and Valenzuela A. A. High frequency characterisation of $YBa_2Cu_3O_{7-x}$ thin films with coplanar resonators, *Superconductor Science and Technology*, **4**, 629–2, 1991

128 Porch, A. Private communication, 1994

4
Microwave measurements

4.1 Introduction

This chapter is mainly concerned with the measurement of the microwave response of a circuit. The equipment that performs these measurements is discussed briefly in Section 4.1, which reviews the capabilities of the network analyser. Following this section a detailed description is given on how to extract the unloaded quality factor of a cavity or resonator from measurements taken on a network analyser. It is well known that the unloaded quality factor can be estimated by simply taking the ratio of the centre frequency to the 3 dB bandwidth of the transmission response of a resonator, and the unloaded Q can be calculated by simply using the insertion loss measured at resonance. However, although this may be adequate for many situations, greater accuracy may be required or reflection measurements from the cavity may only be available. Calculations are therefore given for both one- and two-port measurements. For the two-port case both symmetrical and asymmetrical coupling are considered. Throughout the discussion, indications of how the accuracy of quality factor measurements may be improved are given. The final section of this chapter deals with power dependent measurements. The calculation of the energy stored in a cavity or the current on a resonant transmission line from S (scattering) parameter measurements is given.

4.2 Measurement system

Modern microwave measurement equipment, although expensive, allows accurate measurements to be made with the minimum of effort. Network analysers, signal sources, oscilloscopes and spectrum analysers are available for frequencies up to and beyond 100 GHz. With careful use of these instruments very precise measurements can be made on a microwave system.

The vector network analyser is the workhorse of any microwave laboratory; its functions are shown in Figure 4.2.1. The device under test (DUT) is connected either for transmission or reflection measurements, as shown in Figure 4.2.1. The controls of the analyser are then set in order to define a signal which sweeps from low frequency to high frequency at a certain rate. The resultant transmitted and/or reflected signal is received by the analyser and compared with the original signal. This can measure all four of the *S* parameters of the device, which can be displayed in the various forms shown in Figure 4.2.1. The versatility of the network analysis comes not only from its very high accuracy, which arises from various calibration routines, but also from its ability to visualise the frequency response of the DUT over a wide (or narrow) frequency band. Software in the network analyser not only allows the different displays of information, but also Fourier transforms the data, enabling time domain information to be observed.

- Gain (dB)
- Insertion Loss (dB)
- Insertion Phase (deg)
- S_{12} and S_{21} Transmission Coefficients
- Separation of Transmission Coefficients (Fourier Transform)
- Electrical Length (m)
- Electrical Delay (s)
- Deviation from Linear Phase (deg.)
- Group Delay (s)

- Return Loss (dB)
- S_{11} and S_{22} Reflection Coefficients
- Reflection Coefficient vs. Time (Fourier Transform)
- Impedance (Linear and Smith Chart)
- SWR

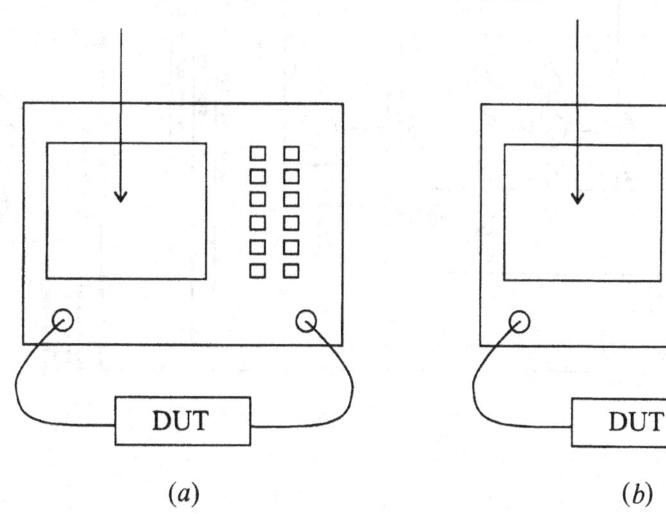

Figure 4.2.1. The functions of a network analyser: (*a*) transmission measurements, (*b*) reflection measurements.

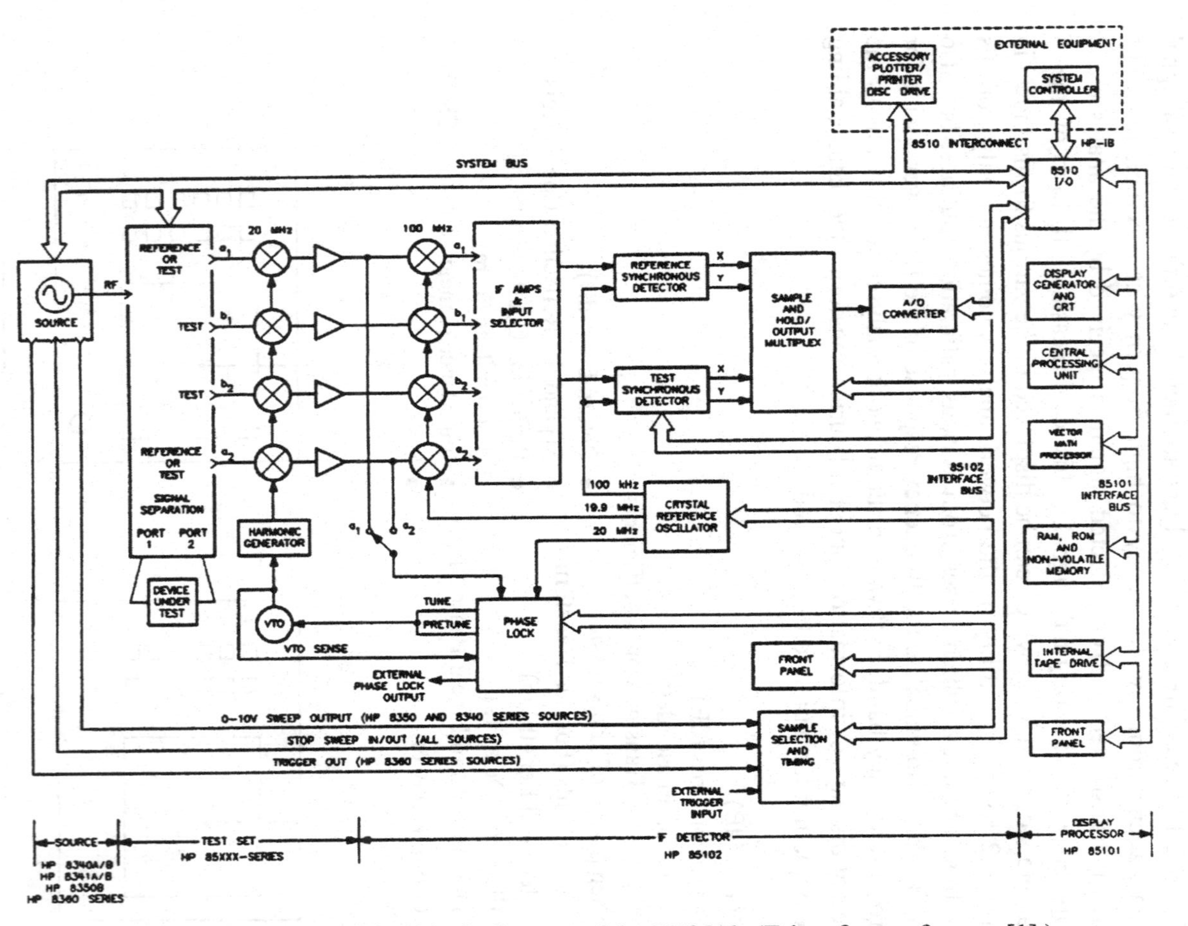

Figure 4.2.2. Simplified block diagram of the HP8510. (Taken from reference [1].)

A simplified block diagram of the Hewlett Packard HP8510 is shown in Figure 4.2.2. The signal source on the left of the diagram is first set to produce a swept frequency, or, typically, it can be set to step through a number of synthesised frequency steps. The signal from the source is applied to the *S* parameter test set, where signal separation occurs. Part of the signal is applied to the DUT and part to a reference channel output (a_1 or a_2). The reflected and/or the transmitted signal from the DUT exits via channels b_1 and b_2. A frequency conversion down to 20 MHz then takes place. This is done by mixing with a harmonic of a voltage tuned oscillator (65–300 MHz), with fine tuning accomplished using a 20 MHz crystal reference. A second down conversion takes place to 100 kHz using internal crystal references. Detection and analogue to digital conversion takes place after this stage. All the frequency conversions are phase coherent and the RF and IF signal paths are carefully matched in terms of electrical length, therefore maintaining both the magnitude and phase relationships of the original signal through to the detection stages. Automatic and fully calibrated IF gain steps maintain the IF signal at optimum levels for detection over a wide dynamic range. The whole system is controlled by a Motorola 68000 microprocessor. Analysis of the resultant data gives the displays shown in Figure 4.2.1.

The accuracy of the network analyser arises from the calibration of the system using precision components; without these the accuracy would be severely limited. The dynamic range of the system after calibration is around 90 dB, with absolute accuracy in signal amplitude of less than 0.1 dB. The frequency resolution can be as low as 1 Hz with the correct signal source, which is adequate to measure the majority of *Q*s from the HTS devices discussed in the preceding chapter.

The network analyser measures the linear frequency response of a device. It also measures this response as a function of input power. What it cannot do is to look at frequency conversion or mixing in devices; a spectrum analyser is required for this. Also, the network analyser cannot look at other non-linear processes such as switching; a fast microwave oscilloscope is useful in this area.

4.3 Measurement of quality factors

The objective of measuring the quality factor of a cavity or resonator is to extract information about the material from which the cavity, or part of the cavity, is constructed. Alternatively, it may be used to determine the parameters required in order to use the cavity for application purposes. Various cavity types are used in order to extract the surface impedance of superconducting samples, and these are discussed in detail in Chapter 3. Whatever the specific type of cavity, it is the measurement of Q and the resonant frequency that is required. Specifically, the unloaded quality factor Q_0 is of interest. A great deal has been written about this subject in the past[2–4] and therefore only the most important information is presented here. Unlike most of the preceding information, it is discussed here in relation to using a swept frequency network analyser,[5] where measurements can easily be performed at a large number of frequency points.

When a cavity resonator is coupled to an external circuit, additional power loss out of the coupling ports occurs. A quality factor can be attributed to this, which is termed the external quality factor Q_e. The combination of the external Q and the unloaded Q gives rise to the loaded quality factor of the cavity Q_l, the quantity which is usually measured using the network analyser. The loaded Q is defined as

$$\frac{1}{Q_l} = \omega \frac{\textit{Energy stored in cavity}}{\textit{Average power loss to external circuit } (P_e)} + \omega \frac{\textit{Energy stored in cavity}}{\textit{Average power loss in cavity } (P_0)} \tag{4.3.1}$$

that is,

$$\frac{1}{Q_l} = \frac{1}{Q_e} + \frac{1}{Q_0} \tag{4.3.2}$$

In order to extract these values of Q from measurements of the reflection and/or transmission coefficients, an equivalent circuit model of a cavity will be considered. Two possible resonant circuits are the series and parallel circuits, shown in Figure 4.3.1.

In general, a cavity has a number of resonances at different frequencies. The circuits in Figure 4.3.1 only give a single resonance. In principle, more resonant circuits could be added in series or parallel, with these circuits representing these other modes. However, the modes can be treated separately provided they are far enough apart in frequency so that they will not interact, that is, there will be no intermode coupling. For the discussion given below, the series resonant circuit is chosen because the derivations are slightly simpler. In fact, both are equivalent and can be converted between each other by, for example, the inclusion of a $\lambda/4$ length of transmission line, as shown in Figure 4.3.1. The coupling to the resonator will be represented by a transformer, as shown in Figure 4.3.2.

Two circuits are shown in Figure 4.3.2. The first is for single-port measurements, where the cavity is only coupled to by a single measurement port and reflection measurements are undertaken. The second is for dual-port measurements, where transmission measurements are usually taken. Both measurement

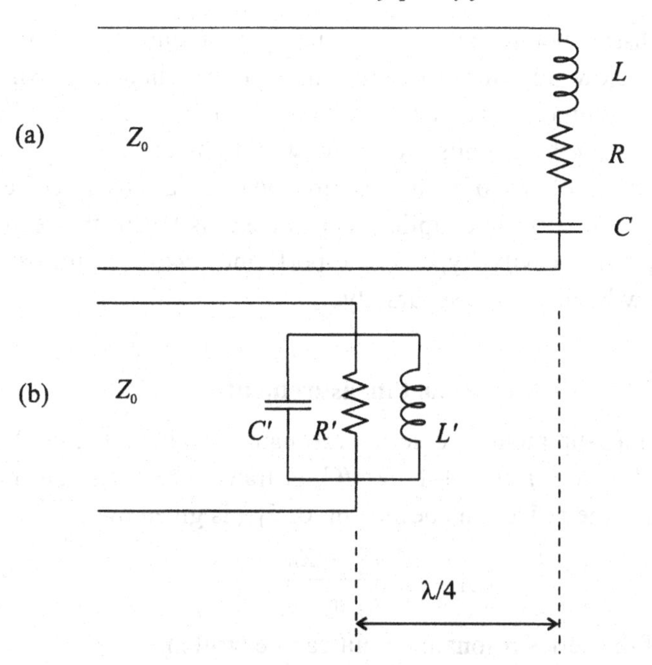

Figure 4.3.1. (*a*) Series and (*b*) parallel resonant circuits.

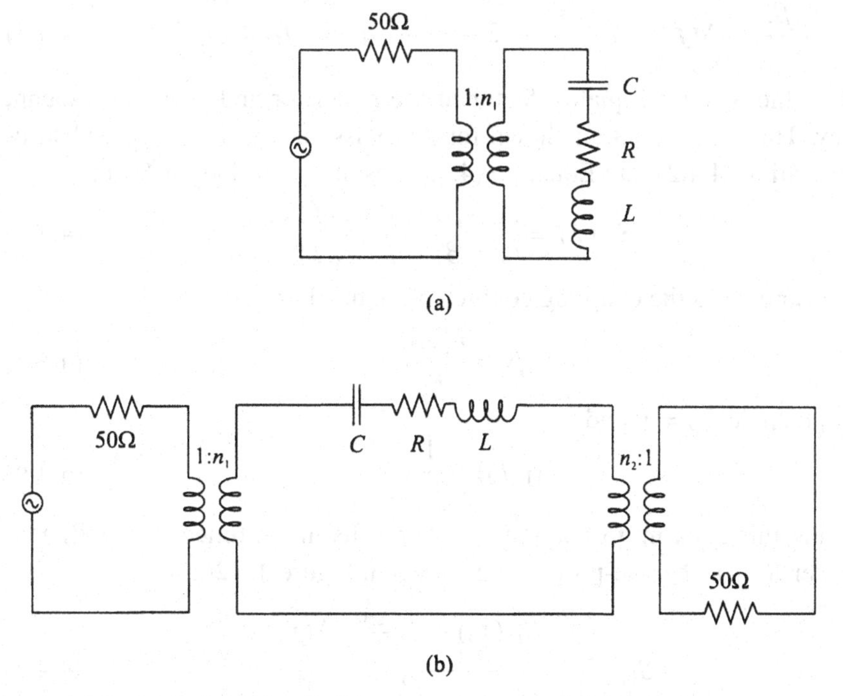

Figure 4.3.2. Equivalent circuit of a cavity resonator: (*a*) for single-port measurements, (*b*) for two-port measurements.

methods are discussed below. The transformers are assumed lossless in the derivations given below, and in most cases this is an excellent approximation. Discussion of measurements when the coupling is not lossless is given by reference [6]. Although the circuits are set up with transformers as a coupling mechanism this does not have to be the case for the calculations given below to be valid. Discussion of specific coupling mechanisms is left to the sections of Chapter 3 on specific cavity types. One-port and two-port measurement techniques will now be discussed separately.

4.4 One-port measurements

For the one-port measurement, the series resonant circuit of Figure 4.3.2(a), with impedance $Z = R + j2\pi fL + 1/j2\pi fC$, is transformed via the transformer to Z/n_1^2. Hence the reflection coefficient, or S_{11}, is given by

$$S_{11} = \frac{Z/n_1^2 - Z_0}{Z/n_1^2 + Z_0} \tag{4.4.1}$$

The impedance of the series resonant circuit can be written

$$Z = R(1 + jQ_0\Delta(f)) \tag{4.4.2}$$

where

$$Q_0 = \frac{2\pi fL}{R}, \quad \Delta(f) = 1 - \frac{f_0^2}{f^2} \approx 2\frac{f - f_0}{f_0} \quad \text{and} \quad f_0 = \frac{1}{2\pi\sqrt{(LC)}} \tag{4.4.3}$$

Here Q_0 is the unloaded quality factor of the resonator and f_0 is the resonant frequency. The approximation shown for $\Delta(f)$ assumes that $f \approx f_0$. Substitution of Equation (4.4.2) into Equation (4.4.1) results in a value for S_{11} of

$$S_{11}(f) = \frac{(1 - \beta) + jQ_0\Delta(f)}{(1 + \beta) + jQ_0\Delta(f)} \tag{4.4.4}$$

where β is known as the coupling coefficient, defined as

$$\beta = \frac{n_1^2 Z_0}{R} \tag{4.4.5}$$

If $f = f_0$ then $\Delta(f_0) = 0$ and

$$S_{11}(f_0) = \frac{1 - \beta}{1 + \beta} \tag{4.4.6}$$

Substituting this back into Equation (4.4.4) results in the reflection coefficient S parameter S_{11} for the one-port circuit shown in Figure 4.3.2(a) as

$$S_{11}(f) = \frac{S_{11}(f_0) + j\dfrac{Q_0}{1 + \beta}\Delta(f)}{1 + j\dfrac{Q_0}{1 + \beta}\Delta(f)} \tag{4.4.7}$$

The external quality factor can be calculated directly from the fundamental definitions given in Equations (4.3.1) and by considering the power dissipated both externally and internally to the resonator:

$$Q_e = Q_0 \frac{P_0}{P_e} = \frac{Q_0}{\beta} \tag{4.4.8}$$

and from Equation (4.3.2):

$$Q_l = \frac{Q_0}{1+\beta} \tag{4.4.9}$$

A network analyser measures the amplitude and phase of the reflection coefficient directly, so it is useful to find the magnitude and phase of Equation (4.4.7):

$$|S_{11}(f)| = \left(\frac{S_{11}^2(f_0) + \frac{Q_0^2}{(1+\beta)^2} \Delta^2(f)}{1 + \frac{Q_0^2}{(1+\beta)^2} \Delta^2(f)} \right)^{1/2} \tag{4.4.10}$$

The phase of $S_{11}(f)$ is given by

$$\phi_{11}(f) = \tan^{-1}\left(\frac{Q_0 \Delta(f)}{1-\beta}\right) - \tan^{-1}\left(\frac{Q_0 \Delta(f)}{1+\beta}\right) \tag{4.4.11}$$

Figures 4.4.1 and 4.4.2 show the magnitude and phase of S_{11} for a number of different values of Q_0 and β; these plots are the same as those which would be seen directly on a network analyser.

In the one-port measurement method, the objective is to measure β and f_0 directly, and then S_{11} at one or more points other than the resonant frequency, from which Δ and hence Q_0 can be calculated. There are a number of ways of accomplishing this, as is detailed below.

It is clear from Equation (4.4.10) that when $f = f_0$, then $|S_{11}(f)|$ is at a minimum value; hence, observation of the position of this minimum value determines f_0. At this resonant frequency, Equation (4.4.6) can, in principle, be used to determine β; however, it is the magnitude of S_{11} that is being measured and hence β can only be determined from

$$\beta = \frac{1 \pm |S_{11}(f_0)|}{1 \mp |S_{11}(f_0)|} \tag{4.4.12}$$

This equation is derived using Equation (4.4.10), with the substitution of β from Equation (4.4.6), and there are two possible solutions for the value of β. As seen from Equation (4.4.12), the values are the inverse of each other, one being greater than unity, the other being less then unity. When $\beta < 1$ the resonant circuit is referred to as undercoupled, when $\beta > 1$, then it is referred

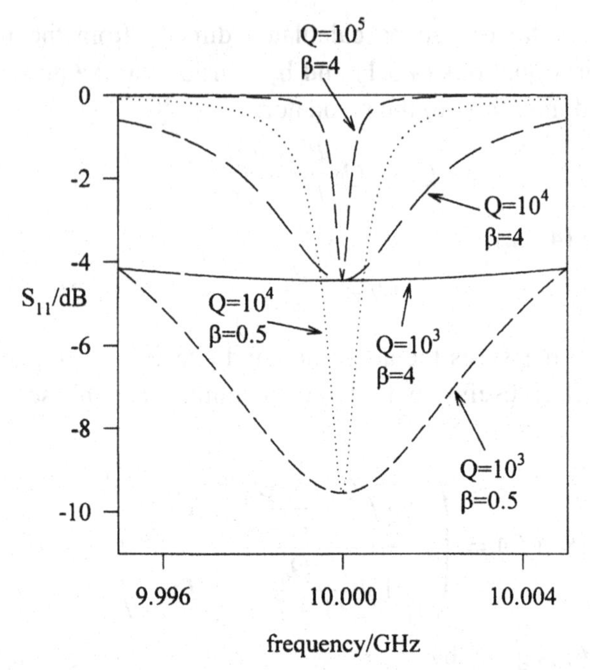

Figure 4.4.1. Magnitude of S_{11} for cavities with resonant frequencies of 10 GHz and various values of Q_0 and β.

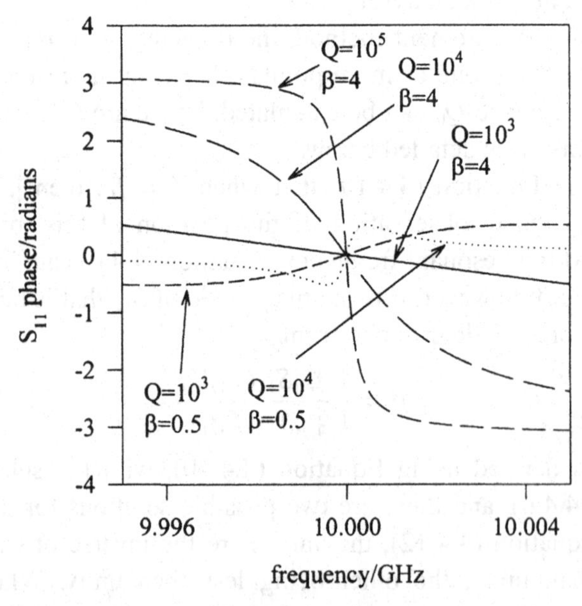

Figure 4.4.2. Phase of S_{11} for cavities with resonant frequencies of 10 GHz and various values of Q_0 and β.

to as overcoupled. In order to determine the value of Q_0 the value of β must be known precisely. Fortunately, there are other methods of determining which of these two values of β is correct for the particular resonance under investigation.

One method of distinguishing between the two different cases in order to find β is to consider viewing the data on a Smith chart, as shown in Figure 4.4.3. Modern network analysers can display Smith charts directly.

If the circle created by the complex reflection coefficient as a function of the frequency of the resonant circuit appears to enclose the centre point ($Z = Z_0$) of the chart, then the resonant circuit is overcoupled. If it passes exactly through the centre it is critically coupled, and if it does not enclose the centre point, then it is undercoupled. This method is useful as a visual check on the coupling, but it does not lend itself to automatic determination by a computer. An alternative method is to consider the gradient of the phase at the resonant frequency. Differentiating Equation (4.4.11) gives

$$\left.\frac{d\phi}{df}\right|_{f=f_0} = \frac{2Q_0}{f_0}\frac{2\beta}{1-\beta^2} \tag{4.4.13}$$

Hence if this gradient is positive, then $\beta < 1$ and the cavity is undercoupled; if the gradient is negative, then $\beta > 1$ and the cavity is overcoupled. β can now be determined by a computer by looking at the phase information around $f = f_0$.

It is now possible to determine Q_0. Equation (4.4.10) can be written in the form

$$Q_0 = \frac{1}{\Delta(f_a)}\left[\frac{|S_{11}(f_a)|^2(1+\beta)^2 - (1-\beta)^2}{1-|S_{11}(f_a)|^2}\right]^{1/2} \tag{4.4.14}$$

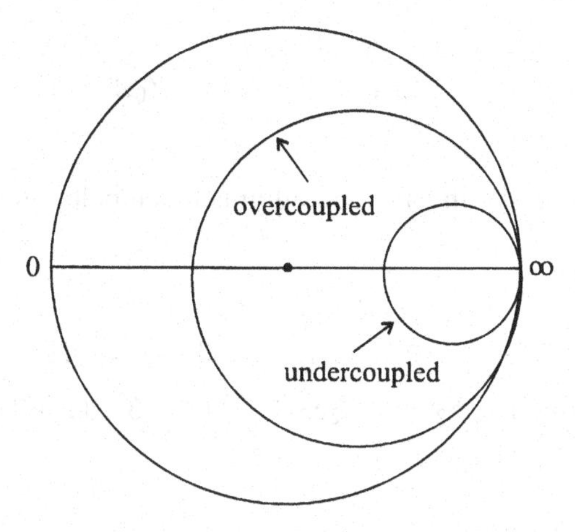

Figure 4.4.3. A Smith chart of two resonant circuits; one is overcoupled, the other is undercoupled.

f_a is some frequency other then the resonant frequency, where the magnitude of S_{11} can be measured accurately. Conventionally, $|S_{11}(f_a)|$ is measured at the frequency which corresponds to $\Delta(f) = 1$, that is, when $f_a = 1.5f_0$. This may not always be the best frequency for accurate measurement, especially for high-Q cavities. However, Equation (4.4.14) can be used more generally for more convenient values of f_a. Using this measurement of $|S_{11}(f_a)|$, the values of f_a, f_0 and the previously measured values of β, Q_0 can be determined. More accurate values can be obtained by using multiple values of f_a and averaging the final value of Q_0 if a computer controlled system is used.

In addition to using the magnitude of S_{11}, the phase of S_{11} can also be used to determine the values of Q_0; this will not be considered here.

4.5 Two-port measurements

For the two-port resonator we require the calculation of all four S parameters. This can be done by calculation of the $ABCD$ matrix[7] for the two-port circuit of Figure 4.3.2(b), and then by converting the $ABCD$ matrix to an S parameter matrix. The $ABCD$ matrix for the circuit in Figure 4.3.2(b) is given by

$$\begin{bmatrix} A & B \\ C & D \end{bmatrix} = \begin{bmatrix} 1/n_1 & 0 \\ 0 & n_1 \end{bmatrix} \begin{bmatrix} 1 & Z \\ 0 & 1 \end{bmatrix} \begin{bmatrix} n_2 & 0 \\ 0 & 1/n_2 \end{bmatrix} = \begin{bmatrix} n_2/n_1 & Z/n_1 n_2 \\ 0 & n_1/n_2 \end{bmatrix} \tag{4.5.1}$$

The resultant $ABCD$ matrix can be translated to the S parameter matrix using the standard expression

$$\begin{bmatrix} S_{11} & S_{12} \\ S_{21} & S_{22} \end{bmatrix} = \frac{1}{Z_0 A + B + Z_0^2 C + Z_0 D} \times \begin{bmatrix} Z_0 A + B - Z_0^2 C - Z_0 D & 2Z_0(AD - BC) \\ 2Z_0 & -Z_0 A + B - Z_0^2 C + Z_0 D \end{bmatrix} \tag{4.5.2}$$

Although not used here, the inverse operation is given below for reference:

$$\begin{bmatrix} A & B \\ C & D \end{bmatrix} = \frac{1}{2S_{21}} \begin{bmatrix} (1 + S_{11})(1 - S_{22}) + S_{12}S_{21} & Z_0[(1 + S_{11})(1 + S_{22}) - S_{12}S_{21}] \\ \frac{1}{Z_0}[(1 - S_{11})(1 - S_{22}) - S_{12}S_{21}] & (1 - S_{11})(1 + S_{22}) + S_{12}S_{21} \end{bmatrix} \tag{4.5.3}$$

Using Equation (4.5.2) gives

$$S_{11}(f) = \frac{(n_2^2 - n_1^2)Z_0 + Z}{(n_1^2 + n_2^2)Z_0 + Z} \tag{4.5.4}$$

$$S_{22}(f) = \frac{(n_1^2 - n_2^2)Z_0 + Z}{(n_1^2 + n_2^2)Z_0 + Z} \tag{4.5.5}$$

and

$$S_{12}(f) = S_{21}(f) = \frac{2Z_0 n_1 n_2}{(n_1^2 + n_2^2)Z_0 + Z} \tag{4.5.6}$$

As described in Section 4.4, the impedance of the series resonant circuit can be written

$$Z = R(1 + jQ_0\Delta(f)) \tag{4.5.7}$$

where

$$Q_0 = \frac{2\pi fL}{R}, \quad \Delta(f) = 1 - \frac{f_0^2}{f^2} \approx 2\frac{f - f_0}{f_0} \quad \text{and} \quad f_0 = \frac{1}{2\pi\sqrt{(LC)}} \tag{4.5.8}$$

The approximation for $\Delta(f)$ assumes that $f \approx f_0$. Substitutions for Z into Equations (4.5.4), (4.5.5) and (4.5.6) result in a value for the S parameters given by

$$S_{11}(f) = \frac{1 + \beta_2 - \beta_1 + jQ_0\Delta(f)}{1 + \beta_1 + \beta_2 + jQ_0\Delta(f)} \tag{4.5.9}$$

$$S_{22}(f) = \frac{1 + \beta_1 - \beta_2 + jQ_0\Delta(f)}{1 + \beta_1 + \beta_2 + jQ_0\Delta(f)} \tag{4.5.10}$$

$$S_{12}(f) = S_{21}(f) = \frac{2\sqrt{(\beta_1\beta_2)}}{1 + \beta_1 + \beta_2 + jQ_0\Delta(f)} \tag{4.5.11}$$

where

$$\beta_1 = \frac{n_1^2 Z_0}{R} \quad \text{and} \quad \beta_2 = \frac{n_2^2 Z_0}{R} \tag{4.5.12}$$

However, at the resonant frequency f_0, these equations reduce to

$$S_{11}(f_0) = \frac{1 + \beta_2 - \beta_1}{1 + \beta_1 + \beta_2} \tag{4.5.13}$$

$$S_{22}(f_0) = \frac{1 + \beta_1 - \beta_2}{1 + \beta_1 + \beta_2} \tag{4.5.14}$$

$$S_{12}(f_0) = \frac{2\sqrt{(\beta_1\beta_2)}}{1 + \beta_1 + \beta_2} \tag{4.5.15}$$

Back substituting into Equations (4.5.4), (4.5.5) and (4.5.6) results in

$$S_{11}(f) = \frac{S_{11}(f_0) + jQ_l\Delta(f)}{1 + jQ_l\Delta(f)} \tag{4.5.16}$$

$$S_{22}(f) = \frac{S_{22}(f_0) + jQ_l\Delta(f)}{1 + jQ_l\Delta(f)} \tag{4.5.17}$$

$$S_{12}(f) = S_{21}(f) = \frac{S_{12}(f_0)}{1 + jQ_l\Delta(f)} \tag{4.5.18}$$

The loaded quality factor has been substituted here in order to simplify the equations. This can be calculated by first considering the external quality factor as shown below. The external quality factor can be calculated directly from the fundamental definitions given in Equations (4.3.1) and (4.3.2) by considering the power dissipated both externally via each of the ports and internally in the resonator:

$$Q_{e1} = Q_0\frac{P_0}{P_{e1}} = \frac{Q_0}{\beta_1} \quad \text{and} \quad Q_{e2} = Q_0\frac{P_0}{P_{e2}} = \frac{Q_0}{\beta_2} \tag{4.5.19}$$

Thus the loaded quality factor can be calculated from Equation (4.3.2), resulting in

$$Q_l = \frac{Q_0}{1 + \beta_1 + \beta_2} \tag{4.5.20}$$

The magnitude and phase of the transmission response can now be calculated as

$$|S_{12}(f)| = \frac{|S_{12}(f_0)|}{(1 + Q_l^2\Delta^2(f))^{1/2}} \tag{4.5.21}$$

$$\phi_{12}(f) = -\tan^{-1}(Q_l\Delta(f)) \tag{4.5.22}$$

Expressions for the reflection coefficients can also be deduced and are similar to those given in Section 4.4. These can be used to determine the quality factor of a cavity as outlined in that section. The following now concentrates on the transmitted response in order to deduce the quality factor of a cavity. It is first instructive to plot the magnitude and phase of the transmission coefficients, and these are shown in Figures 4.5.1 and 4.5.2 for a typical cavity resonating at 10 GHz. Graphs are shown for different values of unloaded quality factor and β.

Using Equations (4.5.21) and (4.5.22) it is quite straightforward to see how Q_l can be deduced from measurements. Rewriting Equations (4.5.21) and (4.5.22) gives

$$Q_l = \frac{1}{|\Delta(f_a)|}\left(\frac{|S_{12}^2(f_0)|}{|S_{12}^2(f_a)|} - 1\right)^{1/2} \tag{4.5.23}$$

$$Q_l = -\frac{\tan(\phi_{12}(f_a))}{\Delta(f_a)} \tag{4.5.24}$$

If some frequency f_a is chosen, conveniently away from the resonant

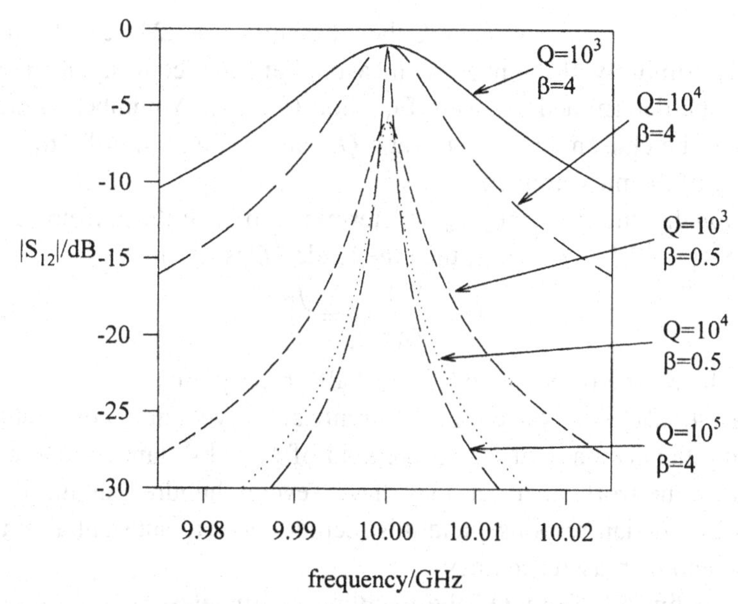

Figure 4.5.1. The magnitude of the transmission coefficient for a cavity resonating at 10 GHz with various values of $\beta = \beta_1 = \beta_2$ and unloaded quality factor.

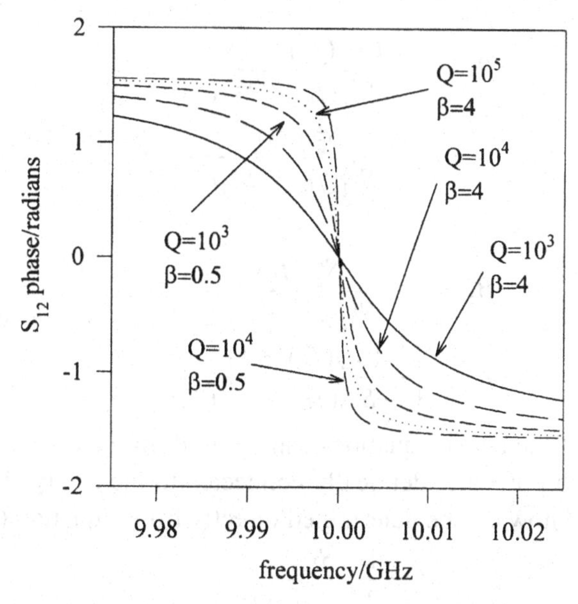

Figure 4.5.2. The phase of the transmission coefficient for a cavity resonating at 10 GHz with various values of $\beta = \beta_1 = \beta_2$ and unloaded quality factor.

frequency, and $S_{12}(f_a)$ is measured, then Equation (4.5.23) can be used to calculate Q_l. Similarly, if the phase is measured at some convenient frequency, then Q_l can be determined by using Equation (4.5.24). A number of different f_a values can be chosen, and the average Q_l can be found in order to increase the accuracy of the measurement.

Conventionally, the frequency f_a is determined to be at the half-power point, such that $|S_{12}^2(f_0)|/|S_{12}^2(f_a)| = 2$; then the loaded Q is given by

$$Q_l = \frac{1}{|\Delta(f_a)|} = \frac{f_0}{B} \tag{4.5.25}$$

where B is the resonator bandwidth at the half-power points.

Equation (4.5.21) is a conventional Lorentzian curve, and a convenient way of improving the accuracy of a measurement of Q_l is by standard curve fitting techniques. A network analyser may have several hundred digitised points across this Lorentzian response, and this method leads to substantial reduction in the noise and increased accuracy.

In order to find Q_0 from Q_l, the coupling coefficients β_1 and β_2 must be found. These can be found directly from Equations (4.5.13)–(4.5.15), giving

$$\beta_1 = \frac{1 - S_{11}(f_0)}{S_{11}(f_0) + S_{22}(f_0)} \tag{4.5.26}$$

$$\beta_2 = \frac{1 - S_{22}(f_0)}{S_{11}(f_0) + S_{22}(f_0)} \tag{4.5.27}$$

or

$$\beta_1 = \frac{(S_{11}(f_0) - 1)^2}{1 - S_{11}^2(f_0) - S_{21}^2(f_0)} \tag{4.5.28}$$

$$\beta_2 = \frac{S_{21}^2(f_0)}{1 - S_{11}^2(f_0) - S_{21}^2(f_0)} \tag{4.5.29}$$

or

$$\beta_1 = \frac{S_{21}^2(f_0)}{1 - S_{22}^2(f_0) - S_{21}^2(f_0)} \tag{4.5.30}$$

$$\beta_2 = \frac{(S_{22}(f_0) - 1)^2}{1 - S_{22}^2(f_0) - S_{21}^2(f_0)} \tag{4.5.31}$$

Any combination of the above equations can be used. In the case of symmetric coupling, when both ports are identically connected to the cavity, then $\beta_1 = \beta_2 = \beta$. The value of β now can be determined directly from Equation (4.5.15):

$$\beta = \frac{1}{2}\frac{S_{12}(f_0)}{1 - S_{12}(f_0)} \tag{4.5.32}$$

In fact, there are two solutions for β which can be determined in a similar manner to the one-port measurements given in Section 4.4; however, one is negative and is rejected as non-physical. It therefore follows from the substitution of Equation (4.5.32) into Equation (4.5.20) that, for a symmetrically coupled cavity,

$$Q_0 = \frac{Q_l}{1 - S_{12}(f_0)} \tag{4.5.33}$$

It is much simpler if the cavity is symmetrically coupled, but care must be taken in the construction if this is to be obtained, thus making sure that $\beta_1 = \beta_2$. If the cavity is weakly coupled and $S_{12}(f_0)$ is small, then $Q_l \approx Q_0$. This is a good situation to aim for if the signal-to-noise of the measurement system is good enough.

4.6 Power dependent measurements

For power dependent measurements it is convenient to be able to calculate the energy stored inside any resonator from measurements of the input power, transmission coefficient and quality factor. This section derives an expression for the energy stored in a resonator and then uses this in order to calculate the total current in a transmission line resonator. The calculation will be performed for transmission measurements.

Starting from the fundamental definition of Q:

$$Q_0 = \omega \frac{\textit{Energy stored in cavity}}{\textit{Average power lost}} \tag{4.6.1}$$

the energy stored can be written

$$\textit{Energy stored} = \frac{Q_0}{\omega}\,\textit{power lost} \tag{4.6.2}$$

The power lost can be written in terms of the difference between the power input and the power reflected and transmitted:

$$\textit{Energy stored} = \frac{Q_0}{\omega}(1 - |S_{11}|^2 - |S_{21}|^2)P_{in} \tag{4.6.3}$$

where P_{in} is the power applied to port 1 of the cavity. However, from Equations (4.5.13) and (4.5.15), assuming symmetrical coupling with $\beta_1 = \beta_2 = \beta$,

$$S_{11} = \frac{1}{1 + 2\beta} \quad \text{and} \quad S_{12} = \frac{2\beta}{1 + 2\beta} \tag{4.6.4}$$

Combining the two equations of Equation (4.6.4) gives (as expected)

$$S_{11} = 1 - S_{12} \tag{4.6.5}$$

Substitution of Equation (4.6.5) into Equation (4.6.3) results in

$$Energy\ stored = \frac{2Q_0}{\omega}(1 - S_{12})S_{12}P_{in} \tag{4.6.6}$$

From Equation (4.5.33), $Q_0(1 - S_{12})$ can be replaced by Q_l, giving

$$Energy\ stored = \frac{2Q_lS_{12}P_{in}}{\omega} \tag{4.6.7}$$

This equation can now be used to work out the energy stored in any cavity resonator from simple measurements made on a network analyser. If a specific cavity is considered, where the expressions for the field variation are known, then the actual levels of the fields can be calculated at any point, including the cavity walls. If the magnetic field is known at the walls, the current distribution in the walls of the cavity can be found. Knowing this current or surface field in absolute terms is important in the measurement of critical current and critical field effects.

This calculation can be taken one stage further if the cavity is assumed to support TEM waves, that is, it is a transmission line cavity. The energy stored can then be equated to the energy stored in the inductance (or capacitance) of the TEM transmission line. Thus

$$\frac{1}{2}LlI^2 = \frac{2Q_lS_{12}P_{in}}{\omega} \tag{4.6.8}$$

where L is the inductance per unit length of the cavity, l is the total length and I is the total current. For a low-loss transmission line we can use

$$Z_0 = \sqrt{\left(\frac{L}{C}\right)} \quad \text{and} \quad c = \frac{1}{\sqrt{(LC)}} \tag{4.6.9}$$

where Z_0 is the transmission line characteristic impedance and c is the velocity of transmission down the line. With this, Equation (4.6.8) becomes

$$I = \sqrt{\left(\frac{4Q_lP_{in}S_{12}}{n\pi Z_0}\right)} \tag{4.6.10}$$

The final assumption to obtain Equation (4.6.10) is that, for resonance, the transmission line length must be an integer (n) multiple of half-wavelengths, requiring that $l = n\lambda/2$.

4.7 References

1 HP8510C *Operating and Programming Manual*, Hewlett Packard, Part No. 08510–90281, Edition 1, 1991
2 Montgomery C. G., Dicke R. H. and Purcell E. M. *Principles of Microwave Circuits*, McGraw-Hill, 1948
3 Ginzton E. L. *Microwave Measurements*, McGraw-Hill, 1957
4 Sucher M. and Fox J. *Handbook of Microwave Measurements*, Wiley, 1964

5 Aitken J. Swept-frequency microwave Q-factor measurements, *Proc. IEE*, **123**(9), 855–62, 1976
6 Malter L. and Brewer G. R. Microwave Q measurements in the presence of series losses, *J. Appl. Phys.*, **20**, 918–25, 1949
7 Ramo S., Whinnery J. R. and van Duzer T. *Fields and Waves in Communication Electronics*, John Wiley, 1984

5
Superconducting filters

5.1 Introduction

This chapter discusses the use of superconductors in the construction of microwave filters. Superconductors can help in two ways. Firstly, the performance of a filter is improved by the use of superconductors in the sense that the insertion loss can be significantly reduced, as well as improving the filter roll off and reducing its bandwidth. Secondly, filters can be miniaturised. The improvement in the performance of filters is generally achieved by using fairly conventional design techniques, and improvement arises due to the reduced dissipation owing to the low surface resistance. However, miniaturisation usually requires a change in the geometry of filters and therefore entirely new types of filter become possible when superconducting materials are used in their construction. Improved performance and miniaturisation are complementary and reducing the size of a filter generally leads to reduced performance, although superconductors allow a much larger reduction in size, when compared with normal metals, whilst still giving improved performance. A third possible reason for using superconductors is to use their special properties such as the change in internal (or kinetic) inductance with microwave power or temperature, or to use their switching capabilities. These functions are discussed further later in this chapter. This chapter is organised under the theme of miniaturisation, where most of the interesting novel work has been in HTS filter development. All the filters discussed, especially those only having a modest amount of miniaturisation, are superior in performance to conventional filters made from normal metals.

Miniaturisation is accomplished by a change in filter geometry; for example, a waveguide filter can usually be redesigned so that the same functions can be performed by a microstrip filter, provided that the losses of the material making up the microstrip are low enough. The move from a three-dimensional structure to a planar (two-dimensional) structure reduces the size of the final filter significantly. It is also possible to reduce the size of planar filters by a change in geometry, and five methods of accomplishing this are given in Table 5.1.1.

Clearly, a reduction of the propagation velocity on transmission lines leads to a reduction in circuit size because of the reduced wavelength for a given frequency. The simplest way to do this is to increase the dielectric constant of the substrate which is shown as the first entry in Table 5.1.1. It is therefore appropriate to look for substrates with high dielectric constants for use with HTS. In fact, $LaAlO_3$ has a reasonably high dielectric constant, giving a reduction in velocity over free space of about 5, and therefore gives significant miniaturisation.

Meandering or coiling of transmission lines uses the surface area available on a substrate more efficiently. However, as the meander or coil becomes tighter, unwanted coupling occurs between neighbouring line sections. This can be reduced by confining the electromagnetic fields more, which can be achieved in the microstrip and stripline by reducing the thickness of the substrate.

The third entry in Table 5.1.1 is internal inductance. This can also be used to reduce the velocity of the wave on a transmission line. This velocity is governed by its capacitance (C) and inductance (L) ($c = 1/\sqrt{(LC)}$). Normally, if the capacitance is increased, say by making the microstrip substrate thinner, then the inductance is proportionally reduced, keeping a constant velocity. However, the internal inductance, which is due to energy storage within the penetration depth of the superconductor, adds an extra inductive term without affecting the capacitance. The effect of the internal inductance can be maximised by reducing the normal external inductance, for example in the microstrip by using thin substrates.

Slow-wave transmission lines are another way of increasing the capacitance on a transmission line without decreasing the inductance (or vice versa). It is accomplished by using discrete inductances and capacitances on the transmission line occupying lengths much less than a wavelength.

The final entry in Table 5.1.1 is the use of lumped element components.

Table 5.1.1. *Methods of miniaturising planar microwave filters*

1. Use of high dielectric constant substrates
2. Meandering or coiling of the transmission lines
3. Use of increased internal inductance
4. Use of slow-wave transmission lines
5. Use of lumped element components

These are discrete inductors and capacitors which, by definition, are used in a regime where they are much less than a wavelength in size. By shrinking these components down in physical size, small filters can be made at high frequencies.

The five methods given above are discussed separately in the following text. However, it must be pointed out that in certain limits some may be considered to be equivalent. For example, as a meander line becomes smaller it can eventually be considered to be a slow-wave structure; this occurs when the meander lengths are much less than a wavelength. The slow-wave transmission line could also be considered to be a lumped element filter. However, it is differentiated from a lumped element filter because of the way it is used in filter design. The slow-wave transmission line can be used as a transmission line element when designing filters and conventional distributed element techniques can be used in the design process, whereas lumped elements need to be considered as separate components in the design of a lumped element filter. The distinction between different miniaturisation techniques is given solely to categorise the discussion below.

Section 5.2 discusses more conventional filters and includes miniaturisation techniques from categories 1 and 2 in Table 5.1.1. Delay line filters are an example of miniaturisation in category 2 and are discussed in a separate chapter. The remaining three miniaturisation methods are discussed in Sections 5.6–5.8.

There is in principle no limit to how small a filter can be produced, and the limitations in performance are determined by the materials used. Low surface resistance allows filters to be reduced in size whilst still maintaining a reasonably low insertion loss. As the filter is reduced in size the current density increases for a given input signal. The current density will eventually reach a critical current of the superconductor; this limit is further complicated by the variation in current distribution within the filter itself. Currents are peaked at the edges of lines within the filter which can cause current limiting at specific points. By careful design such peaking in the current can be reduced, but this is usually accompanied by an increase in the size of the filter. The other loss mechanisms in the filter have little effect. Dielectric loss is dependent upon the relative amount of energy storage within the dielectric and need not change as the filter is reduced in size. Radiation loss in fact decreases as the filter size decreases due to the more confined fields. Other limitations on size reduction may occur due to more practical constraints. The patterning resolution limits the line widths to around, or just less than, a micron in size and the packaging becomes increasingly more difficult because of the tight tolerance required on external components. The mismatch at the connector ports can become the dominant loss mechanism in many low-loss miniature filters. Another practical limitation is substrate thickness, although deposited substrates can in principle overcome this problem. Another potential problem with smaller filters is their sensitivity to external influences, for example as internal inductance becomes more predominant the temperature sensitivity of the centre frequency becomes more of a problem.

The ultimate goal of many of the designs discussed below is not only the design of a high-performance single filter but the design of a filter bank or multiplexer. Here the miniaturisation is much more important when several devices are to be used together.

The last few sections of this chapter look at a miscellaneous set of microwave applications that are closely related to filters. Section 5.9 discusses a number of methods for switching a microwave signal, giving examples of different mechanisms. The entirely different subject of phase shifters is tackled in Section 5.10. Phase shifters are particularly important not only because of their ability to produce a variable phase shift, but also because they are able to produce variable velocity transmission lines. The inclusion of variable velocity transmission lines into filters enables frequency responses with adjustable characteristics to be produced. Two methods for producing this variable delay are discussed in Section 5.10. The chapter is concluded by a brief discussion of a number of other microwave devices which benefit from superconducting materials. Couplers, transformers and power splitters all improve in performance in some way, and all can be miniaturised by careful and novel design. This, of course, applies to other microwave components not specifically discussed in this or other chapters.

5.2 Conventional filters

An enormous amount of literature and software has been written on the design of microwave filters made from normal conductors, and the principles are of course of direct application to superconducting filters. In fact, many groups have designed HTS filters without reference to the superconducting properties, except for a reduced surface resistance; such an approach is valid, although more optimisation must be performed experimentally at low temperature. The design steps for a conventional bandpass filter start with a lowpass lumped element prototype design which is converted to a lumped element bandpass specification. These steps are straightforward and are accomplished in software and/or books.[1–3] The next step is usually to convert the lumped elements to distributed elements or transmission line sections for the filter layout. The layout can then be inserted in a simulation program which can optimise the filter dimensions. Such microwave design packages contain fully characterised (in terms of their S parameters) microwave components such as transmission lines, 'T' sections or coupled lines. The optimised layout can then be produced and the fine tuning done experimentally; or, alternatively, optimisation can be done by full wave simulation software. This allows the final experimental fine tuning to be left out in certain instances. Superconducting filters are designed in exactly the same way except that the extra parameter of temperature dependence must be considered[4,5] in the design. PEM (Section 2.6) can be used for the initial calculations, but the full wave analysis programs must include the effects of temperature dependent internal inductance effects. This can be achieved in part by including the full surface impedance (including the surface reactance) in the programs. However, for thin films of the order of, or less than, a penetration depth, a full calculation of the current within the volume of the superconductor is required. This is discussed in Section 2.7. Such calculations are very time consuming and it is usually appropriate to look at experimental optimisation of the filters.

To illustrate some of the techniques used to reduce the size of a filter consider Figure 5.2.1(a), which shows a simple third-order Butterworth low-pass microstrip filter using 0.5-mm-thick MgO for the substrate. The performance using silver as the conductor is shown in Figure 5.2.2; it is a good filter with low insertion loss in the passband. The filter size can be reduced simply by using $LaAlO_3$ as a substrate which has a higher dielectric constant; the size reduction of the filter is shown in Figure 5.2.1(b). Further size reduction can be

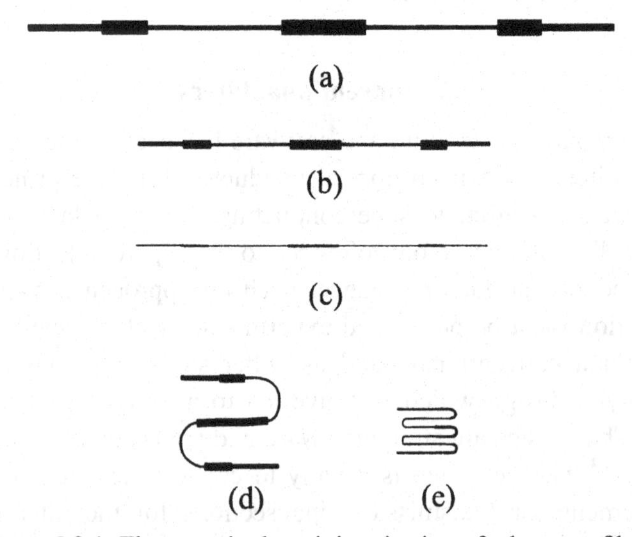

Figure 5.2.1. The steps in the miniaturisation of a lowpass filter.

achieved by thinning the substrate to 0.1 mm (Figure 5.2.1(c)). A more convenient form of the filter is made by meandering the structure, as shown in the last designs in Figure 5.2.1. As the filter reduces in size the performance in terms of stopband insertion loss becomes worse; Figure 5.2.2 shows this deterioration for each of the filters using silver as the conductor. The performance is retrieved if HTS is used in the design, and the results are shown in Figure 5.2.2. The modelling does not take the coupling between sections in the meander line into account. A similar discussion on the methods of size reduction of a bandstop filter is given by Chaloupka;[6] here the stubs in the bandstop design are replaced by discrete lumped element capacitors in order to save space.

The microstrip filter shown in Figure 5.2.3 is of conventional design; it consists of four resonators coupled together by the coupling of the electromagnetic fields over the gaps between the resonators. The filter has a 2.5% bandwidth centred on 9.75 MHz and has been made from both niobium and YBCO.[7]

Figure 5.2.4 shows the frequency response of the YBCO filter of Figure 5.2.3; this filter has a minimum insertion loss of less than 0.5 dB with an approximate 1 dB ripple, whilst the niobium filter has an insertion loss of less than 0.1 dB with negligible ripple. Simulations indicate that a similar gold filter at room temperature would have an insertion loss of about 4.5 dB. The interesting thing about this filter is not the design but the fact that it is produced on a 0.432-mm-thick sapphire substrate using an MgO buffer layer. The use of

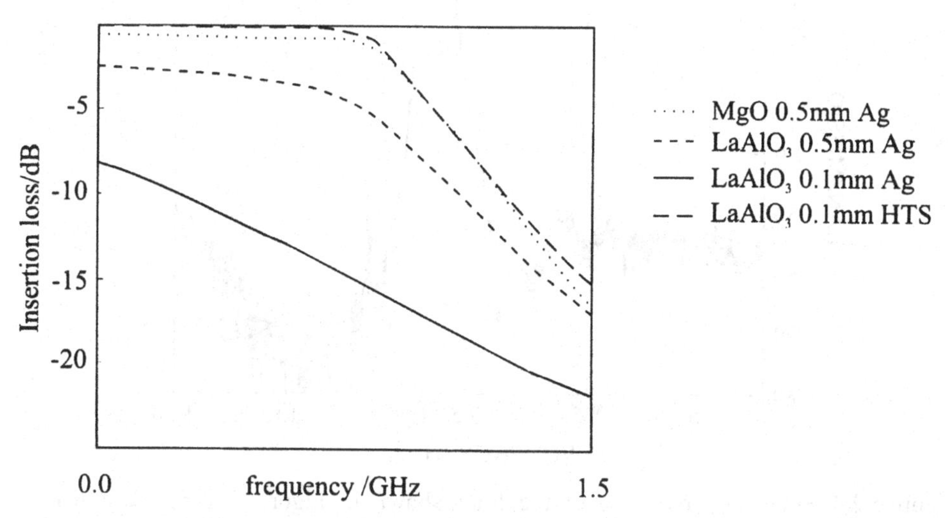

Figure 5.2.2. The predicted performance of the lowpass filters shown in Figure 5.2.1.

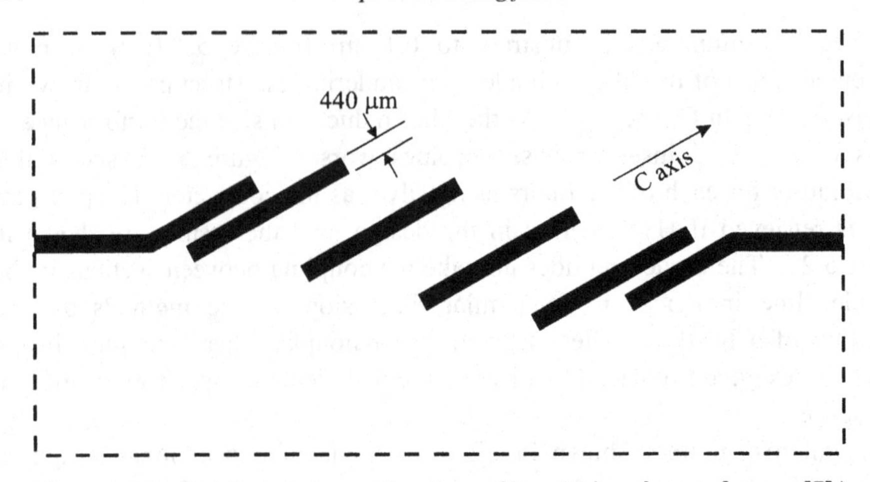

Figure 5.2.3. Distributed element bandpass filter. (Taken from reference [7].)

sapphire has the advantage of being able to reduce the ground plane spacing by lapping the substrate, although having a dielectric constant of only 9.34 gives only a small size reduction.

As with all other filters described in this section, this type of filter can be

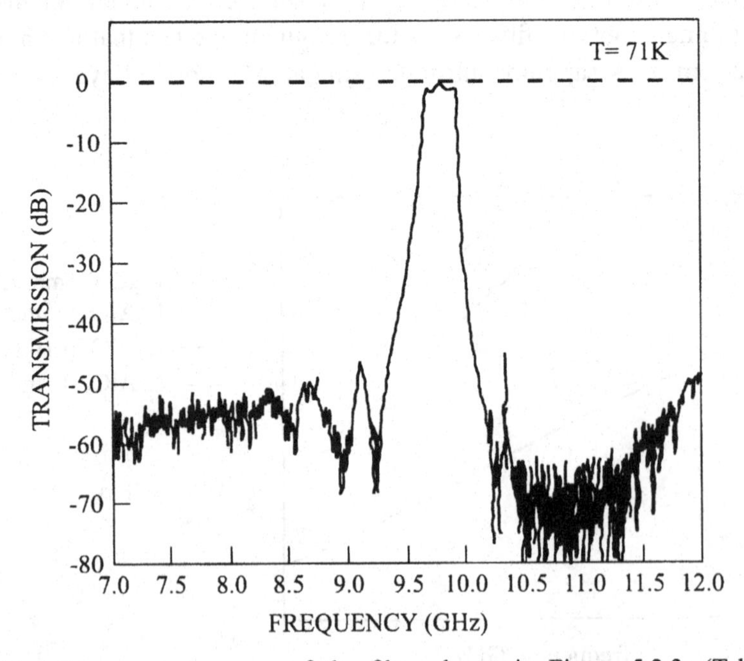

Figure 5.2.4. Frequency response of the filter shown in Figure 5.2.3. (Taken from reference [7].)

used in a filter bank or multiplexer. Such filter banks have applications in electronic warfare systems, as well as communication systems. An example of the architecture of such a system is shown in Figure 5.2.5; it consists of a number of filters connected through 90° hybrid couplers, resulting in the input signal being split into a number of channels. Two filters and hybrid couplers are used per filter section in the diagram. A demonstration of such a filter is based on four channels, each made from YBCO on 0.5-mm-thick, 5 cm-diameter $LaAlO_3$ substrates with gold ground planes.[8] Each filter is fourth order and is based on a Chebyshev design with 0.1 dB ripple. The centre frequencies of the filters are about 4 GHz and have a 50 MHz bandwidth (1.25%) with a 16 MHz guard band between filters. The hybrid couplers are 10% bandwidth and require a termination. This is made from a thin film resistor and a shunt capacitance to ground. The capacitance is resonant with the inductance of the resistor, providing the connection to ground. The resistors are made from molybdenum with a titanium overlayer to prevent oxygenation. The frequency response of the filter bank is shown in Figure 5.2.6. The distortion at the lower frequencies of each passband is caused by the attenuation of the signal at these frequencies from the preceding filter in the series.

The substrate area can be used slightly more effectively by using 'U' or hairpin-shaped resonators, as shown in the filter of Figure 5.2.7.[9–15] This filter works on the same principle as the one discussed above but the half-wavelength resonators are folded. The filter shown in Figure 5.2.7 is designed for a centre frequency of 4 GHz and a 3% bandwidth and is constructed on $LaAlO_3$ measuring 0.0425 × 1.3 × 2.3 cm.

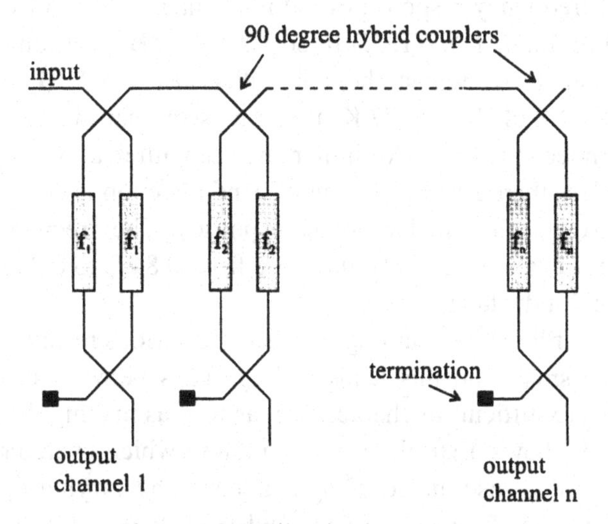

Figure 5.2.5. Filter bank architecture.

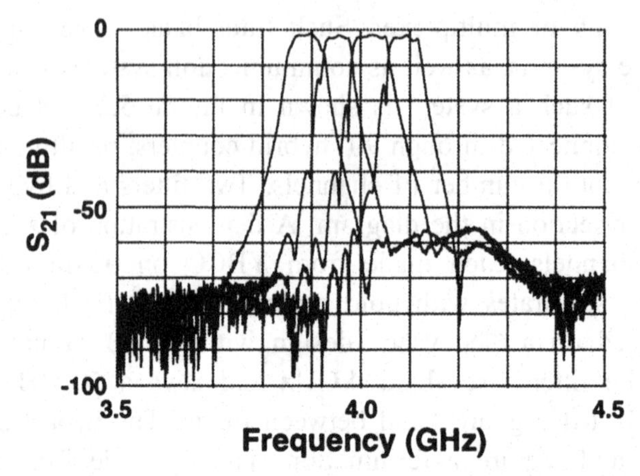

Figure 5.2.6. Frequency response of a HTS filter bank. (Taken from reference [8].)

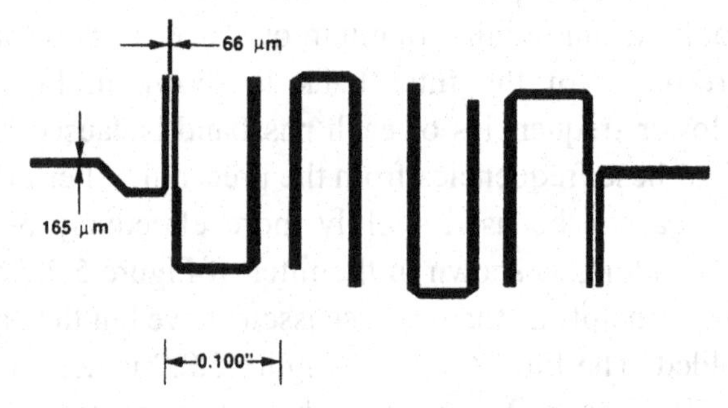

Figure 5.2.7. Hairpin distributed element filter. (Taken from reference [9].)

The measured frequency response of the filter made out of YBCO and gold is shown in Figure 5.2.8. The YBCO filter has a 0.3 dB insertion loss at 77 K and 0.1 dB loss at 13 K, whilst the gold filter has a 5.2 dB loss at room temperature and a 2.8 dB loss at 77 K. It can be seen that the gold filter has a significantly rounded response. A similar niobium filter at 4.2 K showed an insertion loss of less than 0.1 dB. The out of band rejection is better than 50 dB up to 12 GHz except for the harmonic responses. The frequency response shifted by 80 MHz when the temperature rose from $0.84T_c$ to $0.94T_c$ due to the increase in internal inductance.

For narrow band filters the coupling between the various resonant sections of the filter must be small and this leads to large gaps between the resonators. These large gaps are difficult to characterise, as well as making the whole filter large. Figure 5.2.9 shows a stripline filter structure which overcomes some of these problems. In this case the coupling is adjusted by varying the overlap of the resonator elements. There is no passband if all the resonators are in line

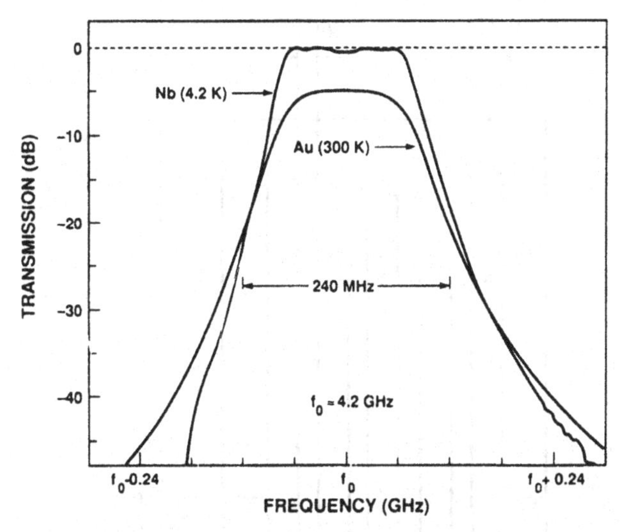

Figure 5.2.8. Frequency response of the filter shown in Figure 5.2.7. (Taken from reference [9].)

and the coupling increases as they become more staggered. This is true only for stripline because the odd and even mode velocities are the same. If microstrip is used then there is inherent coupling even if the resonators are not staggered. The filter shown in Figure 5.2.9 is much smaller than a conventional filter because the resonant sections can be packed more closely.[16–18] A five-pole superconducting filter of this type has been demonstrated with a 2.62% bandwidth centred on a frequency of 2.3 GHz and having a 0.05 dB Chebyshev ripple. The basic design equations are given in reference [18] and the filter was further optimised using Touchstone, a circuit modelling CAD package. The filter is made out of two $LaAlO_3$ substrates sandwiched together, the largest measuring 0.7 × 0.54". Each of the substrates has a ground plane on one side and the filter layout of the other side. Having the filter layout on both substrates helps to minimise the effect of air gaps since the two substrates are pressed together to form the stripline.

The frequency response of the filter is shown in Figure 5.2.10. The insertion loss is less than 0.1 dB at 77 K; this excludes the loss in the connectors which was calibrated out.

Despite the forward coupling caused by the difference in the odd and even velocities, when microstrip is used in the filter shown in Figure 5.2.9, advantages in terms of substrate utilisation can still be gained over the conventional edge coupled filter shown in Figure 5.2.3. For example, a resonator at a mobile radio frequency of about 900 MHz is about 1.7" long when $LaAlO_3$ is used as a substrate, and an edge coupled filter will not

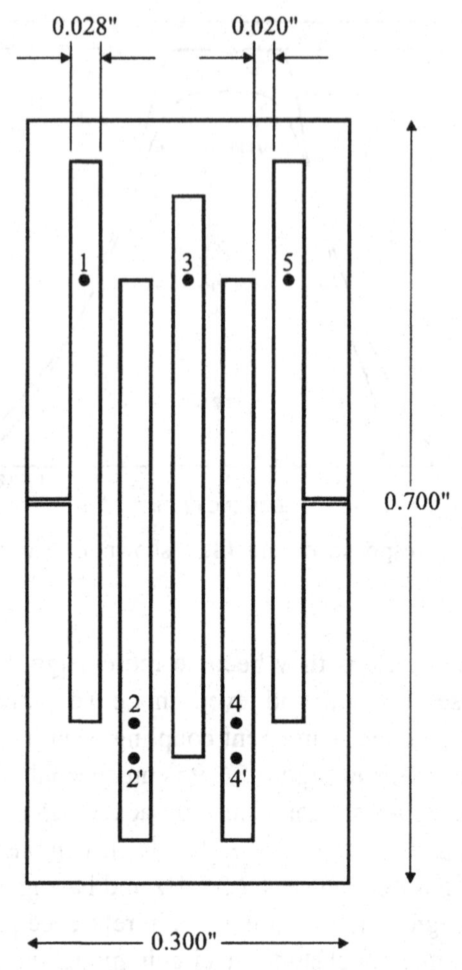

Figure 5.2.9. Staggered resonator distributed element filter. (Taken from reference [18].)

therefore fit on a 2" diameter substrate. However, five- and nine-pole forward coupled filters have been demonstrated on such a substrate.[19] With these filters there is no need to stagger the resonators with respect to each other as there is inherent forward coupling with microstrip even when the resonators are parallel. A brief analysis of such a filter is given by Zhang.[19] Two 900 MHz filters have been demonstrated for mobile radio applications with 2% bandwidth. These filters are made from YBCO on $LaAlO_3$, with resulting insertion losses of 0.1 dB and 0.28 dB for the five-pole and nine-pole devices respectively. Of these values of insertion loss, 0.06 dB is accounted for by the connectors. Another five-pole filter has been designed for higher power handling capabilities by using a lower 10 Ω internal impedance, giving wider

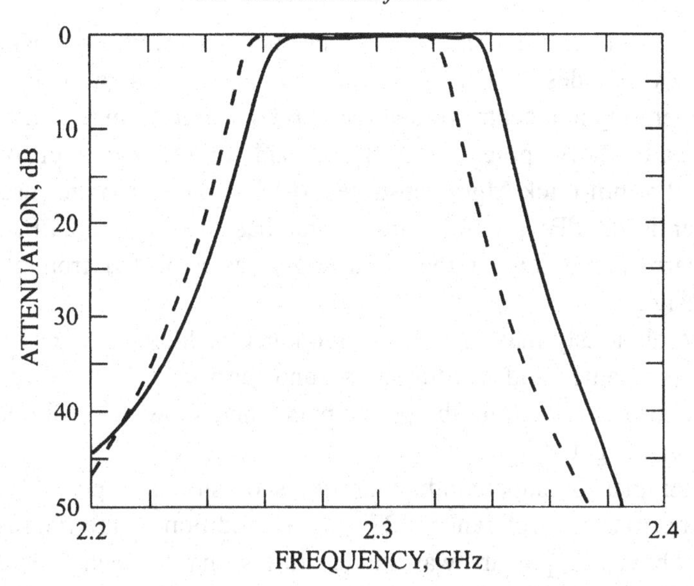

Figure 5.2.10. Frequency response of the filter shown in Figure 5.2.9. (Taken from reference [18].)

lines and a narrower bandwidth of 1.2%. The low impedance produces stronger coupling and the narrower bandwidth requires weaker coupling, making the spacing between the resonators larger. A 900 MHz filter of this specification does not fit on a 2" substrate; however the filter is demonstrated at 2 GHz with a 0.14 dB insertion loss, a 0.2 dB ripple and a return loss better than 14 dB.

Similar results have been obtained with 0.6% filters at 2 GHz, with the insertion loss increasing to 0.4 dB. The power handling capabilities are excellent and are discussed in Section 5.5 below. These filters include both forward and backward couplers in combination.[20,21] Other superconducting filters of this type include three- and seven-pole 2% bandwidth filters[22] and a five-pole 4.3% bandwidth filter.[23] The former has been used in a superconducting diplexer (two-channel multiplexer) to good effect.[24]

Elliptic function filters have the advantage that the attenuation poles are placed at finite frequencies, as opposed to Chebyshev filters where the attenuation poles are placed at infinite frequency. By placing these poles close to the band edges, steep skirts can result. Such filters are designed routinely in conventional waveguide filters using multimode cavity designs. In order to achieve this response, coupling between non-adjacent resonators is required, and such a filter is shown in the section below on dual-mode resonators. Alternatively, a second approach is to extract pole-forming networks from the filter prototype as separate structures for each finite frequency required. For example, this may be done using a shunt resonator tuned to the pole frequency.[25] Such a filter has been designed by Hedges[25] using design techniques by Rhodes.[26] This filter is a four-pole design with two poles extracted operating at a centre frequency of 6 GHz and a bandwidth of 80 MHz (1.3%) with a 0.02 dB ripple. The device is made of YBCO and fits on an inch-square and 0.5-mm-thick MgO substrate. At 77 K the minimum insertion loss of this filter is 0.8 dB; a similar gold filter has a loss of 2.2 dB. The filter exhibits sharp attenuation at the band edges, as expected from the elliptic function design.

There are obviously many more superconducting filters designed and tested by various companies and institutions around the world. The above examples have been chosen to illustrate the points made, and show the different types of filter being investigated.

Other examples of superconducting filters of similar type to those given above can be found in references [27–38]. In addition to the microstrip filter described above, coplanar waveguide filters have been designed and tested.[39–41] Higher-frequency quasi-optical filters have been investigated using HTS and these also show an improvement in performance.[42,43]

5.3 Dual mode filters

Figure 5.3.1 shows a number of dual-mode microstrip resonators. These consist of what is normally considered to be a single-mode resonator, but have a small perturbation in order to split the degenerate mode. All the resonators, shown in Figure 5.3.1 possess two fundamental modes which are at the same frequency.

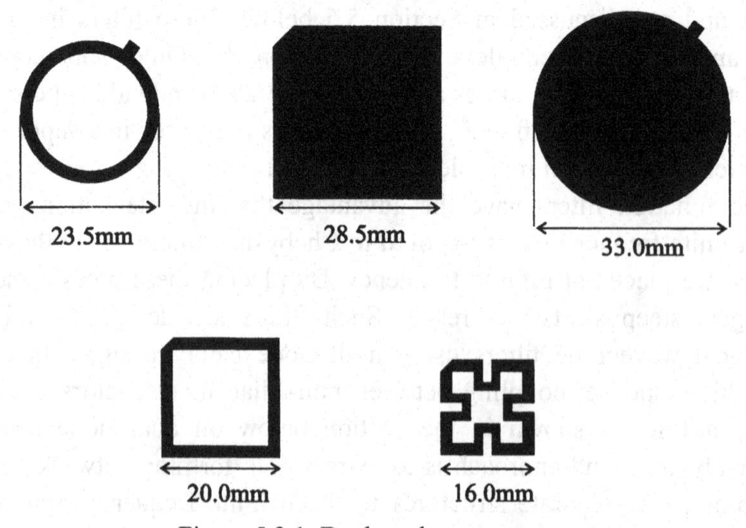

Figure 5.3.1. Dual mode resonators.

The perturbation shifts the frequency of one or both of these modes, separating them in the frequency domain. The separation frequency and the coupling between the modes can be controlled by the size and shape of the perturbation. The perturbations shown are chosen because of their repeatability, symmetry and their ability to tune the centre frequency. Further discussion of dual modes in ring resonators is given in Section 3.7.2. The dual-mode microstrip resonators shown in Figure 5.3.1 all resonate at 1.5 GHz on RT Duroid 6160 ($\varepsilon_r = 10.6$) without the perturbation, and there is therefore a direct size comparison between the various types. The circular patch is the largest, followed by the square patch, the circular loop, the square loop[44] and finally the meander square loop.[45] The latter reduces the size of the square loop by simply meandering the square and, of course, can be shrunk further by increasing the meandering, although coupling between the meander sections eventually becomes a problem.

Figure 5.3.2 shows how these resonators can be made into filters. Transmission lines couple energy into and out of the resonators, with coupling between

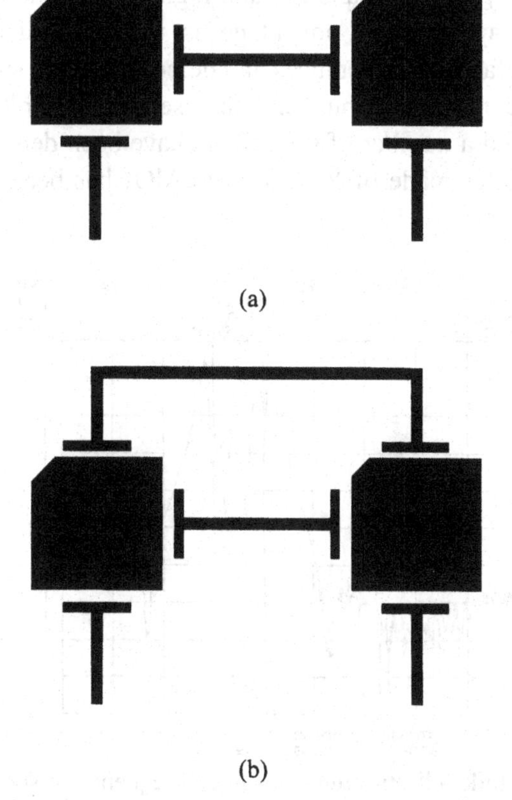

Figure 5.3.2. Dual mode filters: (*a*) Chebyshev, (*b*) elliptic.

the resonators being accomplished in the same fashion. It is possible to cross-couple the modes in order to produce elliptic function filters, as shown in Figure 5.3.2(b). Of course, any of the resonators shown in Figure 5.3.1 can be used in a similar manner. Because each of the resonators shown posseses two modes, the filters shown in Figure 5.3.2 each possesses four poles. The design of these filters is best achieved through electromagnetic full wave simulators.

The advantages of using dual-mode filters over single-mode elements are not immediately clear and only become apparent if considerations are given to size, bandwidth and power handling. The fact that the current is spread over a larger area for the circular and square dual-mode resonators and, to a lesser extent, the other resonators shown in Figure 5.3.1, gives these resonators a good power handling capability. Also, because two modes are used in a single element, this gives, in principle, a size advantage. However, this may only be apparent with narrow bandwidth filters, where the resonant sections of edge coupled filters are weakly coupled and have to be placed far apart. In general, dual-mode filters are of interest in the higher-power, narrow bandwidth regime. The filter can be miniaturised still further by using meander resonators. However, the advantages of the high-power handling and high *Q*s become less as the size is reduced in this way. Another advantage is that the filters are relatively insensitive to manufacturing tolerances and defects in films, such as pin holes.

These filters are not new[46] but lend themselves to implementation using superconductors, and a number of such filters have been demonstrated. A two-pole single-patch filter made of YBCO on $LaAlO_3$ has been demonstrated by

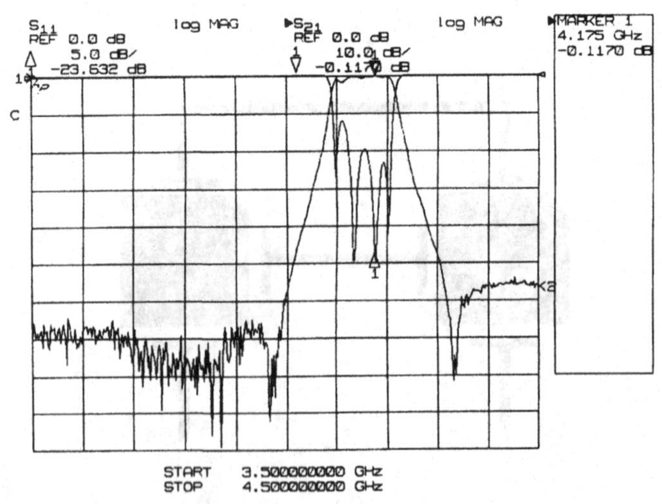

Figure 5.3.3. Dual mode elliptic function filter frequency response. (Taken from reference [50].)

Curtis and Fiedziuszko.[47,48] The filter showed a 3 dB improvement over an identical filter made from normal metal. A four-pole superconducting elliptic function filter, as shown in Figure 5.3.2(b), has an excellent performance, as shown in Figure 5.3.3. The filter has a 2% bandwidth centred on about 4 GHz and a minimum insertion loss less than 0.12 dB. Copper dual-mode filters can also produce excellent results, for example Karacaoglu[49] has demonstrated a two-pole dual-mode ring resonator filter with a bandwidth of 5.5% centred on 5.5 GHz with a minimum insertion loss of 1 dB.

A three-channel superconducting channeliser or multiplexer has been produced using these filters. This is a circulator coupled device operating with circulators at 77 K. Although not made from superconductive material, these circulators have an insertion loss of 0.21 dB over a bandwidth of 500 MHz.[22]

5.4 Cavity and waveguide filters

Three-dimensional filters offer the ultimate in bandwidth and power handling because of the spread of the current over a larger area as compared with the planar filters discussed above. All the three-dimensional resonators discussed in Chapter 3 have been used to design filters made from normal metals and many such filters are used in modern communications and radar systems. Such filters are constructed by coupling a number of resonators together or by using coupled modes in one or more resonators, just as are the dual-mode filters described above.

Some of the resonators described in Chapter 3, such as the coaxial resonator, the helical resonator and the cylindrical cavity, can be made from bulk polycrystalline and thick film materials. Although the surface resistance is not as good as thin films, these resonators can form state-of-the-art performance filters when coupled together. However, the approach taken by most workers is to use thin film material as part of the cavity, for example using dielectric resonators.[51–54] An alternative is to include the resonant sections of the transmission line in a waveguide.[55] Both are three-dimensional structures but are not truly three dimensional since the majority of the current flows on a planar surface.

A high-Q dielectric resonator can be made by suspending a low-loss dielectric centrally inside a cylindrical cavity with normal metal walls, as discussed in Section 3.4. By halving the length of the cavity and placing the dielectric on to a superconducting film a similar performance is achieved if the surface resistance of the film is low enough and optimum modes are chosen. In addition, the mechanical stability of the structure is increased as the dielectric no longer has to be suspended. Resonators can be chained together and linked by coupling holes, as shown in Figure 5.4.1. The tuning of such filters is critical and is usually achieved through the traditional means of tuning screws perturbing the cavity mode frequency.

The performance of four filters designed by Mansour[51,52] is shown in Table 5.4.1. These filters are of the structure shown in Figure 5.4.1. The filters operate at about 4 GHz and each resonator occupies a volume of about 0.5 cubic inches. They are connected by coupling irises. Because of the unacceptable variation in group delay using the resonators only, an additional resonator is added which reduces the variation considerably, as can be seen in the table. The external equalising cavity is connected to the filter via a circulator, although not made from HTS, the circulator operates at cryogenic temperatures.

An alternative method of making a three-dimensional filter is to use a waveguide and insert a planar filter in the centre of the guide. This is known as an E-plane filter. A three-pole HTS filter centred on 34.5 GHz has been

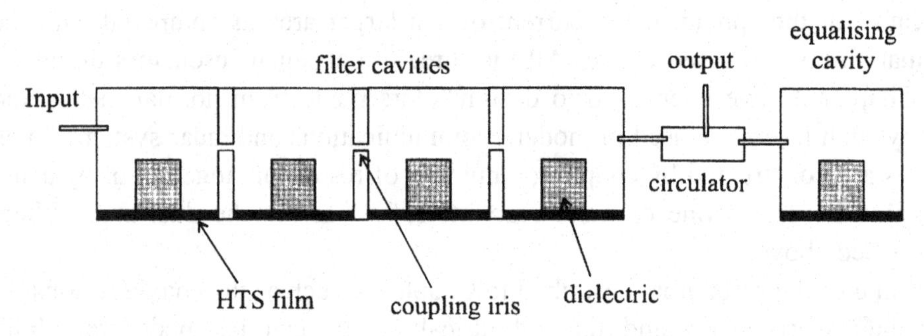

Figure 5.4.1. Eight-pole, externally equalised hybrid dielectric/HTS thin film filter.

Table 5.4.1. *Performance of dielectric resonator filter (after reference [51])*

	Four-pole HTS filter	Four-pole equalised HTS filter	Eight-pole HTS filter	Eight-pole equalised HTS filter
Centre frequency/GHz	3.8	3.8	3.8	3.8
Bandwidth %	1	1	0.8	0.8
Minimum insertion loss/dB	0.12	1.46	0.56	2.21
Loss variation/dB	0.15	0.2	0.25	0.35
Group delay variation over 90% of bandwidth/ns	14	1.5	45	6
Minimum return loss	−20	−23	−14	

described by Mansour.[55] It is made from a combination of bulk polycrystalline HTS for the outer walls of the waveguide and thin film HTS for the inner circuits. The poor performance of this filter, which is especially apparent at these high frequencies, is attributed to the mechanical and electrical qualities of the bulk material.

Most of the filters described in this chapter are based on thin film technology. Thin films have much lower surface resistance than thick film or bulk polycrystalline HTS. However, for low-frequency applications, thick films have a significantly lower surface resistance than normal metals at liquid nitrogen temperatures. Further discussion of this can be found in Appendix 1. In addition, thick films can be deposited on large three-dimensional surfaces. Filters for mobile communication base stations have been produced using thick film material. These are coupled cavity filters, as discussed above, where only the surrounding cylinder is coated with thick film and has an internal split ring resonator which is also made from thick film. The responses of two of these filters designed for mobile communication base stations are shown in Figure 5.4.2. One is for A band operators, the other for B band operators. Further details of mobile communication systems are given in Section 8.4.

Each of the filters has two independent receiving channels, as designated by the communications specification, with the B band device notch levelling out at 52 dB. The insertion loss of the filters is better than 1.5 dB, with a return loss minimum of about 1.5 dB. The two-tone intermodulation distortion (described in Section 5.5) of these filters has been measured, and the two-tone output is −75 dBc for an input level of −10 dBm. In order to achieve these specifications, cavities with *Q*s in excess of 40 000 were used and the A band filter required 16 and 10 poles for the lower and upper bands respectively. The whole filter system is cooled by closed cycle coolers, has a power requirement of 400 watts and weighs approximately 140 lbs.

5.5 Non-linear effects and power handling

Clearly, the power handling of superconducting filters is going to be limited by a critical current of the material making up the filter. Eventually, regions of the superconductor will become normal and the filter response will change. Also, because the current distribution within a filter is non-uniform there will be sections which become normal before others, and thermal effects may become involved as the normal parts of the filter dissipate energy. The effects seen are an increase in the insertion loss of the filter and a distortion of the shape of the filter response. In addition to this effect, another process occurring is the generation of harmonics. Any real material is non-linear to some extent and

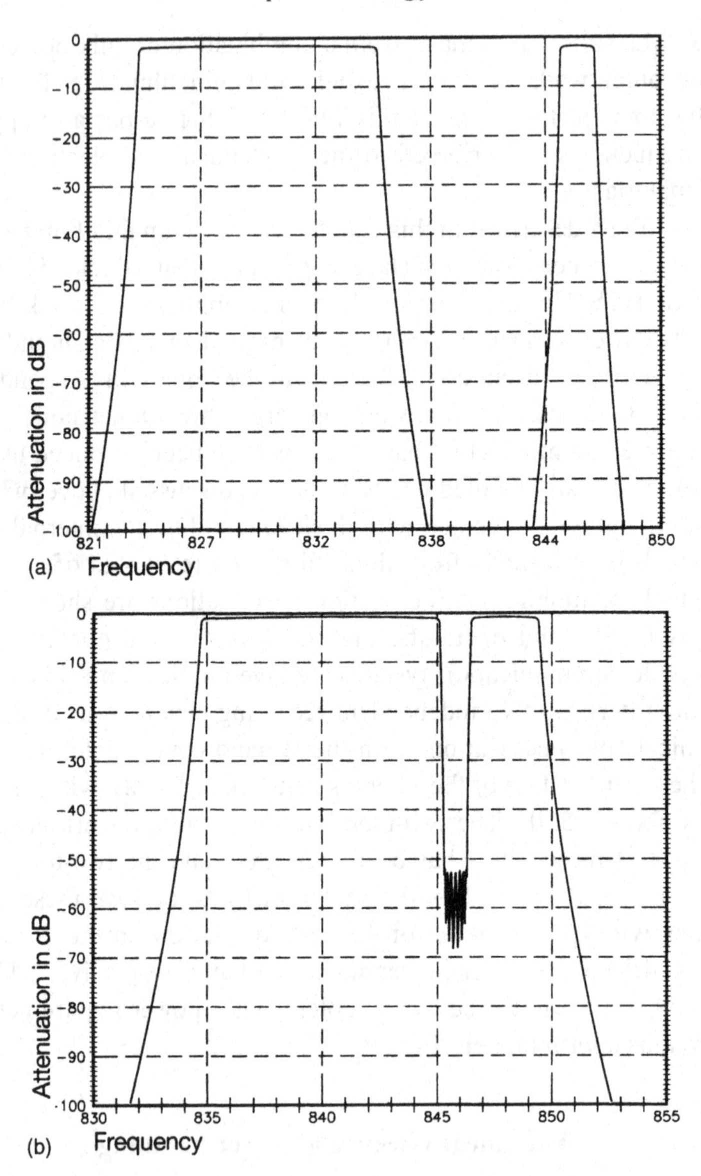

Figure 5.4.2. Filter frequency response of ISC 'A' and 'B' band cellular filter. (Taken from reference [56].)

generates harmonics by the transfer of energy from the fundamental frequency input signal. This occurs in HTS filters both as the current increases above a critical current and, to a much lesser extent, when peak currents are below this current. The effects observed are mainly due to a distortion caused by the harmonic signal mixing down in frequency back into the passband, although some increase in the insertion loss also occurs as energy is transferred to the harmonics. If a signal with two frequencies, ω_1 and ω_2, which are fairly close together, is applied to a filter with non-linear processes operating, then harmonics will be generated. By mixing the third harmonic with the fundamental, a signal with frequency of $2\omega_1 - \omega_2$ can be generated. This frequency component is very close in frequency to the two signals applied to the filter, thus causing distortion and must be assessed if superconducting filters are to be used in high-performance applications.

It is possible to reduce the non-linear effects to a certain extent by the design of the filter. Some methods are: (i) an improvement in the superconducting materials power handling capability; (ii) an increase in the size of the filter, thereby spreading the current over a larger area; (iii) a reduction in the peaking of the current at the edges of the transmission line, for example the use of a microstrip rather than a coplanar line, since the latter has a larger current crowding; alternatively, increase the ground plane spacing; (iv) increasing the film thickness to a large number of penetration depths; (v) increasing the lines' characteristic impedance by using a thicker substrate or a substrate with a large dielectric constant; this must be done without using a narrower line. With careful design these five methods can be used to increase the power handling capability of filters significantly. Demonstration of power handling up to several tens of watts has been achieved with no significant deterioration in the insertion loss of the filter.[57,58]

Two-tone third-order intermodulation can be analysed by considering the relationship between the applied voltage, V, and the current input, I, to the filter.[59] Owing to the non-linear effects, this is no longer linear but can be approximated by the Taylor series

$$I(V) = I(0) + \left(\frac{dI}{dV}\right)_{V=0} \delta V + \frac{1}{2!}\left(\frac{d^2 I}{dV^2}\right)_{V=0} \delta V^2 + \frac{1}{3!}\left(\frac{d^3 I}{dV^3}\right)_{V=0} \delta V^3 + \ldots \tag{5.5.1}$$

Assuming that no bias is applied to the filter, then $I(-V) = I(V)$ and the even terms in Equation (5.5.1) are zero, giving the third-order term as the most dominant non-linear term,

$$I(V) = \frac{1}{R}\delta V + \frac{1}{3!}\left(\frac{d^3 I}{dV^3}\right)_{V=0} \delta V^3 + \ldots \tag{5.5.2}$$

The resistance R has been defined by

$$\frac{1}{R} = \left(\frac{dI}{dV}\right)_{V=0} \tag{5.5.3}$$

The application of two tones to the filter can be simulated by using the substitution

$$\delta V = V_0(\sin(\omega_1 t) + \sin(\omega_2 t)) \tag{5.5.4}$$

giving

$$I(V) = \frac{1}{R} V_0(\sin(\omega_1 t) + \sin(\omega_2 t)) + \frac{1}{3!}\left(\frac{d^3 I}{dV^3}\right)_{V=0} V_0^3(\sin(\omega_1 t) + \sin(\omega_2 t))^3 + \ldots \tag{5.5.5}$$

Taking the cube and sorting out the terms of interest, these are the most dominant fundamental term and the two-tone third-order term, gives

$$I(V) = \frac{1}{R} V_0(\sin(\omega_1 t) + \sin(\omega_2 t)) + \frac{1}{8}\left(\frac{d^3 I}{dV^3}\right)_{V=0} V_0^3 \sin(2\omega_1 t - \omega_2 t) \tag{5.5.6}$$

If alternative mixing products are of interest, then it is possible to retain other terms in this series; however, these will not be considered here. Now consider the amplitude of the two terms in Equation (5.5.6) separately:

$$I_1 = \frac{V_0}{R} \tag{5.5.7}$$

and

$$I_{12} = \frac{1}{8}\left(\frac{d^3 V}{dI^3}\right)_{V=0} V_0^3 \tag{5.5.8}$$

The third derivative of the voltage current relationship, like the resistance, is just a function of the geometry of the filter and the material it is made from. Equation (5.5.7) can be written in a logarithmic form giving

$$20\log(I_1) = 20\log(V_0) - 20\log(R) \tag{5.5.9}$$

Rewriting this in terms of microwave power gives

$$P_{1out} = P_{in} - 20\log(R) \tag{5.5.10}$$

Similarly, using Equation (5.5.8) for the two-tone third-order term:

$$20\log(I_{12}) = 60\log(V_0) - 20\log(8) + 20\log\left(\left(\frac{d^3 V}{dI^3}\right)_{V=0}\right) \tag{5.5.11}$$

and rewriting for the power gives

$$P_{12out} = 3P_{in} - 18.06 + 20\log\left(\left(\frac{d^3 V}{dI^3}\right)_{V=0}\right) \tag{5.5.12}$$

Equations (5.5.11) and (5.5.12) show the power in the fundamental and the power in the third-order intermodulation as a function of the applied input power. The slope of the two-tone intermodulation is, however, much steeper (i.e. 3 as compared with 1 for the fundamental). Consequently, the situation arises where the power in the fundamental becomes equal to the power in the two-tone intermodulation. This can be found by Equations (5.5.11) and (5.5.12), giving

$$P_{in} = -10\log(R) + 9.03 - 10\log\left(\left(\frac{d^3V}{dI^3}\right)_{V=0}\right) \tag{5.5.13}$$

This is known as the third-order two-tone intermodulation intercept and is used as a figure of merit for the non-linearity present in filters. In practice, of course, this intercept can never be measured directly at the power level P_{in} because the additional terms neglected in Equation (5.5.6) become significant before this point is reached. However, it can be measured by plotting the levels of the fundamental and two-tone intermodulation at lower power levels, and then using linear extrapolation to determine the intercept.

Figure 5.5.1 shows some two-tone third-order measurements of a 0.7% bandwidth YBCO microstrip filter on $LaAlO_3$ centred on 2 GHz. The filter is of the forward coupled type, as described above, and the resonant elements have an impedance of 10 Ω in order to increase the power handling capability of the filter. The two-tone third-order intercept at 77 K is 39 dBm, see reference [58]. In certain instances a deviation from the slope of 3 is observed which is probably due to a change in the current distribution within the superconductor

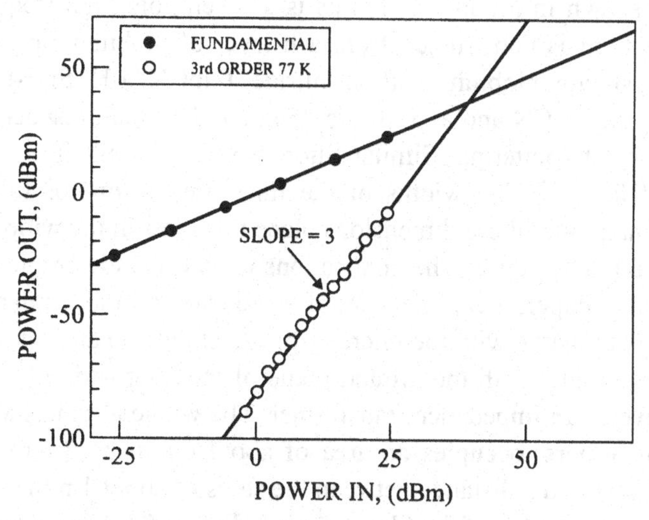

Figure 5.5.1. Two-tone third-order and third harmonic distortions compared with the fundamental. (Taken from reference [58].)

as the power level increases. This effectively alters the third-order derivative in Equation (5.5.12) with an increase in power.

5.6 Internal inductance filters

Another method of reducing the size of a filter is to use the reduction in velocity due to internal or kinetic inductance effects. This velocity reduction arises due to the increase in internal inductance of a transmission line due to additional energy storage within the superconducting film. Clearly, this is only significant if the energy stored in the films is comparable to that stored externally, and for microstrip this means that the dielectric thickness must approach the penetration depth. Predictions for the velocity, loss and impedance for these transmission lines are given in Section 2.3. The quality factor for resonators made using thin films is discussed in Section 3.7.1. As discussed in Section 2.3, it is clear that as the dielectric thickness of a microstrip decreases, the attenuation increases, even though the velocity reduces. Hence, an arbitrary decrease in the size of practical filters cannot be accomplished and careful consideration needs to be given to the balance between the size of the filter and its insertion loss. Early investigations of the possibility of using kinetic inductance effects to produce miniature filters centred on investigation of the properties of delay lines and resonators and it was clear that factors of 30–100 in velocity reduction are possible[60–64] for both high-temperature and low-temperature superconductors.

An example of a filter constructed using the enhancement of internal inductance is shown in Figure 5.6.1; this is a seven-pole high-frequency stub filter.[65,66] The filter is constructed from a trilayer of niobium carbon nitride/hydrogenated silicon/niobium carbon nitride (NbCN/a-H::Si/NbCN). The thicknesses of the NbCN and a-H::Si are 25 nm and 300 nm respectively; both are deposited by rf sputtering. Similar filters have been built from trilayers of NbCN/MgO/NbCN.[65] The widths of the lines are 16 μm for the stub and 8 μm for the main line; these dimensions place this filter in the wide microstrip limit discussed in Chapter 2. The connections to external electronics are made using a coplanar taper; here the central conductor of the coplanar line is linearly tapered to match with the microstrip, whilst the coplanar ground planes are tapered to connect with the ground plane of the microstrip. This structure can also behave as an impedance transformer. The whole circuit, including the coplanar transformers, occupies an area of about 6 mm by 3 mm, with stub lengths of 0.5 mm and a distance between the stubs of about 1 mm.

The frequency response of the filter shown in Figure 5.6.1 is shown in Figure 5.6.2 at a number of different temperatures, together with the predicted

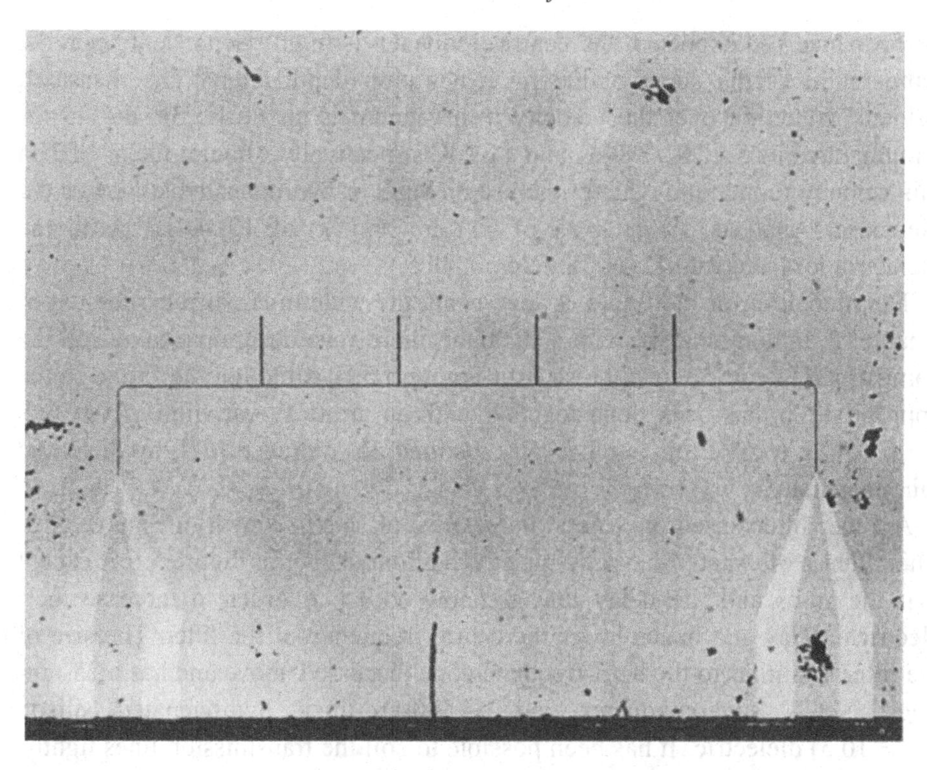

Figure 5.6.1. High-frequency stub filter using kinetic inductance effects for miniaturisation. (Taken from reference [67].)

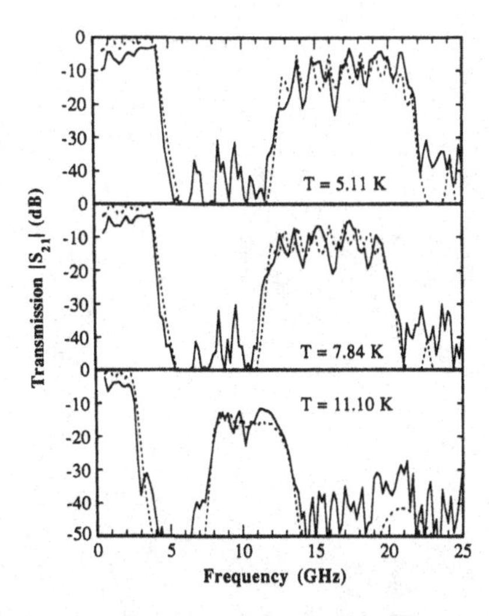

Figure 5.6.2. Frequency response of the high-frequency stub filter shown in Figure 5.6.1 at a number of different temperatures. The dotted line represents the modelled response. (Taken from reference [67].)

performance. As expected, the centre frequency is highly dependent upon the temperature as the superconducting penetration depth alters. The measured velocity reduction over the velocity in a vacuum is about 31, 34 and 45 for temperatures of 5.11 K, 7.84 K and 11.1 K respectively. About a factor of 3 in this velocity reduction is due to the use of MgO or hydrogenated silicon as the dielectric. Analysis yields a λ_0 of 311 nm and T_c of 12.44 K,[65] with the dielectric loss tangent of a-H::Si being 0.0005.

The modelling of the filters is done using the wide microstrip expression of Chapter 2, in combination with a standard microwave design package, and the comparison is good, as can be seen in Figure 5.6.2. Although the fitting to the appropriate λ_0 has been done together with an arbitrary variation of the stub length of up to 5%, this is physically justified, in reference [67], by dielectric inhomogeneity.

Another filter based on kinetic inductance effects is shown in Figure 5.6.3. This filter is of exactly the same type as the filter shown in Figure 5.6.1, except that the stubs and the delay line are now coiled in order to increase their electrical delay and hence lower the centre frequency of the filter. The size of the filter is similar to the high-frequency one discussed above and has a 25-nm-thick NbCN superconductor and a 300-nm-thick hydrogenated silicon ($\varepsilon_r = 10.5$) dielectric. It has been possible to coil the transmission lines tightly, with minimal crosstalk between them; this is due to the very small dielectric

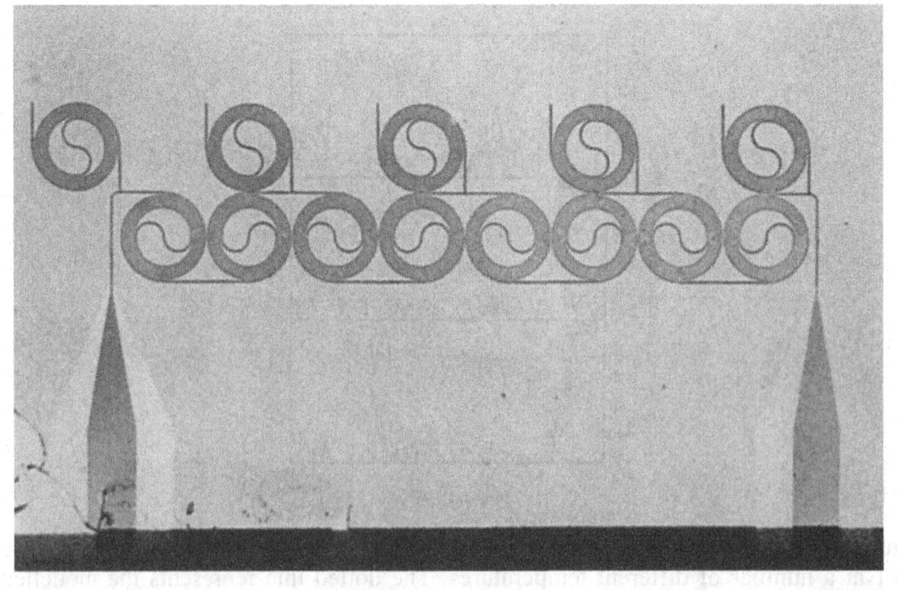

Figure 5.6.3. Low-frequency stub bandpass filter. (Taken from reference [67].)

thickness which helps to confine the field. The lengths of the elements are about a factor of 20 longer than those in the high-frequency filter.

The response of the filter is shown in Figure 5.6.4 at a number of different temperatures, and wideband agreement is obtained using the measured value of penetration depth from the filter properties. The main deviation from the predicted values is due to circuit imperfections, and losses are limited to the loss tangent of the hydrogenated silicon. The performance of this type of circuit is particularly sensitive to circuit imperfections, since it is difficult to deposit good dielectric layers which are constant in thickness over the required area. Any differences in thickness produce unwanted changes in impedance and, in the case of kinetic inductance filters, velocity perturbations. Even small changes can cause unwanted distortion in the passband response. The quality of the dielectric is also important, and the loss tangent presents one of the limiting factors in the performance of the device. Even for the MgO dielectrics, the loss tangent in its thin film form is measured as 0.003, a factor of around 100 higher than some single-crystal values. The superconducting layer must also be of high quality in order to keep the surface resistance low, and must also be kept free of pin holes over a large area. For 50 Ω lines the tracks are narrow and sensitive to degradation due to the patterning process. Wider lines are possible but the impedance reduces, forcing an impedance transformer to be used. However, even with these difficulties, significant progress is being made with kinetic inductance filters, which represent one of the smallest types of superconducting filters possible. The filter in Figure 5.6.3 represents a

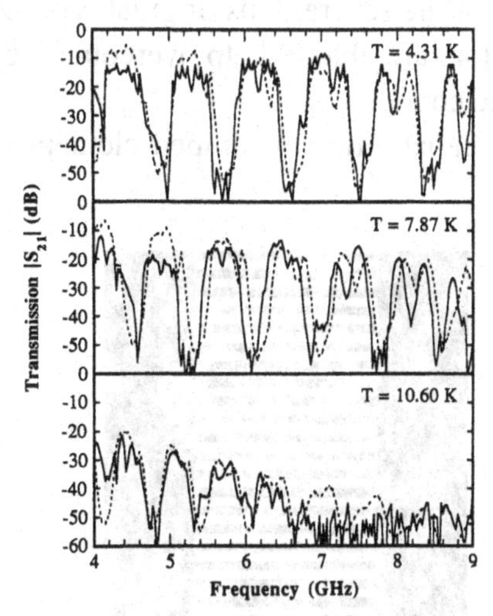

Figure 5.6.4. Frequency response of the low-frequency stub bandpass filter. (Taken from reference [67].)

reduction in size of the order of 100 on a similar filter built from conventional copper microstrip on a substrate with a permittivity of 9.

In contrast to the circuits shown in Figures 5.6.1 and 5.6.3, many conventional microstrip circuits use edge coupling between a number of resonant sections, as discussed in Section 5.2 and shown in Figure 5.2.3. In the case of transmission lines dominated by kinetic inductance, it is difficult to achieve coupling between adjacent transmission lines since the electromagnetic field does not extend a large distance due to the thin dielectric. Submicron patterning would be needed to achieve the required control over line spacing. An alternative is to fabricate microwave filters based on inductive coupling with the lines above one another. Analysis of these structures has been undertaken successfully by a number of authors,[68–70] and demonstrated by inductive coupling to a Josephson junction.[71] The analysis by Pond[72] uses the transverse resonant method to develop a dispersion equation for the two dominant modes of the coupled line, and this is compared with the results of asymmetric coupled line theory to yield equivalent circuit parameters. The results include losses and give design parameters for the coupled structure.

5.7 Lumped element filters

Lumped elements are, by definition, much smaller than the wavelength at which they operate. Hence, at high frequencies, where the wavelength is short, filters based on lumped elements will be physically small. It turns out that where the line widths are limited by the patterning process the centre frequencies of filters are in the several tens of gigahertz range. At these narrow line widths superconductors are able to help overcome the loss associated with the finite resistance of the conductors.

In order to assess the capabilities of lumped element components consider

Figure 5.7.1. A lumped element resonator.

the lumped element resonator shown in Figure 5.7.1. The resonator consists of a number of interdigital fingers forming a capacitor. The central finger connects both sides of the capacitor, acting as an inductor and hence forming a parallel resonant circuit. This circuit can be used to estimate the quality factors available for superconducting lumped element circuits; different geometries will obviously produce different quality factors. The losses in this circuit arise mainly from the inductor because of the high current density on this element.

Simple expressions are available in the literature for the inductance of a planar strip and the capacitance of an interdigital capacitor; together with their associated losses,[73] the Q and the centre frequency of this resonator can easily be estimated. The Q is shown in Figure 5.7.2 as a function of the size of the resonator. This Q is given for the superconductor and copper and the resonator is assumed to be square. This figure is for resonators operating at 10 GHz. In order to maintain this resonant frequency the number of fingers in the capacitor increases as the size of the resonator decreases. This increase is from 7 fingers for the 1-mm-square resonator to 70 for the 200 μm resonator. In order to accomplish this, the width of the fingers is decreased from 85 μm to 1.5 μm. Although this is a very crude estimate, it can be seen that high Qs are possible even with these small-size resonators.

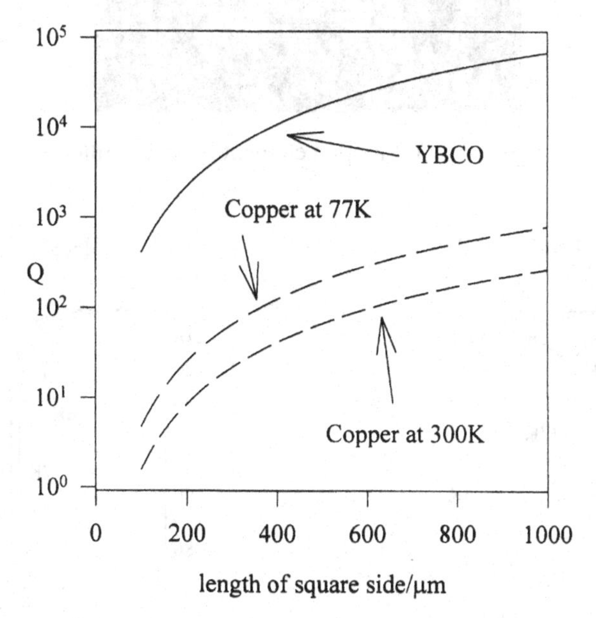

Figure 5.7.2. The quality factor at 10 GHz of a lumped element as a function of resonator size for superconducting and copper materials. (For a discussion of the material parameters see Appendix 1.)

It is easy to use this structure to make a bandstop filter,[74] as shown in Figure 5.7.3. The element shown in Figure 5.7.1 is placed centrally in a coplanar transmission line. The whole YBCO structure fits on a 1-cm-square MgO substrate. In this particular example there are 20 fingers in the interdigital capacitor, each with a length of 1 mm and a width of 10 μm. The coplanar line is 0.41 mm wide with a 0.16 mm gap between the ground plane and the central strip.

The frequency response of the filter is shown in Figure 5.7.4. The bandstop response is centred on about 5 GHz and varies substantially with temperature. This temperature variation is due to the field penetration into the strip inductor

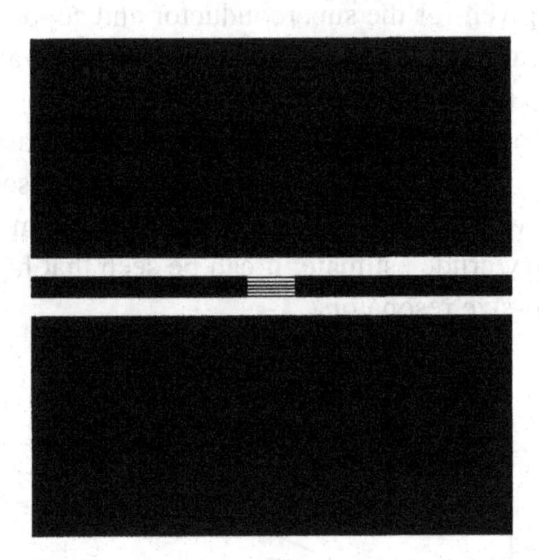

Figure 5.7.3. Lumped element bandstop filter.

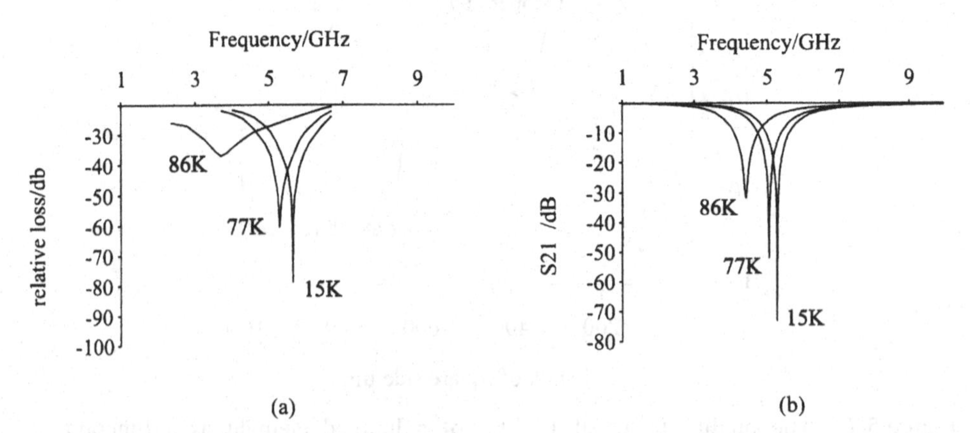

Figure 5.7.4. Bandstop filter performance: (*a*) the insertion loss at a number of different temperatures, and (*b*) the modelled response.

as the superconducting penetration depth alters. Because the film thickness (0.35 μm) is of the order of the penetration depth, these changes cannot be taken into account by conventional methods. Here the volume current distribution in the inductor needs to be calculated. To model the frequency shift the numerical calculation based on the coupling of multiple transmission lines described in Section 2.7 is used, and the results of the calculation are shown in Figure 5.7.4(b). There is a 32% change in resonant frequency between 15 K and 86 K which could be used for tuning the filter by varying the temperature. The stopband performance varies during this temperature range but has a maximum stopband rejection of more than 50 dB. The power dependence of the filter is good, with only a 0.03% change in centre frequency for input power varying from −45 dBm to −10 dBm. The corresponding change in insertion loss is 16% at 15 K and 2% at 77 K.

It is also possible to construct resonators out of this lumped element by leaving capacitive gaps between the resonator element and the coplanar feeding line. Such a resonator made from YBCO on an MgO substrate has a Q ranging from 9400 at 5.734 GHz and 15 K to 1300 at 5.587 GHz and 77 K.[74]

A number of more complex lumped element filters have been produced,[75,76] and one is shown in Figure 5.7.5, with its frequency response being shown in Figure 5.7.6. The size of the filter is 6 mm by 15 mm, and it is a ninth-order Chebyshev filter with a passband between 1.1 and 1.9 GHz. The selectivity is

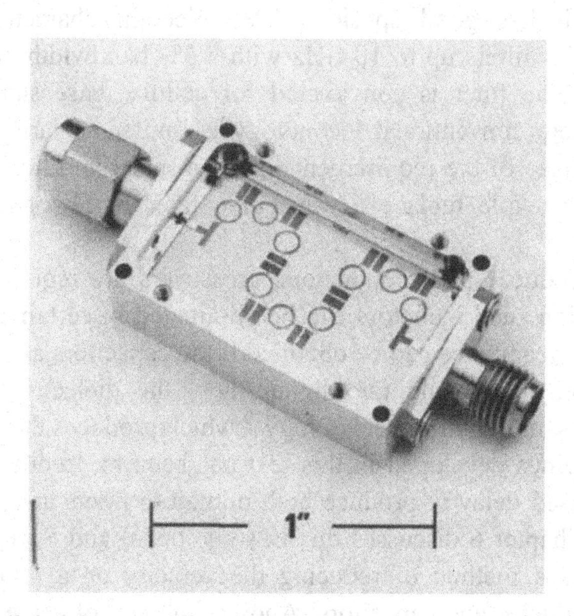

Figure 5.7.5. Cellular base station lumped element bandpass filter. (Taken from reference [77].)

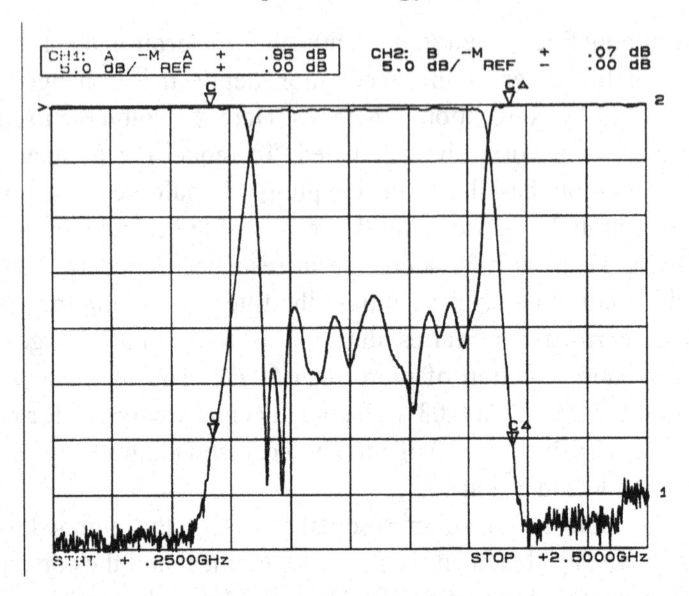

Figure 5.7.6. Frequency response of the lumped element bandpass filter shown in Figure 5.7.5. (Taken from reference [76].)

excellent, with a 0.5 dB bandwidth of 0.84 GHz compared with a 30 dB bandwidth of 1.128 GHz. The minimum insertion loss, including the connectors is 0.2 dB and the return loss is better than 17 dB. As with other lumped element filters it has good spurious free rejection characteristics up to 5 GHz.[76,77] Similar filters up to 10 GHz with a 3% bandwidth have also been demonstrated.[78] The filter is constructed for cellular base station receiving applications, where conventional technology is limited to large-cavity waveguide filters because of the requirement for ultra-selective filter skirts. Such a filter could in principle replace a waveguide filter more than three feet in length.[77]

In order to produce spiral conductors, crossovers are required,[78] although with careful design such inductors can be eliminated in certain cases. Also, a greater value of capacitance can be obtained if the capacitors are in the parallel plate form rather than being interdigital; here the dielectric layer can be deposited by the same thin film technology[79] which produces the HTS itself.

One of the themes developed in this text has been to produce transmission lines with increased delay to produce both miniature filters and signal processing elements. Chapter 6 discusses this in some detail and Section 5.8 of this chapter discusses a method of reducing the velocity on a transmission line using discrete inductor and capacitors. Another method of achieving the same outcome is to construct lumped element allpass filters which have a large group delay; by chaining these filters together long delays can be produced. The applications of these filters are the same as those of the delay lines discussed in Chapter 6. Such an element is shown in Figure 5.7.7,[80] together with its bridged T-equivalent circuit. The design of such a circuit is done by first using approximate expressions for the lumped element components followed by analysis using a full wave CAD package. The structure is constructed from YBCO on $LaAlO_3$ and measures 1.54 mm by 2.78 mm. The design is for a 0.5 ns delay centred on 2.8 GHz. This is in fact achieved over a bandwidth of about 1 GHz. The maximum insertion loss after de-embedding the contacts and

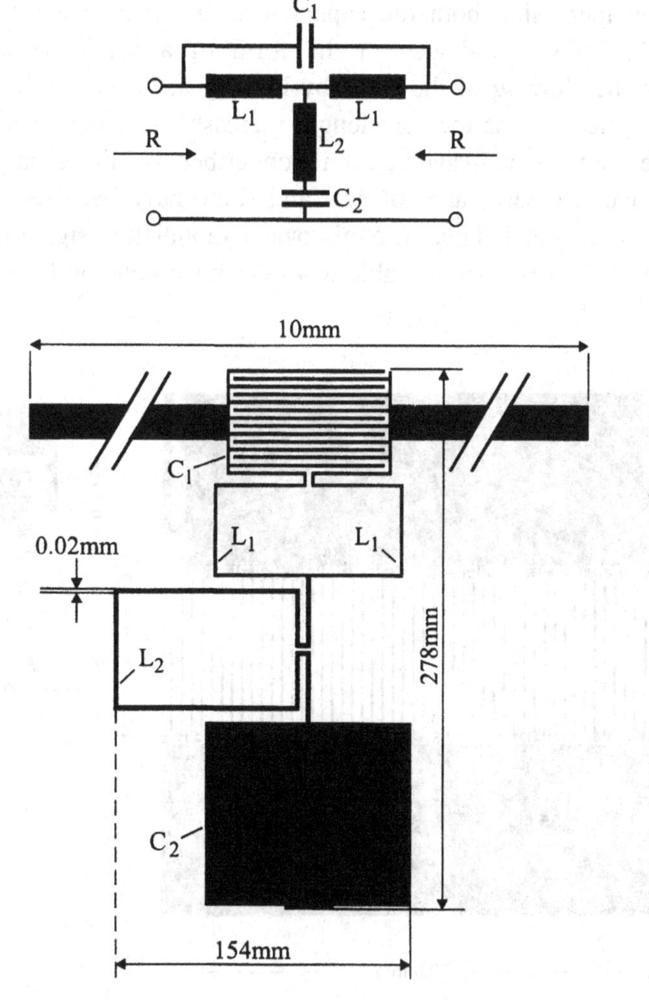

Figure 5.7.7. A lumped element allpass filter with both layout and equivalent circuit. (Taken from reference [80].)

connectors is less than 0.3 dB below 4 GHz. Such a delay in a small length gives an effective velocity of about one hundredth of the speed of light in free space, allowing good flexibility when devices are chained together.

5.8 Filters based on slow-wave transmission lines

A coplanar slow-wave resonator is shown in Figure 5.8.1. It is simply a transmission line formed from discrete inductors and capacitors.[81–83] The inductors are the narrow vertical tracks and the capacitance is gained from the narrow gap between the coplanar ground plane and the central conductor. Making a transmission line in this way allows independent reduction in the velocity by increasing both the capacitance and inductance per unit length. Figure 5.8.1 shows a device in the form of a resonator, convenient for measuring the slowing of the wave by looking at the resonant frequencies. It should be noted that as the wavelength increases to around, and smaller than, the unit cell in the slow-wave line it no longer behaves like a transmission line.

Slow-wave structures are not new and some have been used to match the velocity of the optical signal to a microwave modulating signal in electro-optic modulators.[84,85] Also, considerable work has been done on MMICs in order to

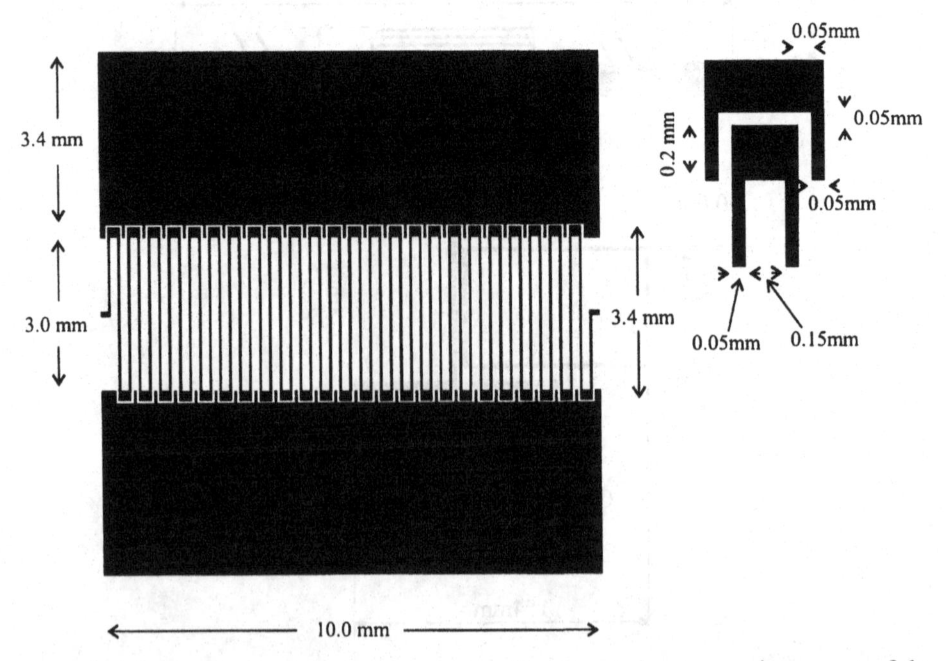

Figure 5.8.1. A coplanar slow-wave resonator. The insert shows an enlargement of the capacitive portion of the structure.

reduce the chip size.[86–88] In this case conventional distributed passive elements are at present the limiting factor on the size of MMICs, as the active devices used are much smaller than a wavelength. However, lumped elements have not generally been used in passive filters because of the associated increase in loss. The first consideration of a superconducting slow-wave structure seems to have been by Gandolfo, in 1968,[89] who produced a lead meander line. This line delayed signals up to 30 ns in a length of about 24 cm at frequencies of about 4 GHz, giving a factor of just less than 40 reduction in velocity over the free space velocity. The response of the device contained ripples and nulls but gave a minimum loss of only 0.5 dB, with the lead superconductor at 4.2 K, compared with 20 dB for copper at the same temperature.

The HTS device shown in Figure 5.8.1 is made from thin film $YBa_2Cu_3O_7$ on a 1-cm-square MgO substrate deposited by laser ablation. The length of the resonator is about 10 mm. The fundamental resonant frequency is 1 GHz at 77 K, providing a velocity reduction factor of 15 over the free space velocity. Figure 5.8.2 shows the frequency response of the resonator. A large number of harmonic resonances are present. They are dispersive, as can be seen by the unequal spacing between the harmonics as the frequency is increased. A cut-off frequency occurs when the wavelength of operation is equal to the length of one period in the structure. The dispersion is complex and is governed not only by the basic response of the inductive and capacitive sections but also by these

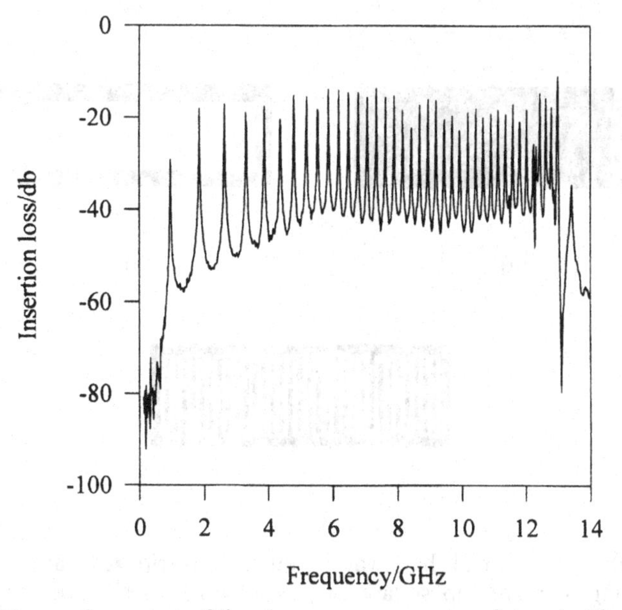

Figure 5.8.2. The performance of the slow-wave resonator shown in Figure 5.8.1.

elements becoming large in terms of wavelength as the frequency increases. In addition, coupling between these elements plays a role.

The geometry discussed above is only an example of one method in which discrete inductors and capacitors can be arranged to produce a transmission line; there are many more and each needs to be examined to see if the velocity can be minimised given the practical constraints of the pattering and the HTS material. The main use of the slow-wave transmission line is not necessarily achieved by making resonators, but by using them as replacements for conventional transmission lines in conventional filters.

Figure 5.8.3 shows how a similar effect can be achieved using microstrip. Consider the standard microstrip resonator shown in Figure 5.8.3(a). The effect of removing the central portion so as to produce the loop of Figure 5.8.3(b) is only small. It effectively turns the standard patch into a loop resonator, and the frequency reduction is small since the width of the patch is small compared with its length. To reduce the frequency of the resonator the loop can be loaded with capacitive fingers as shown in Figure 5.8.3(c). The velocity reduction on this type of transmission line is controlled by the number of fingers within the loop. Copper resonators of this type have shown a 25% reduction in frequency around 4 GHz, with 31 fingers in the loop.[90] Coupled resonators have also been demonstrated,[91] showing that conventional design techniques can be used to design coupled slow-wave lines. A superconducting resonator of this type with

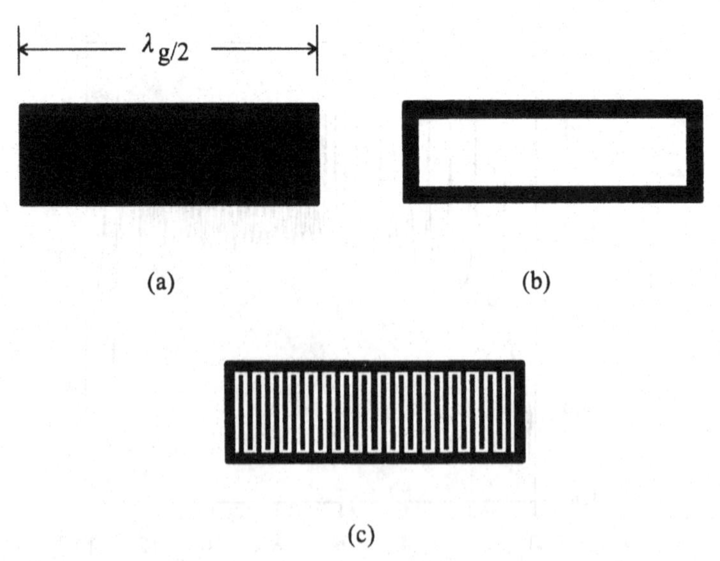

Figure 5.8.3. (*a*) Conventional half-wavelength microstrip resonator. (*b*) Modified resonator forming a microstrip square loop resonator. (*c*) Capacitively loaded loop resonator forming a slow-wave structure.

outside dimensions of 4×1 mm and 195 fingers, each of 10 μm width and 890 μm length, resonates at 10.3 GHz with a Q of 1200 at 77 K. This represents about a 30% reduction in size over the conventional microstrip resonator.

5.8.1 Velocity in periodic structures[92]

It is important in filter design to attempt to characterise the dispersion in these periodic structures. The theory presented below gives general expressions for a transmission line made up of components arranged periodically. All that is required for the calculation is the *ABCD* matrices for each of the elements.[93]

Consider any transmission line made of a periodic structure. The voltages V_1 and currents I_1 at the input of one period are related to the voltages and currents at the output of one period by

$$V_2 = V_1 e^{-j\gamma_p d}$$
$$I_2 = I_1 e^{-j\gamma_p d} \tag{5.8.1}$$

I_1 is defined as the current at the input directed into the section and I_2 is the current at the output directed out of the section. We wish to find γ_p, the propagation constant which is for the whole periodic structure of infinite extent. The voltages and currents are related through the *ABCD* matrix of the section, that is,

$$\begin{bmatrix} V_1 \\ I_1 \end{bmatrix} = \begin{bmatrix} A & B \\ C & D \end{bmatrix} \begin{bmatrix} V_2 \\ I_2 \end{bmatrix} \tag{5.8.2}$$

Here the *ABCD* matrix is made up of the multiplication of the different *ABCD* matrices for the different sections of transmission line which make up the period. For example, for the transmission line shown in Figure 5.8.1 this could be formed by multiplying the *ABCD* matrix of the series inductance by the parallel capacitance. Substituting Equation (5.8.1) into Equation (5.8.2) gives

$$\begin{bmatrix} A & B \\ C & D \end{bmatrix} \begin{bmatrix} V_2 \\ I_2 \end{bmatrix} = K \begin{bmatrix} V_2 \\ I_2 \end{bmatrix} \tag{5.8.3}$$

where $K = e^{j\gamma_p d}$ is the eigenvalue. This is obtained by solving

$$\begin{vmatrix} A - K & B \\ C & D - K \end{vmatrix} = 0 \tag{5.8.4}$$

giving

$$e^{j\gamma_p d} = \frac{A + D}{2} \pm j\sqrt{1 - \frac{(A + D)^2}{2}} \tag{5.8.5}$$

that is,

$$\gamma_p = \frac{1}{d}\cos^{-1}\left(\frac{A+D}{2}\right) \tag{5.8.6}$$

The propagation constant of any periodic structure can now be calculated. First, the *ABCD* matrix of the combined structure can be calculated by multiplying the individual matrices of each of the separate structures and, second, by using Equation (5.8.6). For example, if the periodic structure is made up of two different impedance transmission lines, then the *ABCD* matrix for each line can be multiplied to form the periodic matrix and then Equation (5.8.6) used to obtain the propagation constant and hence the velocity along the structure. Alternatively, if the transmission line is made up of a number of lumped element components, the *ABCD* matrix can be made up of the individual component *ABCD* matrices.

However, this calculation does not explicitly take into account coupling between adjacent components and the effects as the components become large in comparison with the wavelength. Hence, such calculations are only appropriate far away from the cut-off frequency of the slow-wave structures described above.

5.9 Superconducting switches and limiters

Microwave switches are used to highly attenuate a microwave signal by the application of some control signal, and have many applications in the microwave industry. A limiter is similar to a switch except that it is the actual microwave signal rather than a control signal which controls the attenuation; limiters can be used, for example, to protect electronics from the application of large damaging signals. Switches have been used in digital attenuators where different attenuation sections are switched in or out. They can be used in digital phase shifters where different lengths of delay line are switched in to control the phase of a signal. Another application is in switched filterbanks where the frequency response of a filter bank can be changed dynamically in response to external requirements. An example may be in the suppression of narrow band interference; if a bank of narrow band bandstop filters is used on the input stage of a receiver, one or more received interfering signals can be removed by switching in the appropriate filter. The interference may not only come from external sources; for example, when an aircraft uses its own transmitters for communication, this may swamp its own wide band surveillance equipment. Switching in the appropriate filter upon transmission allows the surveillance equipment to continue operation.

Switches must be low loss in their on state and provide high isolation in their off state. Fast switching times are required for many applications, as is a broad bandwidth. However, narrow or medium bandwidth switches are also appropriate for some applications. Switches must have the power handling capabilities for the required application, and there are requirements for high power handling capabilities, particularly for protection applications. Conventionally, PIN diodes or FETs have been used for microwave switches,[94] and these have switching speeds in the nanosecond range and are quite adequate for many applications; superconductors must show improvements over this conventional technology. The advantages of using superconductors are the same as those discussed previously, that is, the low loss of HTS devices and the ability to produce miniature components by various techniques. There are no reasons why conventional PIN diodes or FET switches cannot be used with superconductors and an example is given below.

There are several methods of performing the switching or limiting action with superconductors; these usually involve changing the superconductor from its superconducting state to its normal state. The large difference in surface resistance between the two states produces the required action. The methods which can be used in order to accomplish the transition between states include the application of a magnetic field, an increase in temperature, the use of a bias current and the application of light to the circuit. Application of a magnetic field can be achieved by the close proximity of a wire, with a bias current producing the field. An increase in temperature can be accomplished in a number of ways, including the close proximity of a heater element or the application of light. The application of heat takes the superconductor locally through its transition temperature and is a fairly slow process; this is generally known as the bolometric response of the superconductor. Careful thermal design is required for this type of structure if optimum performance is to be achieved. There is also a much faster non-bolometric response,[95] due to the energy in the control signal splitting the superconducting electron pairs directly. Much research effort has gone into the study of this effect because of the underlying physical mechanisms involved. A number of associated switches have been demonstrated, but they are not necessary for microwave applications.[96–99]

Although the thermal switch is perhaps the simplest in concept, good performance can only be obtained with careful design. Figure 5.9.1 shows a thermal switch,[100] consisting of a narrow section of transmission line between 10 and 100 μm wide and about 100 μm long. In the narrow region the superconductor is thinned to about 10–40 nm by selective etching. This promotes flux flow in the bridge, speeding up the phase transition and increasing the normal state resistance, and thus producing better isolation. A normal metal control line is placed above this area with an interleaving dielectric layer for isolation, and a thermal insulating layer is placed over the top for thermal isolation from the cryogen. In order to speed up the switch the substrate is thinned behind the switch, giving it a lower thermal capacity. Such switches have a switch on time of around 100 ns and a switch off time of around 400 ns. A switchable bandpass filter bank containing three filters uses this switch, and the results are shown in Figure 5.9.2, with each filter being selected in turn. One of the filter banks is constructed out of seven-pole Chebyshev filters with centre frequencies of 9.1, 10 and 10.9 GHz, while the other uses four-pole elliptical dual-mode microstrip filters. Both filters have excellent responses, with low passband insertion loss (< 1 dB) and high stopband rejection (> 50 dB). These filters have to be tuned individually and it is also found that twin-induced dielectric anomalies make reproducibility difficult. This switch has also been demonstrated in a switchable delay line, acting as a phase shifter.[100]

Another example of a switch which also operates as a miniature lumped element bandstop filter is shown in Figure 5.9.3. The structure is described in detail in Section 5.7 in the context of a lumped element filter. However, with the application of a bias current the superconducting inductor is switched into its normal state and the resonance which provides the bandstop function no longer occurs, producing an allpass filter. This filter/switch is only 1 mm square and operates at about 6 GHz. It has almost 40 dB isolation at a temperature of 65 K with a bias current of 50 mA. The bias current applied corresponds closely with the critical current of the inductor section.

Waveguide limiter/filters can also be made using a similar concept by placing a resonant element inside a waveguide. This has been demonstrated at 37 GHz using two slots in a finline inside a waveguide placed a distance $\lambda/4$

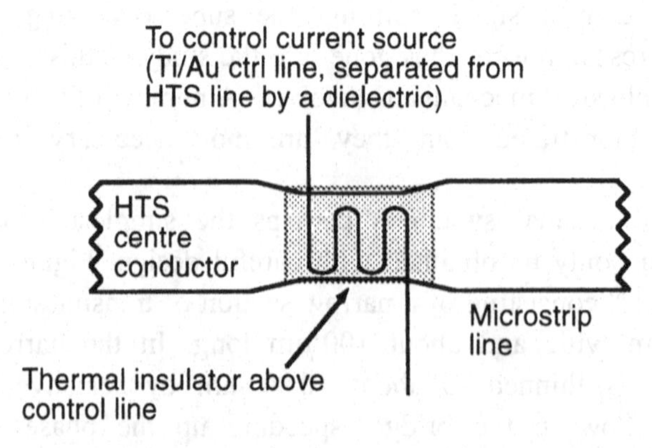

Figure 5.9.1. The structure of the thermal switch. (Taken from reference [100].)

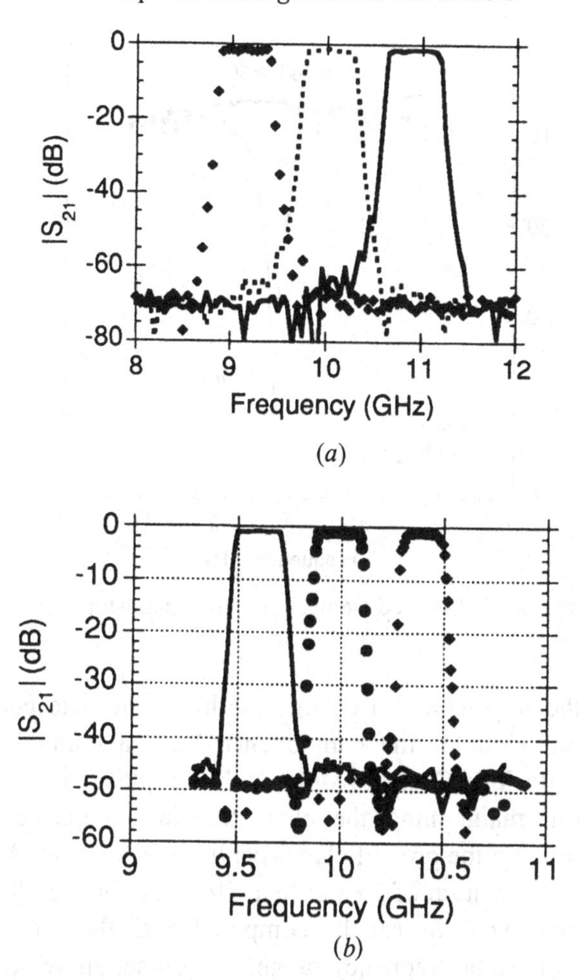

Figure 5.9.2. Composite plot of the switched filter bank, where each of the responses is selected sequentially: (*a*) seven-pole Chebyshev response, (*b*) four-pole elliptic response. (Taken from reference [100].)

apart,[101] with superconducting shorts. When the microwave power down the waveguide is increased, and the superconducting films become dissipative, the slot stubs then provide a band rejection filter. Such a limiter/filter, described in reference [101], has a rejection of 18 dB centred on about 37 GHz, and occupies a band of about 4 GHz. The switching speed of this device is measured as less than 2 ns.

Wide band operation can be achieved by applying the same principles to a superconducting transmission line. A coplanar transmission line is convenient because it is possible to narrow the transmission by using a taper, whilst not altering the characteristic impedance (narrowing the line in a microstrip or

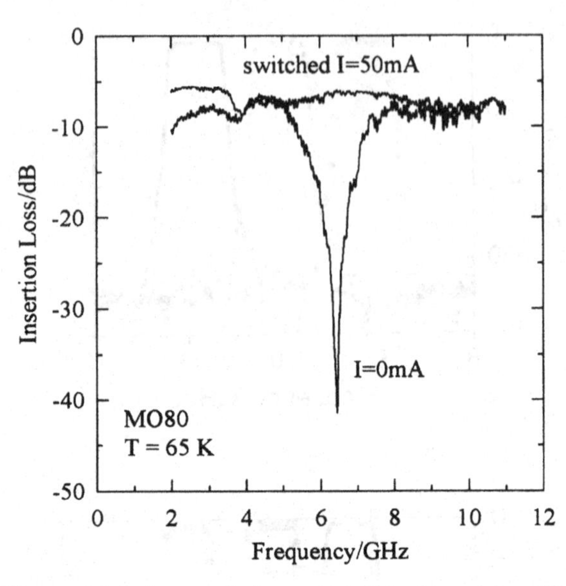

Figure 5.9.3. Lumped element switched bandstop filter.

stripline causes the impedance to change, with the ultimate consequence of narrowing the bandwidth of the component). Such a component has been demonstrated[102] with YBCO on $LaAlO_3$, with a coplanar line tapering from 150 μm to 10 μm, maintaining the characteristic impedance of 50 Ω by simultaneously reducing the ground plane–centre strip spacing. At 77 K with a 15 V bias voltage an isolation of greater than 20 dB is obtained for frequencies from 5 GHz to 30 GHz. This can be compared with the off-state losses of around 2 dB over the same frequency range. Of course, all switches operating by turning a section of the component into their normal state can also be used as limiters where the input microwave signal turns the superconductor to normal.

A very similar structure, that is, a tapered coplanar line, can be used for a superconducting switch with an optical control signal. A spot of light shone on to the narrow section of the coplanar line produces switching due to the bolometric or non-bolometric response. Such a switch has been demonstrated to switch on faster than 78 ps,[103] although the off transition is longer and is in the nanosecond to microsecond range, depending upon the film thickness and laser energy. This is because the film and substrate are required to cool down once the optical signal has been removed.

There are a number of advantages in using optical control. With conventional current biasing the bandwidth of the circuits may be reduced due to the additional circuitry required to be in close proximity with the microwave filter circuits. Each bias line requires a lowpass filter in order to isolate the bias line from the microwave transmission line, but with optical control the isolation is inherent in the technique. This is particularly important where a large number of switches are involved, for example in filter banks. It is far simpler to have a large number of optical fibres as the control circuit rather than a large number of bias lines all being applied to the microwave circuit. The problem of microwave coupling to these bias circuits is prohibitive and will increase the size of the component considerably. In the optical case, fibre optic cable can be used to deliver the optical control signals. These give good thermal isolation to the outside world and connect to standard optical sources.

Another alternative for the optical control of superconducting devices is to use conventional semiconductor components which are sensitive to the optical signal, and then place these strategically within the device. Such a technique has been demonstrated in a bank of up to 32 notch filters in series.[104,105] A diagram of a section of the device, which corresponds to one of the notch filters, is shown in Figure 5.9.4.

The fundamental building block of the device is a folded half-way resonator, with open circuits at the side away from the transmission line. Over the open circuits is placed a piece of gallium arsenide which acts as the switching element. The connection with the device is by a direct ohmic contact to metallised pads. When there is no incident light on the GaAs it acts as a dielectric; however, when light (880 nm) is incident, carriers are generated and it effectively shorts out the resonator at this point.

When this occurs, the resonator becomes a full wavelength resonator and the frequency of the filter is moved to half the resonant frequency, moving out of the band of the filter, and it is effectively turned off. The time for this to occur is about 10 μs for the reject to pass state and 350 μs for the pass to reject state. This is quite a long time, but the use of silicon can speed this up so that both on and off times are less than 2 μs. The resonators have Qs of about 3000–

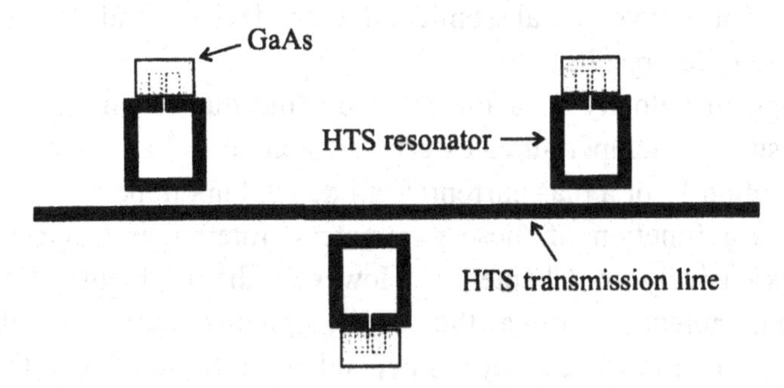

Figure 5.9.4. HTS bandstop filter element using semiconductor switches.

4000 at 77 K in the low X-band frequency region. Three filters are placed in series in order to produce a three-pole notch filter with a 3 dB bandwidth of 70 MHz and a 30 dB bandwidth of 20 MHz. For a 32-channel filter bank, 32 of these notch filters (and hence 96 resonators) are placed in series, each with its own optical fibre control line. The total loss when all the filters are in the pass state is only 3 dB.

5.10 Superconducting phase shifters

A number of sophisticated applications become possible if the velocity of a wave on a microwave transmission line can be adjusted by the application of a control signal. It is possible to alter the phase of a signal dynamically and shift the resonant frequencies of resonators. If these resonators are used as the basic building blocks of a filter, then the frequency response of the filter can be changed in response to external conditions. Specific applications include tunable antenna phased arrays, where the direction of reception of a signal can be controlled, tunable antennas and scanning of filter passbands in frequency-agile systems.

There are many ways of producing change in the velocity of a signal on a superconducting transmission line; these include changes in temperature, the application of external magnetic fields, the application of bias currents, or increasing the microwave power. It is also possible using optical irradiation. In order to increase these effects it is possible to construct transmission lines from an array of junctions, although the power handling of the transmission line will be degraded in this case. Alternatively, by the choice of specific substrates or the inclusion of specific dielectric materials into the microwave device, it is possible to modulate the dielectric permittivity or permeability[106] by the application of an electric or magnetic field. Such phase shifters must be compared in performance with conventional room-temperature devices based upon ferrites or semiconductor devices such as PIN diodes.[107] Of course, combinations of conventional semiconductor devices and HTSs can also produce the required results.

The change in velocity of a transmission line due to changes in external parameters such as temperature, external magnetic fields, microwave power, optical irradiation[108] or a bias current[109] all depend upon the change in internal inductance as a function of these parameters. Internal inductance has been discussed extensively in Chapter 1. However, the problem with all these methods is the increase in loss as the control signal is applied, and this may be the limiting factor in devices that are dependent on these effects; this depends upon the amount of velocity change required for a particular application. The effect of a bias current is particularly interesting as this is the traditional way of controlling microwave devices, although optical tuning is also important if large arrays of devices are to be used. Both these methods have been used to switch devices, as discussed in Section 5.9. The tuning of the lumped element switch in Section 5.9 showed a 32% change in resonant frequency from 15 K to 86 K.

The change in internal inductance as a function of bias current can be predicted by the Ginzburg–Landau theory and is given approximately by[110]

$$L_i = \mu_0 \lambda^2 \left(1 + \frac{4I^2}{9I_c^2}\right) \tag{5.10.1}$$

Here I is the applied current and I_c is the critical current of the superconductor; the applied current is assumed to be small compared with the critical current. This equation has been shown to be consistent with measurements of the delay times of short pulses on superconducting striplines.[110]

An alternative to producing a continuous phase shift by altering the velocity of a transmission line is to produce a digital phase shifter. This can be accomplished by switching in different lengths of superconducting transmission lines, each with a different phase delay.[111] Superconducting switches can be used, as discussed in Section 5.2.

5.10.1 Junction phase shifters

A variable transmission line velocity can be accomplished by using Josephson junctions made from superconductors. Figure 5.10.1 shows a basic microstrip transmission line made from an array of SQUIDs. The velocity on the transmission line is governed by the conventional inductance plus a contribution from the SQUID loop inductances. The inductances of the SQUID loops can be varied by controlling the flux which penetrates the loops containing the junctions. The change in velocity can be increased by increasing the loop diameter and reducing the substrate thickness; increasing the number of

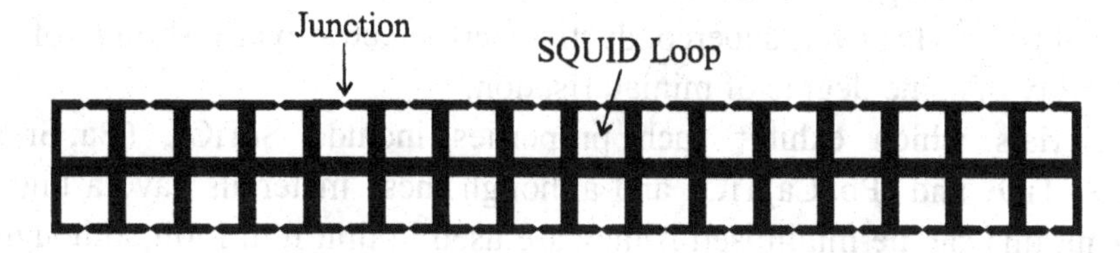

Figure 5.10.1. A microstrip transmission line made from an array of SQUIDs.

SQUIDs also increases the phase shift. Not all SQUIDs in the device need to function optimally, since the final response is a combination of the performance of all devices and a perfect yield is not important. The advantage of this type of phase shifter is that it is low loss with a continuous phase shift, potentially up to very high frequencies. In addition, the application of the control signal to shift the phase does not affect the loss of the device, in contrast to the kinetic inductance devices discussed above. However, because of the use of low critical current junctions, the power handling capability of such a phase shifter is severely limited.

A number of devices have been made from such SQUIDs in the form of both resonators and phase shifters with niobium[112] and YBCO junctions.[113,114] The YBCO devices were made from step edge junctions on $LaAlO_3$ and exhibited low loss (< 1 db) and a phase shift of around 20°. The measurements were made using a resonator with a centre frequency of about 10 GHz; this gave a frequency shift of 0.01 GHz for an applied field of $\Phi_0/6$. Here Φ_0 is the flux quantum. As expected, the results show a poor power dependence, indicating that these devices are not practicable for high-power applications.

An alternative method of performing a phase shift using junctions is to use a series array of junctions, where the change in inductance is controlled solely by the Josephson inductance.[115] This, however, has the disadvantage that the phase shift is accompanied by higher losses.

5.10.2 Ferroelectric phase shifters

The permittivity of some ferroelectric materials can be altered by the application of an electric field. If these materials are used as the dielectric in the construction of a transmission line, then a phase shifter immediately results. The loss along the transmission line must be kept to a minimum and the phase change must be large per unit length of line; this implies that the loss tangent of the ferroelectric must not be too large. Also, it does not necessarily follow that dielectric materials with good, appropriate properties at low frequencies have the same properties at microwave frequencies. The use of ferroelectrics is not new; they have been studied since the early 1960s,[116] and room-temperature X-band waveguide phase shifters have been demonstrated with up to 360° phase shift.[117] However, superconducting ferroelectric phase shifters offer the possibility of some degree of miniaturisation.

Materials which exhibit such properties include $SrTiO_3$, $(Ba, Sr)TiO_3$, $(Pb, Sr)TiO_3$ and $(Pb, Ca)TiO_3$ and although these materials have a high loss tangent, this can be minimised if they are used in thin film form, although this is at the expense of phase shift per unit length. A number of measurements of the properties of these materials have been made at both high and low frequencies. Low-frequency measurements are accomplished by simply making a capacitor with the ferroelectric as a spacer and with superconducting electrodes. Interdigital capacitors[118,119] are useful in minimising the number of film depositions, although parallel plate capacitors[120] are more sensitive. The dielectric constants of these materials are a strong function of temperature in addition to material quality. The dielectric constant of a thin film (~ 300 nm thick) of $Ba_xSr_{1-x}TiO_3$ ($x = 0.08$) has been observed to change from about 200 to 255,[118] with fields from 0 to 6.6 MV/m at a frequency of 100 kHz. The loss tangent of the same material is about 0.03. Measurements on a parallel plate capacitor using 800 nm thin films of similar material[120] ($x = 0.5$) at 1 MHz and 77 K show a change in permittivity from 180 to 370 for applied fields of 6.2 MV/m to −3.7 MV/m, with a loss tangent of 0.36.

For microwave application purposes it is more appropriate to look at these effects at higher frequencies and this can be accomplished with both transmission lines[121,122] and resonators.[123] Figure 5.10.2 shows the cross-section of three coplanar transmission lines with strontium titanate (STO) thin films deposited with different geometries. The coplanar guide is composed of a short exponential taper from the larger size where the connectors are fitted. Application of a bias voltage between the ground plane and the guide produces a phase change down the guide.[124] Such a construction produces the phase shift shown in Figure 5.10.3. Based on best results given in reference [124], a hypothetical phase shifter 2.2 cm long gives continuous tuning to 90° with 3.9 dB loss.

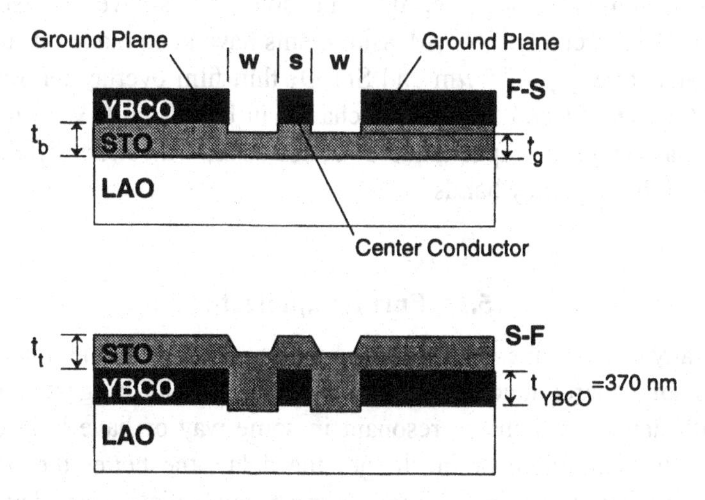

Figure 5.10.2. Cross-section of three coplanar transmission lines. (Taken from reference [124].)

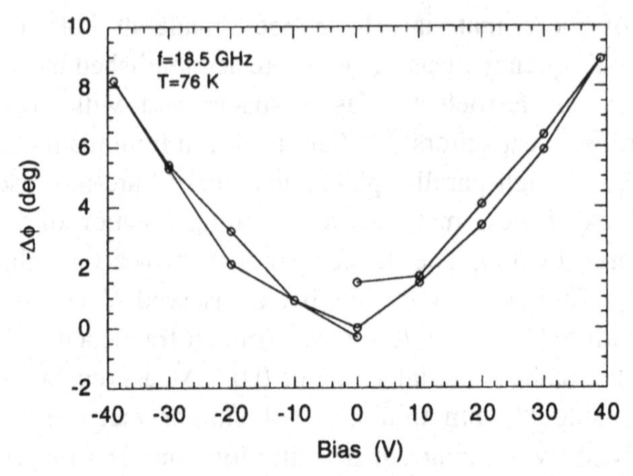

Figure 5.10.3. Phase change as a function of applied bias voltage for two $SrTiO_3$ thin film coplanar transmission lines ($S = 26\ \mu m$, $W = 50\ \mu m$). (Taken from reference [124].)

Another measurement technique is the use of a capacitive loaded resonator. A planar capacitor can be placed at the centre of an open-ended coplanar resonator; this has the effect of adding additional capacitance to the structure and the resonant frequency reduces. This, however, does not happen with the even resonant modes, where there is a current node at the central position where the capacitor is located. These harmonics are thus unaffected by the presence of the capacitance and can be used to calibrate the losses. A bias needs to be applied to the resonator via chokes in order not to affect the resonant response, although certain removable probes have been shown not to affect the Q appreciably.[125,126] Measurements have been made on such a device with a capacitive gap of 5 μm and $SrTiO_3$ thin film overlay on YBCO grown on $LaAlO_3$ at 6, 13 and 20 GHz. A change in bias from 0 V/m to 12 MV/m produced a change in capacitance of about 68%, with little difference across the three high-frequency bands.

5.11 Further applications

Almost any microwave device can be improved in some manner by the inclusion of superconductors in its construction. The biggest improvements arise with devices which are resonant in some way or have long delays. The narrower the bandwidth or the longer the delay, the better the improvement with superconductors. In addition, almost any microwave device can be miniaturised in some way, whilst still keeping or improving its performance, when superconductors are used. It does, of course, depend upon the final application whether the use of superconductors is warranted, although modern high-performance applications are becoming increasingly more demanding. Active devices are also of importance in microwave systems and devices such as mixers, amplifiers, analogue to digital converters, track and hold circuits and three-terminal devices can all be integrated into microwave systems. Active devices are not covered in this text but some examples of their integration into signal processing systems are given in Chapter 8. The remainder of this chapter is now devoted to describing some additional passive microwave devices which have been successfully designed for enhanced performance by the inclusion of superconductors.

5.11.1 Couplers

Couplers represent one of the most important microwave components and there are many different designs both in waveguide and planar circuits. With careful thought most of the designs can be miniaturised in some way. As an example, a forward wave coupler is shown in Figure 5.11.1. The different odd and even mode velocities in this structure cause forward coupling, and they can be designed with a quadrature output and 3 dB coupling.[127] The coupling is controlled by the geometry of the structure and increases as the distance between the lines is reduced. Figure 5.11.1(b) shows a serpentine coupler[128] where the structure actually increases the coupling, enabling shorter lengths to

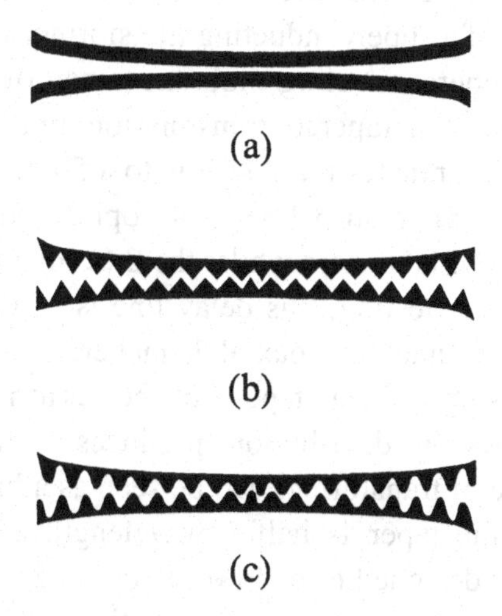

Figure 5.11.1. Some miniaturised couplers: (*a*) smoother coupler, (*b*) serpentine coupler, (*c*) rounded serpentine coupler. (After reference [129].)

be used for a given forward coupling. The peaked structures in the serpentine coupler can be rounded (Figure 5.11.1(c)) to some extent with little reduction in performance; this can be important since there is a high current density at the peaks which can cause non-linear effects when superconductors are used on their implementation.

Couplers of the type shown in Figure 5.11.1 have been constructed from normal metals on alumina of 18.6 mm length and performed well at frequencies between 12 and 18 GHz.[129] Isolation of −20 dB was observed. The serpentine coupler can achieve a 30% reduction in size over the smooth coupler due to its increased coupling constant.

5.11.2 Transformers

A microwave transformer is used for converting from a circuit with one impedance level to another. This is usually the impedance of a specific device or circuit to the 50 Ω system impedance level. Devices which require transformation include filters, active devices[130] and some of the delay lines and filters discussed in Chapter 6. Transformers can be produced from distributed or lumped elements, or a mixture of both,[131] and the reasons for making superconducting ones are the same as those for filters. An additional reason may be that the rest of the circuit is superconducting and an interface is required; there need not necessarily be a performance or miniaturisation reason in this case. Microwave transformers can take many forms and the design principles are well documented in many microwave texts, so these will not be discussed here. However, an example of a superconducting transformer is given below.

An example of a superconducting niobium transformer has been discussed by McGinnis.[132] This is a tapered transmission line transformer with two functions. The first is to transform a 2 Ω line to a 50 Ω system impedance; the second is to convert a microstrip line to a coplanar line. The coplanar line connects to 50 Ω coaxial launchers whilst the 2 Ω microstrip line is connected to a short delay line. For testing, this delay line is connected through another identical transformer to another coaxial launcher. It is possible to construct such a transformer using a linear taper but the performance is poor. A taper with a Dolph–Chebyshev distribution produces a much more optimised transformer. This type of transformer can be seen as a highpass filter, such that when the length of the taper is half a wavelength a low-frequency cut-off occurs. This cut-off is designed to be lower than the 5 GHz specification of the transformer. The upper limit of the specification is 15 GHz. The design is 15 mm long and is split into three sections as follows:

1. The 2 Ω microstrip consists of an 80-μm-wide, 300-nm-thick niobium strip on the surface of a SiO_2 dielectric layer, 1 μm thick. The ground plane below this is niobium and is again 300 nm thick, deposited on to a silicon substrate. The impedance of this section is gradually increased to 8 Ω by making the microstrip line wider.
2. When the microstrip has a characteristic impedance of 8 Ω a transition from microstrip to coplanar takes over. In this region the ground plane is split directly underneath the upper conductor and the split gradually becomes wider. Expressions for the impedance as the conversion from microstrip takes place are not available in the literature, so interpolation of the expressions for coplanar line and microstrip are required. During this period the upper signal line increases in width and the ground planes separate faster, so the structure gradually turns into a coplanar line. This takes over at an impedance of 23 Ω.
3. The final section is a tapered coplanar line which goes from 23 Ω to the required 50 Ω. The construction is not a true coplanar line because the SiO_2 layer still separates the ground plane. However, the thickness of this layer (1 μm) is much smaller than the spacing between the ground plane and the signal line, making the transmission line effectively a coplanar line.

The construction of this transformer is quite complex and the design can be done initially through the use of the analytical expressions available in the literature, although microwave CAD packages help in the optimisation and checking.

The results of the transmitted response show good agreement with the design predictions, with a good low-loss frequency response over the whole band of 5–15 GHz. However, a number of dropouts occur across this band and are attributed to substrate resonances.

5.11.3 Splitter/combiner

Figure 5.11.2 shows a superconducting splitter or combiner.[133] It is a broad band device with an impedance transformation based on a seven-section Chebyshev design. As each section has a length of about a quarter-wavelength of transmission line at the centre frequency, the transmission lines are fairly long. To fit it on a reasonably sized substrate requires fairly narrow meandered lines; hence the performance improves when using superconductors. The device is designed to operate over the frequency range 0.5–4.5 GHz.

The performance of the device is good over the full frequency band, showing a 3.3 dB insertion loss, of which 3 dB is the loss due to the splitting of the signal. The remaining 0.3 dB is due to losses in the device and unwanted

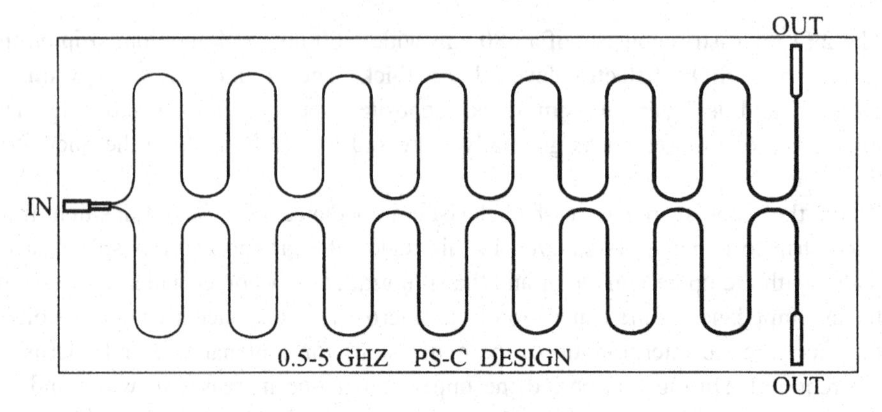

Figure 5.11.2. Superconducting splitter/combiner. (Taken from reference [133].)

reflections. A similar gold circuit showed a 7 dB loss, which would make the output of a combined signal less than the sum of the input signals.

5.12 References

1 Matthaei G. L., Young L. and Jones E. T. *Microwave Filters, Impedance Matching Networks and Coupling Structures*, Dedham MA Artech House, 1980

2 Fusco V. F. *Microwave Circuits – Analysis and computer aided design*, Prentice-Hall International, 1987

3 Collin R. E. *Foundations for Microwave Engineering*, McGraw-Hill, 1992

4 Vendik O. and Kollberg E. Software models HTSC microstrip and coplanar lines, *Microwaves and RF*, pp. 118–21 July, 1993

5 Gevorgian S., Carlsson E., Gal'chenko S., Kaparkov D. and Vendik I. Critical limitations in low-loss narrow band HTSC filter design, *European Microwave Conference*, pp. 522–7, 1994

6 Chaloupka H. High temperature superconductors – A material for miniaturised or high performance microwave components, *Freqenz*, **44**(5), 141–4, 1990

7 Liang G. C., Withers R. S., Cole B. F., Garrison S. M., Johansson M. E., Ruby W. S. and Lyons W. G. High temperature superconductive delay lines and filters on sapphire and thinned $LaAlO_3$ substrates, *IEEE Trans. on Applied Superconductivity*, **3**(3), 3037–41, 1993

8 Talisa S. H., Janocko M. A., Meier D. L., Moskowitz C., Grassel R. L., Talvacchio J., LePage P., Hira G., Buck D. C., Pieseski S. J., Brown J. C. and Wagner G. R. High temperature superconducting four channel filterbanks, *IEEE Trans. on Applied Superconductivity*, **5**(2), 2079–82, 1995

9 Lyons W. G., Bonetti R. R., Williams A. E., Mankiewich P. M., O'Malley M. L., Hamm J. M., Anderson A. C., Withers R. S., Meulenberg A. and Howard R. E. High T_c superconductive microwave filters, *IEEE Trans. on Magnetics*, **MAG-27**, 2537–9, 1991

10 Nagai Y., Itoh K., Suzuki N. and Michikami O. 1.5 GHz bandpass microstrip filters

fabricated using EuBaCuO superconducting films, *Jap. J. Appl. Phys.*, **32**, L260–L263, 1993
11 Enokihara A., Setsune K., Wasa K., Sagawa M. and Makimoto M. High T_c bandpass filter using miniaturised microstrip hair pin resonators, *Electronics Lett.*, **28**(20), 1925–6, 1992
12 Hedges S. J., Humphreys R. G., Chew N. G. and Goodyear S. W. Development of planar microwave filters from double-sided YBCO thin films on MgO substrates, *Electron. Lett.*, **27**, 2311–13, 1991
13 Lyons W. G. *et al.* High temperature superconductive passive microwave devices, *IEEE Device Research Conference, Santa Barbara CA*, June, 1990
14 Mankiewich P. M. *et al.* Reproducible technique for fabrication of thin films of high transition temperature superconductors, *Appl. Phys. Lett.*, **51**, 1753, 1987
15 Lyons W. G. and Withers R. S. Passive microwave device applications of High T_c superconducting thin films, *Microwave Journal*, p. 85, November, 1990
16 Matthaei G. L. and Hey-Shipton G. L. Novel staggered resonator array superconducting 2.3 GHz bandpass filter, *IEEE MTT-S Int. Microwave Symp. Digest*, **3**, 1269–72, 1993
17 Matthaei G. L. and Hey-Shipton G. L. Concerning the use of high temperature superconductivity in planar microwave filters, *IEEE Trans. on Microwave Theory and Techniques*, **42**(7), 1287–93, 1994
18 Matthaei G. L. and Hey-Shipton G. L. Novel staggered resonator array superconducting 2.3 GHz bandpass filter, *IEEE Trans. on Microwave Theory and Techniques*, **41**(12), 2345–52, 1993
19 Zhang D., Liang G.-C., Shih C. F., Withers R. S., Johansson M. E. and Cruz A. D. Compact forward coupled superconducting microstrip filters for cellular communications, *IEEE Trans. on Applied Superconductivity*, **5**(2), 2656–9, 1995
20 Liang G.-C., Zhang D., Shih C.-F., Johansson M. E., Withers R. S., Anderson A. C. and Oates D. E. High power high temperature superconducting microstrip filters for cellular base station applications, *IEEE Trans. on Applied Superconductivity*, **5**(2), 2652–5, 1995
21 Liang G.-C., Zhang D., Shih C. F., Withers R. S., Johansson M. E., Ruby W., Cole B. F., Krivoruchko M. and Oates D. E. High power HTS microstrip filters for wireless communication, *IEEE MTT-S Digest*, pp. 183–6, 1994
22 Mansour R. R. Design of superconductive multiplexers using single mode and dual mode filters, *IEEE Trans. on Microwave Theory and Techniques*, **42**(7), 1411–18, 1994
23 Schmidt M. S., Forse R. J., Hammond R. B., Eddy M. M. and Olson W. L. Measured performance at 77 K of superconducting microstrip resonators and filters, *IEEE Trans. on Microwave Theory and Techniques*, **39**(9), 1475–9, 1991
24 Mansour R. R., Rammo F. and Dokas V. Design of hybrid coupled multiplexers and diplexers using asymmetrical superconducting filters, *IEEE MTT-S Digest*, pp. 1281–4, 1993
25 Hedges S. J. and Humphreys R. G. Extracted pole planar elliptic function filters, *Proc.* ESA/ESTEC *Workshop on Space Applications of High Temperature Superconductors*, pp. 97–106, Noordwijk, April, 1993
26 Rhodes J. and Cameron R. Generalised extracted pole synthesis technique with application to low loss Te_{011} mode filters, *IEEE Trans. on Microwave Theory and Techniques*, **28**(9), 1018–28, 1980
27 Suginosita F., Imai K., Yazawa N., Suzuki K., Fujino S., Takenaka T. and Nakao K. 13.3 GHz YBCO microstrip bandpass filter, *Electron. Lett.*, **28**(4), 355–7, 1992

28 Newman H. S., Chrisey D. B., Horwitz J. S., Weaver B. D. and Reeves M. E. Microwave devices using $YBa_2Cu_3O_{7-\delta}$ films made by pulsed laser deposition, *IEEE Trans. on Magnetics*, **27**(2), 2540–3, 1991
29 Talisa S. H., Janocko M. A., Jones C. K., McAvoy B. R., Talvacchio J., Wagner G. R., Moskowitz C., Buck D. C., Billing J. and Brown R. Microwave superconducting filters, *IEEE Trans. on Magnetics*, **27**(2), 2544–7, 1991
30 Kang K.-Y., Lee S. Y., Han S. G. and Ahn D. Microwave multipole lowpass and bandpass filters fabricated by high T_c superconducting thin films, *IEEE Trans. on Applied Superconductivity*, **5**(2), 2671–4, 1995
31 Oh B., Kim H. T., Choi Y. H., Moon S. H., Hur P. H., Kim M., Lee S. Y. and Denisov A. G. A compact two pole X-band high temperature superconducting microstrip filter, *IEEE Trans. on Applied Superconductivity*, **5**(2), 2667–70, 1995
32 Talisa S. H., Janocko M. A., Moskowitz C., Talvacchio J., Billing J. F., Brown R., Buck D. C., Jones C. K., McAvoy B. R., Wagner G. R. and Watt D. H. Low- and high-temperature superconducting microwave filters, *IEEE Trans. on Microwave Theory and Techniques*, **39**(9), 1448–53, 1991
33 Jing D., Shao K., Cao C. H., Zhang L. X., Jiao G., Zhang Z. J., Li S. Q., Guo C. N., Yang B. C., Wang X. P., Xiong G. C. and Lian Q. J. 10 GHz bandpass YBCO superconducting microstrip filter, *Superconductor Science and Technology*, **7**, 792–4, 1994
34 Zahopoulos C., Sridhar S., Bautista J. J., Ortiz G. and Lanagan M. Performance of a high T_c superconducting ultra low-loss microwave stripline filter, *Appl. Phys. Lett.*, **58**(9), 977–9, 1991
35 Rensch D. B., Josefowicz J. Y., Macdonald P., Nieh C. W., Hoefer W. and Krajenbrink F. Fabrication and characterisation of high T_c superconducting X-band resonators and bandpass filters, *IEEE Trans. on Magnetics*, **MAG-27**, 2553–6, 1991
36 Kuhn M., Horn R., Klinger M. and Hinken J. H. HTSC microstrip UHF filter at 2 GHz, *Superconducting Science and Technology*, pp. 471–2, 1991
37 Matthaei G. L. and Hey-Shipton G. L. High temperature superconducting 8.45 GHz bandpass filter for the deep space network, *IEEE MTT-S Int. Microwave Symp. Digest*, **3**, 1273–6, 1993
38 Bonetti R. R. and Williams A. E. Preliminary design steps for thin film superconducting filters, *IEEE MTT-S Int. Microwave Symp. Digest*, **1**, 273–5, 1990
39 Weigel R., Nalezinski M., Valenzuela A. A. and Russer P. Narrow band YBCO superconducting parallel coupled coplanar waveguide bandpass filter at 10 GHz, *IEEE MTT-S Int. Microwave Symp. Digest*, **3**, 1285–8, 1993
40 Swanson D. G. and Forse R. J. An HTS end-coupled CPW filter at 35 GHz, *IEEE MTT-S Symp. Digest*, pp. 199–202, 1994
41 Chew W., Riley A. L., Rascoe D. L., Hunt B. D., Foote M. C., Cooley T. W. and Bajuk L. J. Design and performance of a high T_c superconductor coplanar waveguide filter, *IEEE Trans. on Microwave Theory and Techniques*, **39**(9), 1455–61, 1991
42 Zhang D., Matloubian M., Kim T. W., Fetterman H. R., Chou K., Prakash S., Deshpandey C. V., Bunshah R. F. and Daly K. Quasi-optical millimetre wave bandpass filters using high T_c superconductors, *IEEE Trans. on Microwave Theory and Techniques*, **39**(9), 1493–7, 1991
43 Zhang D., Rahmat-Samii Y., Fetterman H. R., Prakash S., Bunshah R. F., Eddy M. and Nilsson J. L. Application of high T_c superconductors as frequency selective

surfaces: Experiment and theory, *IEEE Trans. on Microwave Theory and Techniques*, **41**(6/7), 1033–6, 1993

44 Hong J. S. and Lancaster M. J. Bandpass characteristic of new dual-mode microstrip square loop resonators, *Electron. Lett.*, **31**(11), 891–2, 1995

45 Hong J. S. and Lancaster M. J. Microstrip bandpass filter using degenerate modes of a novel meander loop resonator. Accepted for publication in *IEEE Microwave and Guided Wave Letters*

46 Wolff I. Microstrip bandpass filter using degenerate modes of a microstrip ring resonator, *Electron. Lett.*, **8**(12), 302–3, 1972

47 Curtis A. and Fiedziuszko S. J. Miniature dual mode microstrip filters, *IEEE MTT-S Int. Microwave Symp. Digest*, **2**, 443–6, 1991

48 Curtis A. and Fiedziuszko S. J. Dual mode microstrip filters, *Appl. Microwave*, pp. 83–6, Autumn, 1991

49 Karacaoglu U., Robertson I. D. and Guglielmi M. An improved dual-mode microstrip ring resonator filter with simple geometry, *European Microwave Conf.*, pp. 472–7, 1994

50 Mansour R. R. Design of superconductive multiplexers using single mode and dual mode filters, *IEEE Trans. on Microwave Theory and Techniques*, **42**(7), 1411–18, 1994

51 Mansour R. R., Dokas V., Thomson G., Tang W.-C. and Kudsia C. M. A C-Band superconductive input multiplexer for communication satellites, *IEEE Trans. on Microwave Theory and Techniques*, **42**(12), 2472–9, 1994

52 Mansour R. R. and Dokas V. C-band externally equalised superconductive input channel filters, *IEEE MTT-S Digest*, pp. 187–90, 1994

53 Curtis J. A., Fiedziuszko S. J. and Holme S. C. Hybrid dielectric/HTS resonators and their applications, *IEEE MTT-S Int. Microwave Symp. Digest*, pp. 447–50, 1991

54 Kogami Y. *et al.* Low loss bandpass filter using dielectric rod resonator orientated axially in a high-T_c superconducting cylinder, *IEEE MTT-S Int. Microwave Symp. Digest*, pp. 1345–8, 1991

55 Mansour R. F. and Zybura A. Superconducting millimetre-wave E-plane filters, *IEEE Trans. on Microwave Theory and Techniques*, **39**(9), 1488–92, 1991

56 Illinois Superconductor Corporation IS80AE Filter and IS80BE Filter, Specification Sheets, 1995

57 Liang G.-C., Zhang D., Shih C.-F., Johansson M. E., Withers R. S., Anderson A. C. and Oates D. E. High power high temperature superconducting microstrip filters for cellular base station applications, *IEEE Trans. on Applied Superconductivity*, **5**(2), 2652–5, 1995

58 Liang G.-C., Zhang D., Shih C. F., Withers R. S., Johansson M. E., Ruby W., Cole B. F., Krivoruchko M. and Oates D. E. High power HTS microstrip filters for wireless communication, *IEEE MTT-S Digest*, pp. 183–6, 1994

59 Shen Z.-Y. and Wilker, C. Raising the power handling capacity of HTS circuits, *Microwaves and RF*, April, 1994

60 Hohenwarter G. K. G., Martens J. S., Bayer J. B., Nordman J. E. and McGinnis D. P. Design of variable phase velocity kinetic inductance delay lines and their measured characteristics when fabricated by a simple Nb based process, *IEEE Trans. on Magnetics*, **25**(2), 1100–3, 1989

61 Rauch W., Valenzuela A. A., Sölkner G., Behner H., Gieres G. and Gornik E. Microstrip transmission line resonator with epitaxial $YBa_2Cu_3O_{7-x}/NdAlO_3/YBa_2Cu_3O_{7-x}$ trilayer, *Electron. Lett.*, **28**(6), 579–80, 1992

62 Pond J. M., Claassen J. H. and Carter W. L. Kinetic inductance microstrip delay lines, *IEEE Trans. on Magnetics*, **23**, 903–7, 1987
63 Pond J. M., Claassen J. H. and Carter W. L. Measurements and modelling of kinetic inductance microstrip delay lines, *IEEE Trans. on Microwave Theory and Techniques*, **35**, 1256–62, 1987
64 Pond J. M., Carroll K. R., Horwitz J. S. and Chrisey D. B. $YBa_2Cu_3O_{7-\delta}/LaAlO_3/YBa_2Cu_3O_{7-\delta}$ trilayer transmission lines for measuring the superconducting penetration depth, *IEEE Trans. on Applied Superconductivity*, **3**(1), 1438–41, 1993
65 Carroll K. R., Pond J. M. and Cukauskas E. J. Microwave losses in kinetic inductance devices fabricated from NbCN/MgO/NbCN trilayers, *IEEE Trans. on Superconductivity*, **3**(1), 2808–11, 1993
66 Pond J. M., Carroll K. R. and Cukauskas E. J. Ultra-compact microwave filters using kinetic inductance microstrip, *IEEE Trans. on Magnetics*, **27**(2), 2696–9, 1991
67 Carroll K. R., Pond J. M. and Cukauskas E. J. Superconducting kinetic inductance microwave filters, *IEEE Trans. on Applied Superconductivity*, **3**(1), 8–16, 1993
68 Owen C. S. and Scalapino D. J. Inductive coupling of Josephson junctions to external circuits, *J. Appl. Phys.*, **41**, 2047–56, 1970
69 Yoshida K., Kudo K., Hossain Md. S., Watanabe K., Enpuku K. and Yamafuji K. Microwave applications of inductively coupled superconducting striplines, *IEEE Trans. on Applied Superconductivity*, **3**(1), 2788–91, 1993
70 Vedavathy T. S. and Tripathi A. Stacked superconductor structure for microwave transmission, *Applied Superconductivity*, **2**(2), 111–15, 1994
71 Yoshida K., Nagatsuma T., Kumataks S. and Enpuku K. Millimetre wave emission from Josephson oscillator through the thin film junction electrode, *IEEE Trans. on Magnetics*, **23**, 1283–6, 1987
72 Pond J. M., Weaver P. F. and Kaufman I. Propagation and circuit characteristics of inductively coupled superconducting microstrip, *IEEE Trans. on Microwave Theory and Techniques*, **38**(11), 1635–43, 1990
73 Wadell B. C. *Transmission Line Design Handbook*, Artech House Inc., Norwood MA USA, 1991
74 Lancaster M. J., Li J., Porch A. and Chew N. G. High temperature superconductor lumped element resonator, *Electron. Lett.*, **29**(19), 1728–9, 1993
75 Ye S. and Mansour R. R. Design of manifold-coupled multiplexers using superconductive lumped element filters, *IEEE MTT-S Int. Symp. Digest*, pp. 191–4, 1994
76 Hammond R. B., Hey-Shipton G. L. and Matthaei G. L. Designing with superconductors, *IEEE Spectrum*, April, 1993
77 Madden J. Microwave HTS application notes, *Superconductor Technologies Application Note*, June, 1994
78 Swanson D. G., Forse R. and Nilsson B. J. L. A 10 GHz lumped element high temperature superconductor filter *IEEE MTT-S Digest*, pp. 1191–3, 1990
79 Bhushan M., Green J. B. and Anderson A. C. Low-loss lumped element capacitors for superconductive integrated circuits, *IEEE Trans. on Magnetics*, **25**(2), 1989
80 Patzelt T., Aschermann B., Chaloupka H., Jagodzinski U. and Roas B. High-temperature superconductive lumped-element microwave all-pass sections, *Electron. Lett.*, **29**(17), 1578–9, Aug., 1993
81 Gorur A., Karpuz C. and Lancaster M. J. Modified coplanar meander transmission line for MMICs, *Electron. Lett.*, **30**(16), 1317–18, 1994

82 Lancaster M. J., Huang F., Cheung H., Li J. C. L., Gorur A., Porch A., Avenhaus B., Woodall P., Wellhofer F., Humphreys R. G. and Chew N. G. Miniaturisation of microwave filters using high temperature superconductors, *HTSED Workshop May 26–28, Whistler, Canada*, 1994

83 Lancaster M. J., Li J. C., Gorur A., Porch A., Avenhaus B., Wellhofer F., Gough C. E., Humphreys R. G., Chew N. G. and Woodall P. Miniaturisation of high temperature superconducting filters, *EUCAS 4–9 Oct. 1993 Göttingen Germany DGM Informationsgesellschaft mbH*, pp. 1473–6, 1993

84 Jaeger N. A. F. and Lee Z. K. F. Slow wave electrode for use in compound semiconductor electrooptic modulators, *IEEE J. Quantum Electronics*, **28**(8), 1778–84, 1992

85 Dagli N. and Spickermann R. Millimetre wave coplanar slow wave structure on GaAs suitable for use in electro-optic modulators, *Electron. Lett.*, **29**(9), 774–5, 1993

86 Seki S. and Hasegawa H. Cross-tie slow wave coplanar waveguide on semi-insulating GaAs substrates, *Electron. Lett.*, **17**(25), 940–1, 1981

87 Fukuoka Y., Yi-Chi S. and Itoch T. Analysis of slow-wave coplanar waveguide for monolithic integrated circuits, *IEEE Trans. Microwave Theory and Techniques*, **MTT-31**(7), 567–73, 1983

88 Pettenpaul E., Kapusta H., Weisgerber A., Mampe H., Luginsland J. and Wolff I. CAD models of lumped elements on GaAs up to 18 GHz, *IEEE Trans. Microwave Theory and Techniques* **MTT-36**(2), 294–304, 1988

89 Gandolfo D. A., Boornard A. and Morris L. C. Superconductive microwave meander lines, *J. Appl. Phys.*, **39**(6), 2657–60, 1968

90 Hong J. S. and Lancaster M. J. Capacitively loaded microstrip loop resonator, *Electron. Lett.*, **30**(18), 1494–5, 1994

91 Hong J. S. and Lancaster M. J. Edge coupled microstrip loop resonators with capacitive loading, *IEEE Microwave and Guided Wave Letters*, **5**(3), 87–9, 1995

92 Ishimaru A. *Electromagnetic Wave Propagation, Radiation, and Scattering*, Prentice Hall, 1991

93 Ramo S., Whinnery J. R. and van Duzer T. *Fields and Waves in Communication Electronics*, John Wiley, 1984

94 Shu Y.-H., Navarro J. A. and Chang K. Electronically switchable and tunable coplanar waveguide–slotline bandpass filters, *IEEE Trans. on Microwave Theory and Techniques*, **39**(3), 548–54, 1991

95 Tabib-Azar M., Bhasin K. B. and Romanofsy R. R. Comparative study of bolometric and nonbolometric switching elements for microwave phase shifters, *NASA Tech. Memo 104435*, 1991

96 Gupta D., Donaldson W. R., Kortkamp K. and Kadin A. M. Optically triggered switching of optically thick YBCO films, *IEEE Trans. on Applied Superconductivity*, **3**(1), 2895–8, 1993

97 Leuthold A. C., Wakai R. T., Hohenwarter G. K. G. and Nordman J. E. Characterisation of a simple thin superconductivity switch, *IEEE Trans. on Applied Superconductivity*, **4**(3), 181–2, 1994

98 Frenkel A., Venkatesan T., Lin C., Wu X. D. and Inam A. Dynamic electrical response of YBCO, *J. Appl. Phys.*, **67**, 3767–75, 1990

99 Karasik B. S., Milostnaya I. I., Zorin M. A., Elantev A. I., Gol'tsman G. N. and Gershenzon E. M. High speed current switching of homogenous YBaCuO films between superconducting and resistive states, *IEEE Trans. on Applied Superconductivity*, **5**(2), 3042–5, 1995

100 Martens J. S., Hietala V. M., Ginley D. S., Tigges C. P., Plut T. A., Truman J. K., Track E. K., Young K. H. and Young R. T. HTS-based switched filter banks and delay lines, *IEEE Trans. on Applied Superconductivity*, **3**, 2824–7, 1993
101 Keis V. N., Kozyrev A. B., Samoilova T. B. and Vendik O. G. High speed microwave filter-limiter based on high-T_c superconducting films, *Electron. Lett.*, **29**(6), 546–7, 1993
102 Poulin D. G., Hegmann F. A., Lachapelle J., Moffat S. H. and Preston J. S. A superconducting microwave switch, *IEEE Trans. on Applied Superconductivity*, **5**(2), 3046–8, 1995
103 Cao W.-L., Liu Y.-Q., Lee C. H., Mao S. M., Bhattacharya S., Xi X. X., Venkatesan T., Shen Z.-Y., Pang P., Kountz D. J. and Holstein W. L. Picosecond superconductor opening switch, *IEEE Trans. on Applied Superconductivity*, **3**, 2848–51, 1993
104 Soares E. R., Raihn K. F., Fenzi N. O. and Matthaei G. L. Optical switching of HTS band reject resonators, *IEEE Trans. on Applied Superconductivity*, **5**(2), 2276–8, 1995
105 Fenzi N. O., Raihn K. F., Negrete G. V., Soares E. R. and Matthaei G. L. An optically switched bank of HTS bandstop filters, *IEEE MTT-S Digest*, pp. 195–8, 1994
106 Dionne G. F., Oates D. E. and Temme D. H. YBCO/Ferrite low loss microwave phase shifter, *IEEE Trans. on Applied Superconductivity*, **5**(2), 2083–6, 1995
107 White J. F. Review of microwave phase shifters, *IEEE Trans. on Microwave Theory and Techniques*, **MTT22**, 658–74, 1974
108 Tsindlekht M., Golosovsky M., Chayet H., Davidov D., Chocron S. and Brodskii B. Frequency modulation of superconducting parallel plate microwave resonator by laser irradiation, submitted to *Appl. Phys. Lett.*
109 Hohenwarter G. K. G., Martens J. S., Beyer J. B., Nordman J. E. and McGinnis D. P. Design of variable phase inductance delay lines and their measured characteristics when fabricated by a simple Nb based process, *IEEE Trans. on Magnetics*, **25**(2), 1100–03, 1989
110 Cho S., Erlig H., Kain A. Z., Fetterman H. R., Liang G.-C., Johansson M. E., Cole B. F. and Withers R. S. Electrical tuning of the kinetic inductance of high temperature superconductors, *Appl. Phys. Lett.*, **65**(26), 3389–91, 1994
111 Heiden V. O., Bhasin K. B. and Long J. K. Emerging applications of high temperature superconductors for space communications, *NASA Tech. Memo. 103629*, prepared for the *World Congress on Superconductivity. Houston Texas Sept. 10–13*, 1990
112 Durand D. J., Carpenter J., Ladizinsky E., Lee L., Jackson C., Silver A. H. and Smith A. D. A distributed Josephson inductance phase shifter, *IEEE Trans. on , Applied Superconductivity*, **2**(1), 33–8, 1992
113 Jackson C. M., Kobayashi J. H., Guillory E. B., Pettiette-Hall C. L. and Burch J. F. Monolithic HTS microwave phase shifter and other devices, *J. Superconductivity*, **5**(4), 419–24, 1992
114 Takemoto-Kobayashi J. H., Jackson C. M., Pettiette-Hall C. L. and Burch J. F. High T_c superconducting monolithic phase shifter, *IEEE Trans. on Applied Superconductivity*, pp. 39–44, 1992
115 Chen G. and Beasley M. R. Shock wave generation and pulse sharpening on a series array Josephson function transmission line, *IEEE Trans. on Applied Superconductivity*, **1**, 140–4, 1991

116 *US Army Final Report Contract No. DA 36-039-AMC-02349(E)* Microwave ferroelectric phase shifters and switches, US Army electronics laboratories, Fort Monmouth NJ, 1964
117 Varadan V. K., Ghodgaonkar D. K., Varadan V. V., Kelly J. F. and Glikerdas P. Ceramic phase shifters for electronically steerable antenna systems, *Microwave Journal*, p. 116, January, 1992
118 Wu H.-Y., Zhang Z., Barnes F., Jackson C. M., Kain A. and Cuchiaro J. D. Voltage tunable capacitors using high temperature superconductors and ferroelectrics, *IEEE Trans. on Applied Superconductivity*, **3**(3), 156–60, 1994
119 Naziripour A., Outzourhit A., Trefny J. U., Zhang Z.-H., Barnes F., Cleckler J. and Herman A. M. Fabrication of $Ba_{1-x}Sr_xTiO_3$ tunable capacitors with $Tl_2Ba_2Ca_1Cu_2O_x$ electrodes, *Physica*, C**233**, 387–94, 1994
120 Miranda F. A., Mueller C. H., Cubbage C. D., Bhasin K. B., Singh R. K. and Harkness S. D. HTS/ferroelectric thin films for tunable microwave components, *IEEE Trans. on Applied Superconductivity*, **5**(2), 3191–4, 1995
121 Carroll K. R., Pond J. M., Chrisey D. B., Horwitz J. S. and Leuchtner R. E. Microwave measurement of the dielectric constant of SrO.5BaO.5TiO3 ferroelectric films, *Appl. Phys. Lett.*, **62**, 1845–47, 1993
122 Jackson C. M., Kobayashi J. H., Lee A., Pettiette-Hall C., Burch J. F. and Hu R. Novel monolithic phase shifter combining ferroelectrics and high temperature superconductors, *Microwave Opt. Technol. Lett.*, **5**, 722–6, 1992
123 Hermann A. M., Price J. C., Scott J. E., Yandrofski R. M., Naziripour A., Galt D., Paranthaman H. M., Tello R., Cuchario J. and Ahrenkiel R. K. Tunable microwave resonators utilising thin high temperature superconducting films and ferroelectrics, *Bull. Amer. Phys. Soc.*, **38**, 689, 1993
124 DeGroot D. C., Beall J. A., Marks R. B. and Rudman D. A. Tunable microwave properties of $YBa_2Cu_3O_{7-\delta}/SrTiO_3$ thin film transmission lines, *IEEE Trans. on Applied Superconductivity*, **5**(2), 2272–5, 1995
125 Galt D., Price J. C., Beall J. A. and Harvey T. E. Ferroelectric thin film characterisation using superconducting microstrip resonators, *IEEE Trans. on Applied Superconductivity*, **5**(2), 2575–8, 1995
126 Galt D., Proce J. C., Beall J. A. and Ono R. H. Characterisation of a tunable thin film microwave $YBa_2Cu_3O_{7-x}/SrTiO_3$ coplanar capacitor, *Appl. Phys. Lett.*, **63**(22), 3078–80, 1993
127 Ikalainen P. K. and Mathaei G. L. Wideband forward-coupling microstrip hybrids with high directivity, *IEEE Trans. on Microwave Theory and Techniques*, **MTT-35**, 719–25, 1987
128 De Ronde F. C. Wideband high directivity MIC proximity couplers by planar means, *IEEE MTT-S Int. Microwave Symposium Digest*, pp. 480–2, 1980
129 Uysal S., Turner C. W. and Watkins J. Nonuniform transmission line codirectional couplers for hybrid mimic and superconductive applications, *IEEE Trans. on Microwave Theory and Techniques*, **42**(3), 407–14, 1994
130 Olsson H. K. Novel microstrip transformer for superconducting microelectronics, *Electron. Lett.*, **23**(21), 1152–3, 1987
131 Zhu L. and Linne'r L. J. P. Mixed lumped and distributed network applied to superconducting thin-film broadband impedance transforming, *IEEE Trans. on Applied Superconducting*, **3**(4), 3067–73, 1993
132 McGinnis D. P. and Beyer J. B. A broad-band microwave superconducting thin film transformer, *IEEE Trans. on Microwave Theory and Techniques*, **36**(11), 1521–5, 1988

133 Shen Z.-Y., Wilker C., Pang P. S. W., Laubacher D. B., Holstein W. L., Face D. W., Kountz D. J., Matthews A. L., Meriwether J. M. and Carter C. F. High T_c superconducting device development at Du Pont, *IEEE Antennas and Propagation Society Int. Symp. Digest*, **2**, 970–3, 1992

6
Superconducting delay lines

6.1 Introduction

Delay lines are an important application of superconductors and one in which they have a distinct advantage over other technologies. A delay line consists of a long transmission line, usually microstrip, stripline or coplanar line, which is deposited on to one or more substrates. This chapter is split into two main sections. Section 6.3 discusses delay lines which just delay the signal over a wide bandwidth; in principle the output should be a replica of the input signal but delayed by a certain time. However, the more interesting aspect of delay line technology is discussed in Section 6.5 and the following sections. Here, delay lines are described which perform a filtering or signal processing function and have applications both in electronic warfare and communications.

An idea of the capabilities of superconducting delay lines can be gained by considering the wide microstrip. Figure 6.1.1[1] shows the attenuation of a wide microstrip calculated from Equation (2.3.6), together with a number of other methods of delaying a signal. It can be seen that it is possible to obtain hundreds of nanoseconds of delay for only several decibels of loss for a superconducting wide microstrip at 10 GHz. Superconducting delay lines offer one of the lowest loss transmission media, even approaching the attenuation of the atmosphere. The transmission along optical fibre is also very low loss and in fact a whole technology related to signal processing using optical fibre has been developed.[2] Surface acoustic wave (SAW) devices are also very low loss in terms of delay per unit time. Losses in copper structures such as coaxial cable, waveguide or stripline are also shown in Figure 6.1.1. Further, more detailed, comparison of different types of technology are given in Section 6.7.

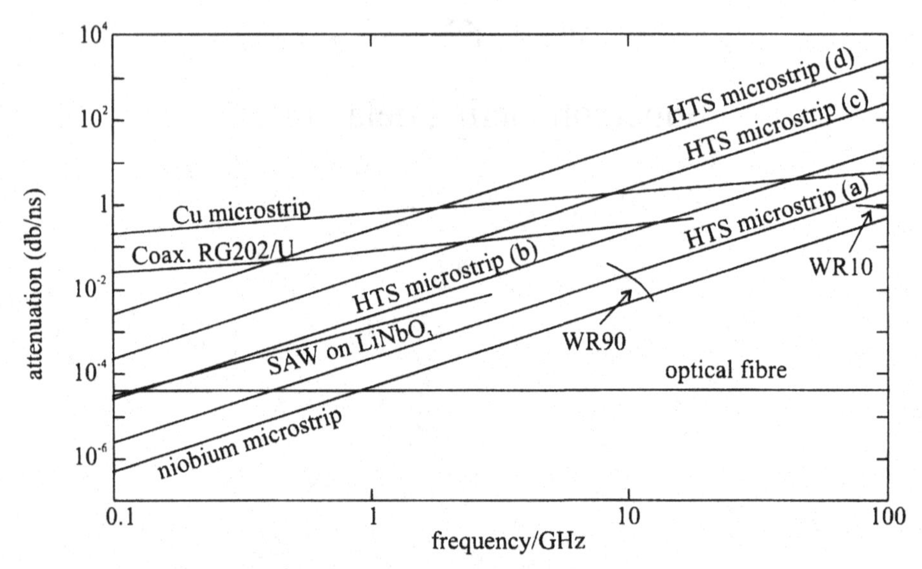

Figure 6.1.1. Attenuation per nanosecond as a function of frequency for various types of transmission media. The HTS is for a wide microstrip at 77 K with $R_s = 100\ \mu\Omega$ (at 10 GHz) and is 10 μm thick, except for (*d*) which is 0.35 μm thick; all are on a MgO substrate. The thickness of the substrate is (*a*) 1 mm, (*b*) 0.1 mm, (*c*) 0.01 mm and (*d*) 1 μm. The copper microstrip is on the same substrate with a thickness of 1 mm, and is at 77 K with a surface resistance at 10 GHz of 8.7 mΩ. The niobium microstrip is the same size as the copper, with a surface resistance of 20 $\mu\Omega$ at 10 GHz and 4.2 K. Losses for coaxial cable (RG402U), waveguide (WR10 and WR90), surface acoustic waves[3] and optical fibres are also shown.

6.2 Substrates for delay lines

Section 2.4 discusses substrates and transmission media for superconducting filters and a number of different types of material are tabulated in Appendix 2. However, microstrip and stripline delay lines have an additional requirement over the more conventional filters in that the thickness of the substrate controls the insertion loss and the cross-coupling between adjacent lines. The thinner the substrate, the closer the lines can be packed without coupling between them. This results in longer delays on a given area substrate. However, as the substrates become thinner, the microstrip lines become narrower in order to maintain an impedance which is reasonably well matched to the system impedance; the current density of the narrower lines then increases, resulting in a high insertion loss. Figure 6.1.1 gives some indication of the delays obtainable for particular thicknesses of films. The choice of a particular substrate and its thickness depends upon the requirements of the system in which the delay line is to be inserted. Shorter delays generally allow for less stringent requirements on the substrate. Substrates for both low-temperature and high-temperature superconductors are discussed briefly below, with more information being given in Appendix 2. A great deal of work has been done on the substrates for low-temperature superconducting devices.[4]

The best substrate thickness tends to be around several tens of microns, generally too thick for good epitaxial deposition and too thin for polishing substrates to reduce their thickness; both methods have been attempted. It is possible to produce thick film dielectrics several tens of microns thick[5] using thick dielectric films; however, processing is difficult with superconducting layers already present. It is possible to reduce the thickness of conventional substrates by lapping; for example, a delay of 44 ns has been produced on a $LaAlO_3$ substrate by reducing its thickness to 127 μm.[6] Sapphire offers an alternative, since it is rugged and large area substrates can be obtained. Delay lines fabricated on 430 μm-thick sapphire have been demonstrated, with good performance.[7]

6.3 Delay lines

This section deals with delay lines which do not attempt to process signals over a limited bandwidth, that is, they should ideally just delay a signal over a very wide band. A later section deals with delay lines which use the delaying process to filter the input signal. One specific example of the application of these delay lines is discussed in the section on systems (Section 8.7), that is, the instantaneous frequency measurement system. Many other examples are available, including beam-steering antenna arrays and many electronic warfare (EW) and satellite communication systems. For example, in EW systems, signals are required to be delayed a short time whilst fast pre-processing of the input occurs. Depending upon the outcome of this pre-processing, the now delayed signal can be routed to the appropriate signal processor. The time scales here are generally a few hundred nanoseconds. Similarly, in satellite communications systems, signals are required to be delayed whilst switching takes place.

Table 6.3.1 shows some examples of delay lines for these applications. The delays range from early devices with a 1 ns delay to devices with delays of 100 ns. The insertion loss is also variable, but fractions of a decibel per nanosecond of delay are generally produced, as can be seen from Figure 6.1.1. The majority of the delay lines given in Table 6.3.1 are constructed from a double spiral in the microstrip or stripline.

One novel design is a coplanar design constructed out of small unit cells,[14] which is shown in Figure 6.3.1. Each of the small cells seen in Figure 6.3.1 is

Table 6.3.1. *Some superconducting delay lines*

Author	Company	Delay/ns	*IL*/(dB/ns)	Construction
Bourne[8]	—	1.09	0.005	1 × 1 cm $LaAlO_3$
Höfer[9]	Daimler–Benz	3	—	—
Track[10]	STI	8	2.5 @ 20 GHz 20 K	2" dia. $LaAlO_3$
Liang[7]	Conductus and MIT	9	0.17 @ 6 GHz, 77 K	Sapphire
Shen[11]	Du-Pont	11	0.3 @ 8 GHz 77 K	1" × 1" $LaAlO_3$
Talisa[12]	Westinghouse	22.5	0.22 @ 20 GHz, 77 K	2" dia. $LaAlO_3$
Liang[6]	Conductus and MIT	44	0.36 @ 6 GHz 77 K	5 cm dia. $LaAlO_3$
Hohenwarter[13]	Hypres	45	—	—
Fenzi[14]	STI	100	0.08 @ 6 GHz, 77 K	2" dia. $LaAlO_3$, 4 in series

Figure 6.3.1. 25 ns coplanar delay line based on the unit cell concept. (Taken from reference [14].)

modelled numerically, accounting for the cross-coupling between adjacent lines. Then the cells are cascaded. Further optimisation gives the desired loss, dispersion and size of the final device. As would be expected, the layout of the cells with respect to each other is crucial, but it is possible to find a layout which produces low interaction. The device in Figure 6.3.1 has 475 cells fitted on to a 2"-diameter $LaAlO_3$ substrate. As with other coplanar structures, it is important to maintain a consistent ground plane if other modes of the structure are not to be stimulated. In the design shown in Figure 6.3.1, this is achieved by bonding crossover wires between the ground planes over the top of the signal line. The number is important; enough are required to suppress the slotline modes, but too many will increase the loss. A considerable number of crossovers is required.

The performance of the delay line illustrated in Figure 6.3.1 is shown in Figure 6.3.2. The line has low loss over the 1–7 GHz band shown, and it has a low ripple. In many delay line designs a ripple is observed on the magnitude response; this is either due to coupling between the line sections or reflections at the connector ports or bends in the line. Although this magnitude response is

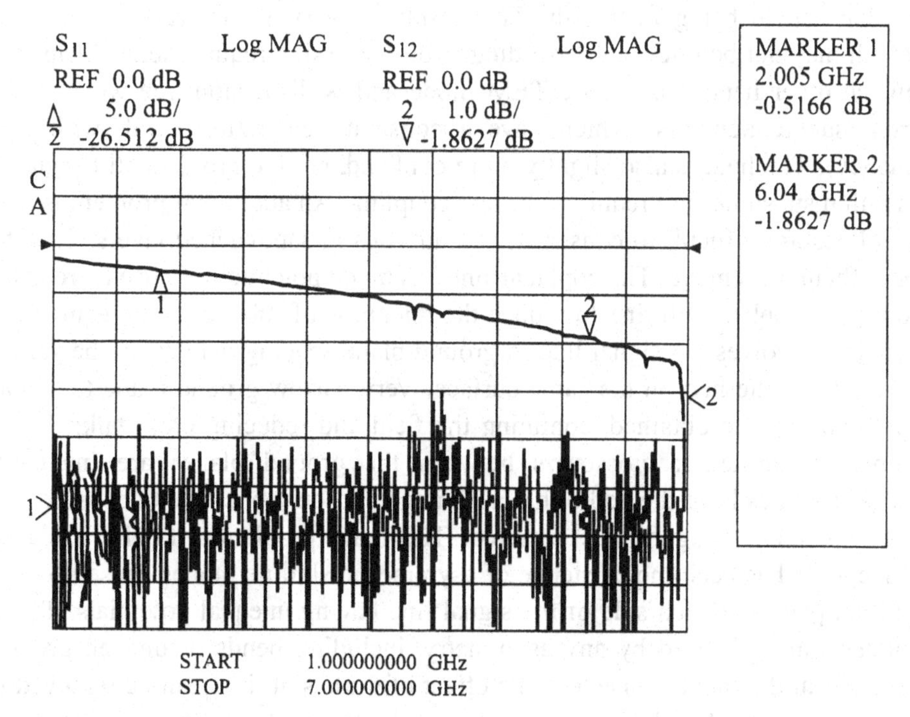

Figure 6.3.2. Performance of the 25 ns coplanar delay line shown in Figure 6.3.1. (Taken from reference [14].)

a good indication of the performance of the delay line it is also important to observe the group delay and/or the phase response of the device. It is possible in principle to obtain a good flat magnitude response, but a non-linear phase and a non-constant delay over the bandwidth, resulting in a dispersive delay line which would be useless for application purposes.

The delay line given as an example has been chosen because of its novelty in design. It may not necessarily be the optimum design, and more conventional double spiral designs may prove better for certain applications.

6.4 Choice of transmission line type

The choice of microwave transmission line type is important for delay lines and each has its own advantages and disadvantages. The microstrip line is the microwave industries' standard and is usually used if all other factors are equal. The disadvantages of the microstrip are the extension of the field within the dielectric substrate beyond the line, resulting in coupling to adjacent lines; this can be improved substantially if the substrates are thinned. Also, microstrip is not a pure TEM transmission line and, as the frequency increases, more of the electric field becomes confined between the microstrip and the ground plane, as opposed to being in the air. This results in a small decrease in velocity, which may not be relevant, depending upon the design requirements. Stripline, on the other hand, has a pure TEM mode and is dispersion-free when made from superconductors (which have a constant penetration depth with frequency). The field is also slightly more confined, but the problem still remains that thin substrates are required for low coupling. An additional problem is that striplines are difficult to construct, and air gaps in sandwiched layers prove to be difficult to remove. The coplanar line has much potential because there is no longer a problem in trying to reduce the thickness of substrate. The equivalent problem involves the signal line to ground plane spacing, which can be varied since it is patterned on the same surface. Very narrow ground plane to signal line gaps can be obtained, confining the field and reducing cross-talk; this is improved further if crossovers between the ground planes are included. Coplanar lines have the additional advantage that only one-sided deposition of the superconducting film is required. The main problem with the coplanar waveguide has been the existence of asymmetric slotline modes caused by the ground planes at each side of the signal line having unequal potentials. These modes can be caused by any asymmetry, including bends or uneven ground contacts at the input connector. The effect of these slot lines can be removed if crossovers are placed between the ground planes at appropriate distances. These can, however, increase the attenuation and cause reflections along the line, and care must be taken in implementing them. The current distribution on a coplanar line is more peaked at the edges than in the microstrip or stripline and these higher current densities result in higher losses. The problems of dispersion for microstrip, as mentioned above, also apply to the coplanar line. The velocity in a stripline is just given by $c_0/\sqrt{\varepsilon_r}$. In a microstrip it is slightly higher than this because a small amount of the field is in the air. For the coplanar waveguide, the velocity is approximately $c_0/\sqrt{(\varepsilon_r + 1)/2}$. This is larger than the stripline velocities if the dielectric constant of the substrate is the same. It could be increased by placing another substrate over the top of the coplanar line, reducing the velocity to that of a stripline.

6.5 Delay line filters

The great advantage of the delay line comes when it is used as a filter, and it can provide a number of valuable signal processing functions. The delay line filter is based on a transversal filter or finite impulse response filter as shown in Figure 6.5.1. The filter consists of a number of discrete delay periods τ running in series. At specific points along this delay chain the signal is tapped off. Each of the tapped signals is multiplied by the weighting value w and then they are all summed to create the output of the filter. The filter response is controlled by the values of τ and w.

The great advantage of this type of filter is in the control over the phase response. The phase can be accurately tuned to some design criteria over the whole bandwidth of the filter. Other types of microwave filter do not have this flexibility of both amplitude and phase control. One specific filter response which is very useful is where the impulse response is a chirp or linear frequency modulated signal, that is, the application of a sharp pulse to the filter produces a signal which starts at one frequency and gradually changes in

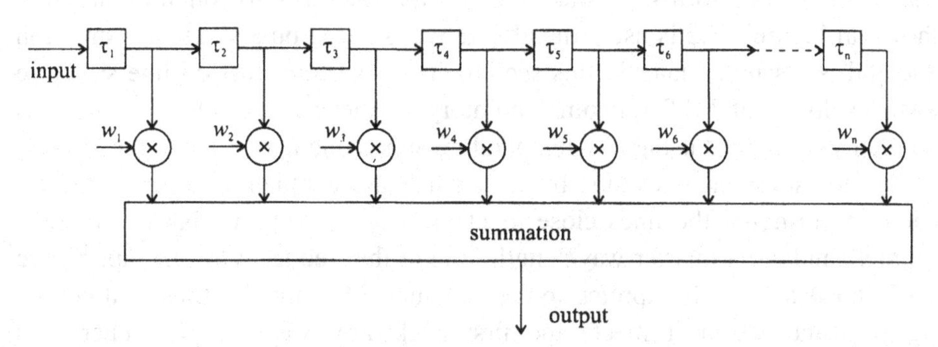

Figure 6.5.1. The principle of the transversal filter.

frequency as time progresses. This is produced by slowly increasing or decreasing the delay times τ along the length of the filter. The value of the weights may be either constant if a chirp of constant amplitude is required, or it may vary if the chirp is required to have a weighted or modulated response.

An example of the application of such a filter is in a pulse compression radar system. Here an impulse is applied to a chirp filter to produce the chirp signal, which is then amplified and transmitted. The received signal, which has been reflected off the target, is applied to another chirp filter. This second chirp filter has a reversed-frequency characteristic, that is, if the impulse response of the first has an increasing frequency chirp with time then the one on the receive side decreases the frequency with time. The result of the returned chirp going through the second filter is that it is compressed back to a pulse. This filter is known as a matched filter and it helps to optimise the signal-to-noise of the radar system. Signal-to-noise is increased by having as much energy in the transmitted signal as possible. This could be obtained by having a very large single pulse, but amplifiers are then required to deal with this large signal. A large energy is achieved in the pulse compression radar case by spreading the signal over time. The amount of energy is proportional to the product of the length of the chirp in time and the frequency bandwidth of the chirp, and is called the TB product. The improvement in signal-to-noise is directionally proportional to the TB product, which is also known as the processing gain of the system. Pulse compression radar also has good range resolution because timing can be taken from the time difference between the input impulse and the resulting impulse from the receiver. It can be seen that such systems require filters having as large a delay as possible which are also accompanied by a wide bandwidth. There are many other applications of delay line filters, including convolvers, correlators, programmable filters and spectrum analysers; these are discussed in Chapter 8.

The transversal filter has been demonstrated using superconducting transmission lines in two forms. In the first, the double transmission line structure shown in Figure 6.5.2 is used; the alternative is based on a single transmission line and is discussed later in this section. The dual transmission line structure was developed at MIT Lincoln laboratory in the early 1980s[15–17] and was originally developed using low-temperature superconductors. The filter consists of two transmission lines side-by-side which are coupled together at certain points by bringing the lines close together. The couplers are backward wave couplers and are a quarter-wavelength long at the frequency of the tap. Figure 6.5.2 shows an impulse applied to the structure. This impulse travels along the top line and, when it meets the first backward wave coupler, energy is transferred to the second line in the frequency band of the coupler. This travels towards the output of the device along the lower transmission line. The input impulse continues along the upper line, transferring energy at each coupler. The resultant output is shown in Figure 6.5.2. In this case it is a chirp signal starting at high frequency and decreasing in frequency with time. This is also depicted in a time–frequency graph in the diagram. It should be noted that the device is really a four-port device and that the other two terminals can be used to produce a chirp which increases in frequency with time. One of the potential problems with such a structure is multiple reflections. Signals travelling towards the output on the lower line also couple back up to the upper line; this effect can be taken into account in the design procedure outlined in Section 6.6. In order for the delay line to fit into a reasonable area, it is coiled into a spiral.

A number of filters of this type have been produced at MIT from niobium thin films. These filters are usually made from stripline with either silicon or sapphire substrates, usually around 125 μm thick. For 50 Ω lines on sapphire, a 42 μm line width is required. Filters with less than 50 Ω characteristic impedance with larger line widths have the advantage of lower sensitivity to width variations, lower conductor losses as the wide microstrip limit is reached, and variation in the effective constant is minimised as the effect of voids between the dielectric layers is reduced. The disadvantage of using lines other than 50 Ω is that impedance transformers are required. The delay of the filter is

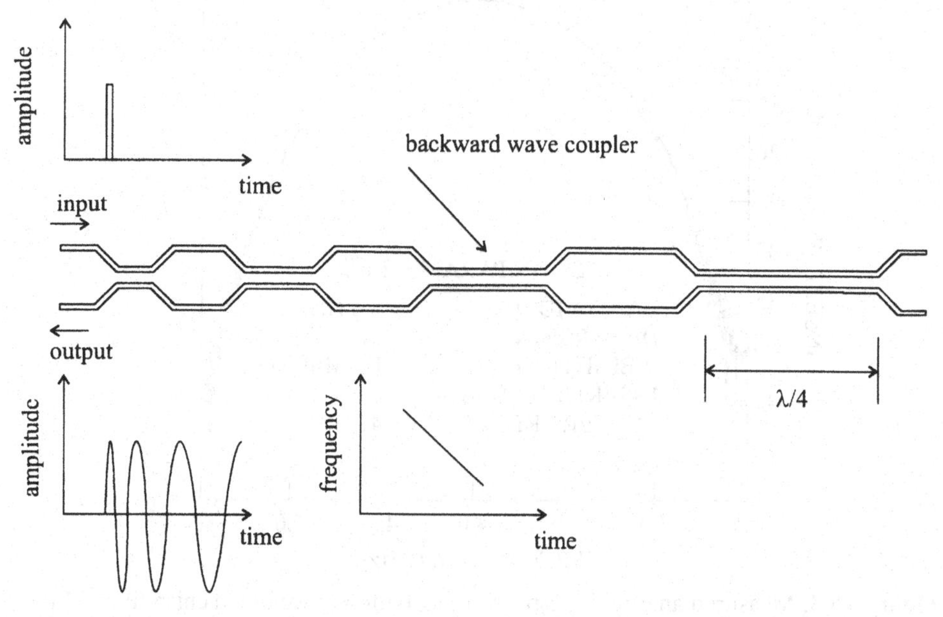

Figure 6.5.2. The double transmission line delay line filter.

maximised by coiling the dual transmission line into a tight spiral without the coupling between the spirals affecting the filter response. With this method, several tens of nanoseconds of delay are possible. Higher delays have been demonstrated by stacking up a number of substrates.[18] Packaging is important in all high-frequency structures and great care must be taken in their design and implementation.[18]

The development of these devices has taken place over many years and has led to a mature technology, with very accurate phase and amplitude response filters being produced.[19] The magnitude and phase response of a chirp filter using a double niobium delay line are shown in Figures 6.5.3 and 6.5.4. The filter is designed to have 37.5 ns of dispersive delay over a bandwidth of 2.6 GHz centred on 4 GHz. This gives a time–bandwidth product of 98. The design is Hamming weighted; a similar filter (but flat weighted) is described in reference [20].

The agreement between the experimental results and the design is excellent for both phase and amplitude. In Figure 6.5.4 the quadratic phase response has been removed in order to highlight the deviation from quadratic. This is measured to be 3.1° rms when compared with a predicted error of 3.5°. Similar

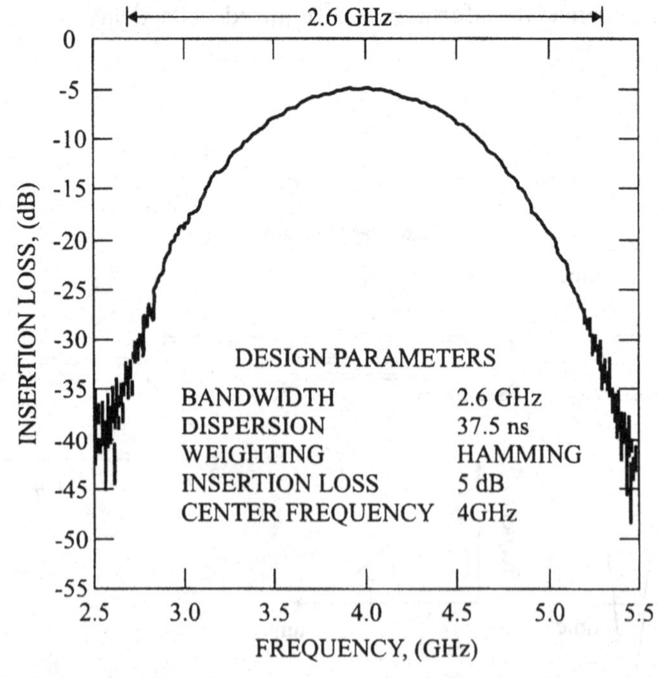

Figure 6.5.3. Measured amplitude response of a Hamming weighted chirp filter. (Taken from reference [20].)

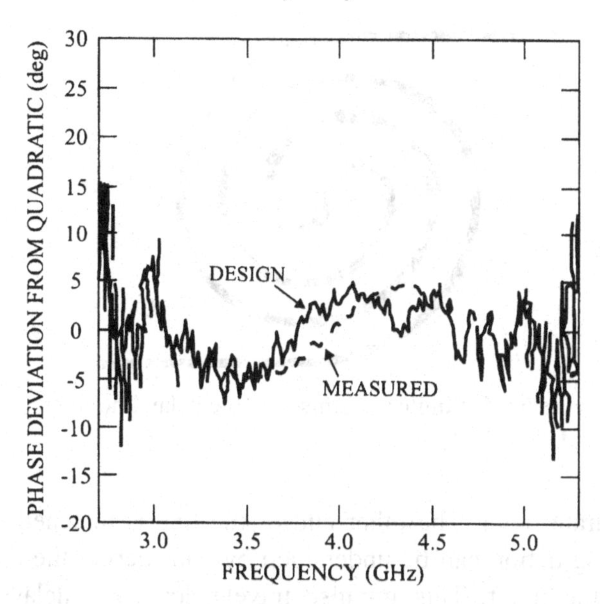

Figure 6.5.4. Design model and measured phase response of a Hamming weighted chirp filter. (Taken from reference [20].)

filters have been constructed with a 6 GHz bandwidth and a 42 ns delay, giving a time–bandwidth product of 252, see reference [20].

A number of HTS delay line filters in the form of the double transmission lines have been produced,[21–24] and good agreement with the design parameters has been obtained. Lyons gives an example of a chirp filter with an 11.5 ns delay using 48 backward wave couplers. The bandwidth is 2.6 GHz, giving a time–bandwidth product of 30. The structure is constructed on a 50-mm-diameter wafer of $LaAlO_3$, with YBCO deposited on the surface. The line width is 120 μm, giving a characteristic impedance of 40 Ω, which is converted to 50 Ω by a tapered line impedance transformer to bring the total delay to 14 ns.

6.5.1 Reflective delay line filters

A second type of delay line filter is based on a single delay line with sections of varying impedance. Figure 6.5.5 shows an example of such a filter. It can be seen that the impedance of the microstrip line varies along the line length. It is this variation which causes the filtering action. Although the width appears to vary smoothly in the diagram it actually varies in small steps. The input and output of the filter occur at the same port, and a directional coupler is required

Figure 6.5.5. Single transmission line delay line filter.

in order to separate them. The other end of the line is matched with a 50 Ω load. The filtering action can be understood by considering the application of an impulse to the input. This impulse travels down the delay line and is reflected at each of the impedance steps. Some of the reflections emerge at the input/output port whilst others are multiply reflected. The impulses emerging constitute the impulse response of the filter which can be converted to the frequency response by a Fourier transform. A design method for this and the double delay line filter is given in Section 6.6.

The advantages of this filter over the double delay line filter are clear. Only a single delay line is required, so that a significantly longer delay can be produced on any given substrate. Also, because of the synthesis technique, which is described below, the harmonic responses of the filter are suppressed. With the absence of the third harmonic, the bandwidth of the filter can be increased above 100%. This could be important for some applications where large time–bandwidth product devices are required, or simply when processing is required over very large bandwidths. The disadvantage of the filter is that a wideband directional coupler is required to separate the input from the output. The obvious choice is a superconducting directional coupler which occupies the same substrate; some superconducting directional couplers are discussed in Chapter 5.

A number of HTS[25,26] and copper[27,28] filters of this type have been produced. The frequency response of the filter given in Figure 6.5.5 is shown in Figure 6.5.6. The filter is designed to give a linear phase response over a 4 GHz bandwidth centred on 10 GHz.[26] The filter is made from 0.35-μm-thick YBCO on a 1"-square, 300-μm thick MgO substrate. Double-sided deposition of the superconductor is required. With this design, delays of around only 1 ns have been demonstrated. The designed and measured responses are shown in Figure

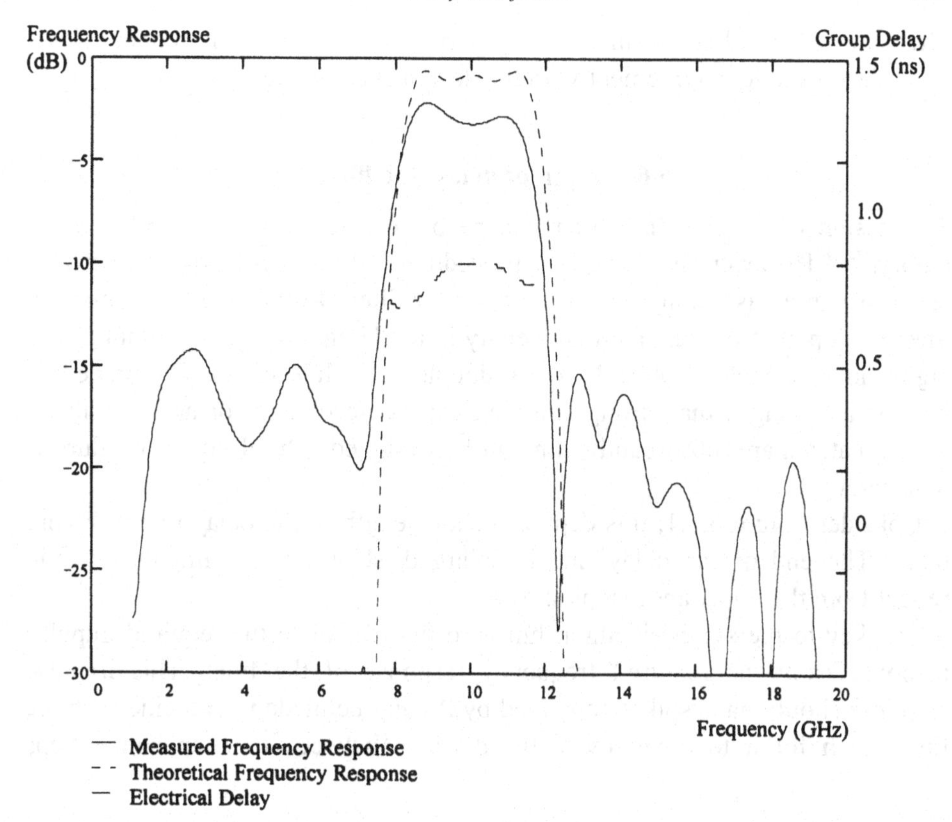

Figure 6.5.6. Frequency response of the HTS linear phase delay line filter shown in Figure 6.5.5.[26]

6.5.6. Good agreement is observed across the passband but the sidelobes are high in the stopband. This is due to coupling in the spiral and reflections at the connectors. It has been found that coplanar lines with not too great a bend radius (in order not to stimulate the slotline mode) can be used for the reduction of cross-talk. A coplanar device with similar specifications as the filter discussed above have shown improved performance in terms of a reduction of the level of sidelobe close to the passband.[26] Copper filters have been constructed mainly to demonstrate the synthesis technique described below. However, some interesting filters have been developed. Chirp or quadratic filters have been demonstrated with a centre frequency of 1 GHz, a bandwidth of 1.33 GHz, and a centre frequency delay of about 11 ns, with a dispersion of 12 ns/GHz.[28] Another copper device, mainly for the demonstration of phase equalisation, was designed to shift phase by −90° between 0.1 and 0.9 GHz and +90° between 1.1 and 1.9 GHz, with a delay of 8 ns. This filter showed excellent results.[27] In general, the results on the copper delay lines show excellent agreement with the design specification.

6.6 Design of delay line filters

The design of a delay line filter can be performed by using coupled mode theory.[16,29] However, an alternative procedure is discussed below. This synthesis procedure has mainly been developed for filters with a single transmission line developed at Birmingham University in the United Kingdom, although the algorithm is equally applicable to the double delay line filters. The procedure has the advantages that strong coupling can be accounted for and that filters with greater than 100% bandwidth can be designed. The design procedure is non-iterative.

Consider Figure 6.6.1; this depicts a short length of the delay line in Figure 6.5.5. The end of the delay line is enlarged, showing how impulses would reflect from the impedance steps shown.

The key to the synthesis algorithm is to first calculate the required impulse response from the required frequency response of the filter. This impulse response should have spikes separated by Δt and be made to coincide with the time taken for it to travel twice the distance between the impedance steps

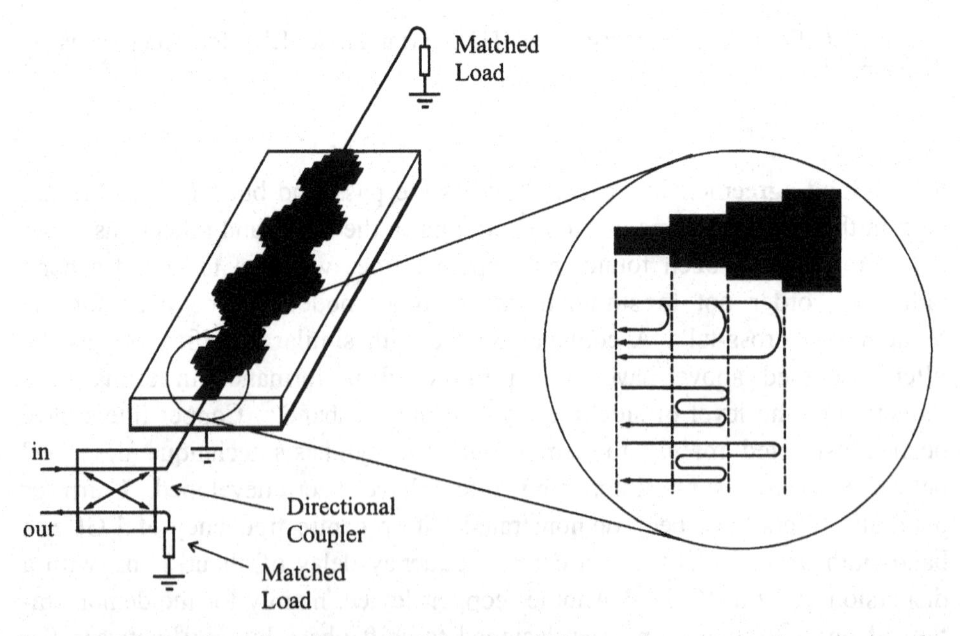

Figure 6.6.1. Section of a microstrip delay line with impedance steps along its length. The insert shows how an impulse would be reflected at the start of the line.

shown in Figure 6.6.1. The calculation of the required reflection coefficients of the impedance steps to give the desired impulse response proceeds as follows.

If an impulse is applied to the transmission line in Figure 6.6.1, then the first impulse returned from the first reflector or impedance change coincides with the first spike in the required impulse response. Thus it is possible to calculate the required reflection coefficient to meet this requirement. The second spike in the required impulse response is made up of a transmission through the first impedance step, reflection from the second impedance step and transmission back through the first impedance step. Thus knowing the reflection, and hence the transmission coefficient of the first step, the required reflection coefficient of the second step can be calculated. The third spike in the required impulse response is more complicated and is made up of an impulse which is reflected from the third impulse step plus a contribution due to one multiple reflection between the first and second steps. The only unknown is the third step reflection coefficient, which can be found. The synthesis procedure continues along these lines, solving for all the reflection coefficients. Once the reflection coefficients are known, then the impedance, and hence the width of the lines, can be found.

It can be seen that this procedure takes multiple reflections into account explicitly and there is no constraint on the strength of the reflection as in the coupled mode synthesis technique. However, it should be noted that the impulse response of the device is infinite, whereas for this procedure the impulse response in finite. There are extra outputs caused by extra multiple reflections which are unaccounted for and which occur after the final spike in the required impulse response. Because these outputs are caused by a large number of multiple reflections, the resultant effect is small. If the procedure is used for the coupled transmission line, then the spikes are a quarter-wavelength apart, but for the single-line type there may be many impedance changes per wavelength; thus the edges of the filter resemble a quantised sine wave. Harmonic responses are eliminated, allowing for filters with greater than 100% bandwidth. These subwavelength impedance steps also allow equal spacing of the changes in impedance along the line, even in chirp devices. This allows simpler implementation of the algorithm. With large changes in width of the transmission line, parasitic capacitances and inductances are produced which need to be taken into account. With small changes in width, these parasitic elements do not need to be considered.

The following gives the detailed design equations for the method and closely mimics reference [27]. As stated above, the filter specification is usually in terms of a frequency response. This frequency response must be Fourier transformed to give an impulse response. This impulse response can be 'windowed' in order to give a finite length impulse response for the synthesis algorithm. If time is quantised by using $t = i\Delta t$, then at $t = 0$ the input impulse applied to the filter can be represented by

$$a_{0,i} = 1 \quad i = 0$$
$$a_{0,i} = 0 \quad \text{otherwise} \tag{6.6.1}$$

The output of the filter will be designated $b_{0,i}$. In order to represent the amplitude of the wave incident upon a particular impedance change, the first subscript is used, $a_{0,i}$, being at the input, $a_{m,i}$ being at the mth reflector, as shown in Figure 6.6.2. Similarly, the waves travelling to the left in Figure 6.6.2 are represented by $b_{m,i}$, m, referring to the mth reflector or impedance step.

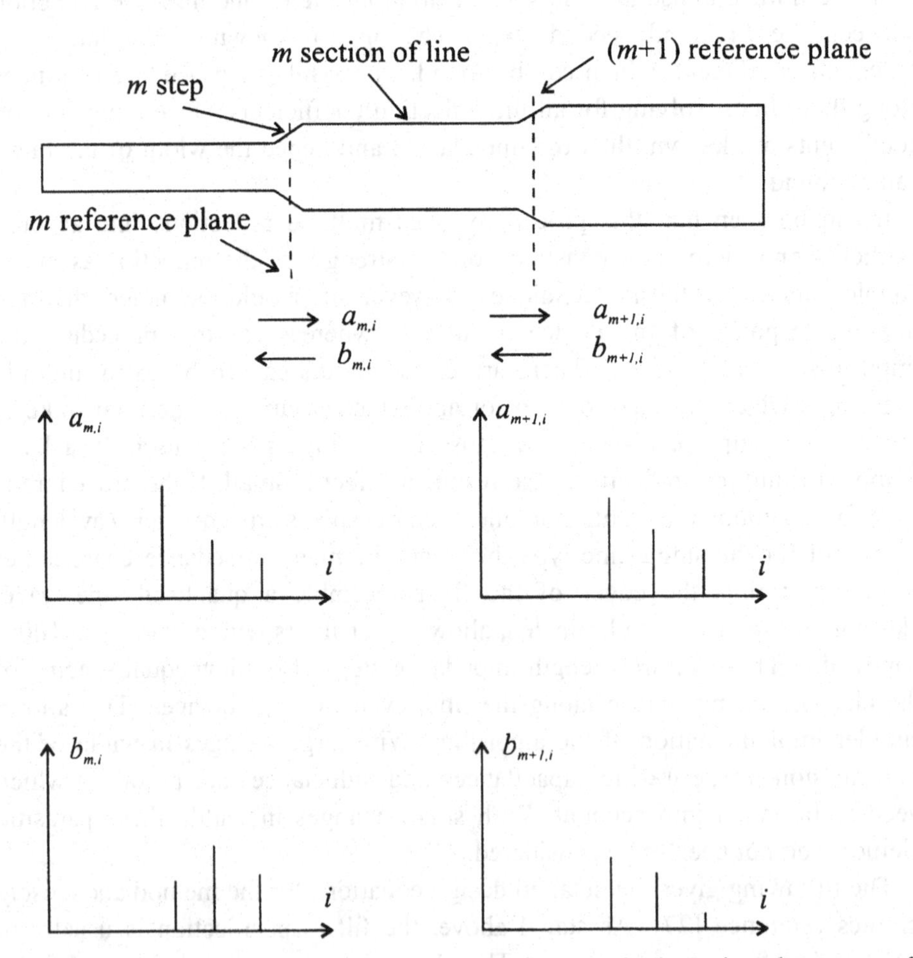

Figure 6.6.2. The mth transmission line section showing the input impulses $a_{m,i}$ and the output impulses.

Thus the reflection coefficient at the mth impedance step is

$$r_m = \frac{b_{m,m}}{a_{m,m}} \tag{6.6.2}$$

since at the propagation delay $m\Delta t$, the first impulse is reflected from the mth section, and after this multiple reflections contribute to the output at the mth reflector. The algorithm described above can be written as

$$a_{m+1,i} = \frac{\eta_m}{\rho_m}[a_{m,i-1} - r_m b_{m,i-1}] \tag{6.6.3}$$

$$b_{m+1,i} = \frac{1}{\eta_m \rho_m}[b_{m,i+1} - r_m a_{m,i+1}] \tag{6.6.4}$$

where

$$\rho_m^2 = 1 - r_m^2 \tag{6.6.5}$$

η_m is the magnitude of the transmission coefficient through each section of line and can be used to represent any conductor or dielectric losses and is assumed to be independent of frequency, although the frequency dependence can be taken into account by modification of the algorithm.[30] Amplitudes have been normalised to equal the square root of power, so ρ_m is real and the step can be suitably designed. The solution process continues along the length of line until all the impedance steps have been included and the procedure stops. At first glance the memory storage requirements seem large, but with careful programming they can be reduced to one-dimensional arrays for $a_{m,i}$ and $b_{m,i}$.[27] A simple example of the design of this filter is given in reference [27].

The errors associated with the implementation of the weighting function and the truncation of the synthesis algorithm can be assessed by analysing the resultant reflectivities or impedance steps. The analysis processes are easy and accurate, and can be done either by transmission matrices[27] or *ABCD* matrices (both are equivalent). In both cases each section of the transmission line is represented by a matrix; to obtain the frequency response of the whole line, the matrices for each section are just multiplied together. The impulse and frequency response of the filter can be calculated from the resultant matrix. After this process, if the match to the original frequency response is not adequate, then either the weighting function can be altered or the number of reflectors increased.

6.7 Comparison with other technologies

There are a number of processing technologies which can potentially compete with superconducting delay lines, including digital, surface acoustic wave (SAW),[31,32] acousto optic (AO), optical,[33] magneto static,[34] acoustic charge transport devices (ACT),[35] and charge coupled devices (CCD).[36] All these technologies possess the ability to delay a signal in order to construct some form of transversal filter.

One of the most important criteria is the loss per unit delay. This has already been discussed in Section 6.1, and the losses in units of dB/ns for various transmission media are shown in Figure 6.1.1. It can be seen that, as far as loss is concerned, superconductive delay lines are an excellent choice of delay mechanism, although at the higher frequencies the loss increases rapidly due to the f^2 increase in surface resistance. Although Figure 6.1.1 shows the attenuation, it gives no indication of the size of the filters produced by means of the various technologies. Clearly, the waveguide, although having low loss, is large and impracticable for delay lines; this also applies to coaxial cable, especially for the lower frequencies. SAW devices are clearly important because of the very low velocity of the acoustic wave, and hence they are very small in size. As discussed above, superconductors can also produce small devices by spiralling or meandering transmission lines on the substrate surface.

Figure 6.7.1 shows another assessment of some of the technologies, this time not only based on their delay capability but also on their potential bandwidths.[37] This diagram does not necessarily represent any demonstrated performance of the technologies but it shows what may be achievable from fundamental considerations. As expected, the superconductive technology has the highest bandwidth capability, followed by SAW and then the CCD devices. Also included are the equivalent performance of a Cray 2 computer with two Giga Operations Per Second (GOPS) and a dedicated parallel processor operating at 10 GOPS.[37]

SAW technology was developed mainly in the 1970s and is now a mature technology and an excellent choice for producing both delay line and other filters in the frequency region of about 10 MHz–1 GHz. The great advantage of SAW is the long delay times available. This is due to the low velocity ($\sim$ 3400 m/s for Y cut $LiNbO_3$) of the acoustic waves. This delay can be as high as several hundred microseconds, and bandwidths of around several gigahertz are possible. A SAW filter is produced by having interdigital electrodes on the surface of a piezoelectric crystal. Application of an alternating voltage produces the surface waves which are picked up and converted back to an electrical signal by a similar transducer. The filtering action arises owing to the design of the transducers or interaction of the surface waves with metallic diffractors or reflectors placed between the transducers. As well as the fundamental loss shown in Figure 6.1.1 limiting the upper frequency, the size of the interdigital transducers becomes very small and patterning is required to be submicron in the low gigahertz frequency region. There are also inherent

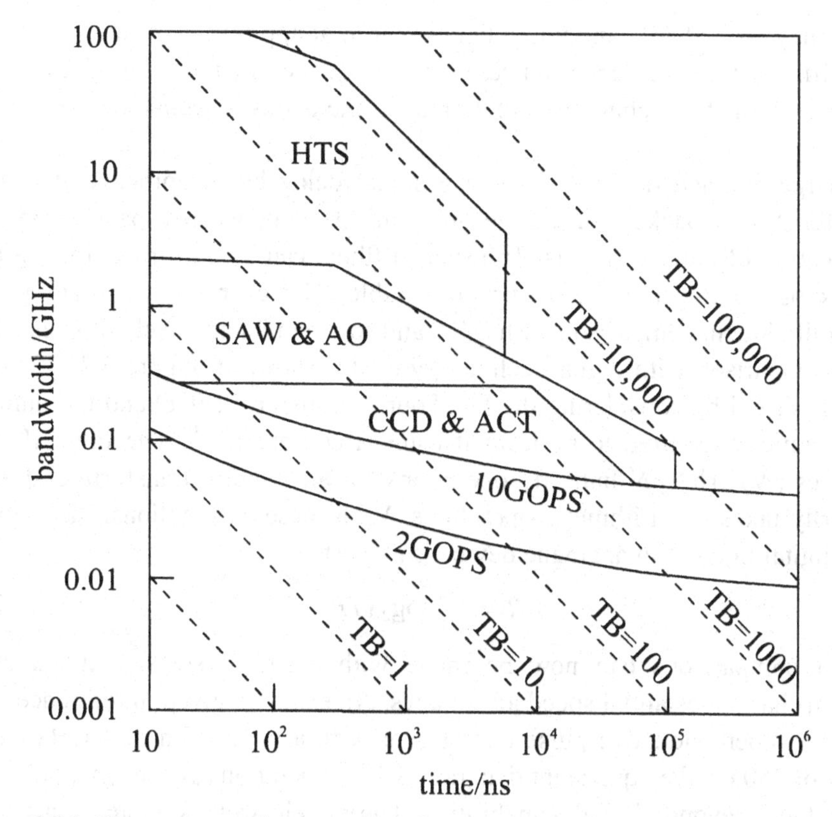

Figure 6.7.1. Near-term projections for signal processing capabilities for various technologies.[37]

additional losses associated with SAW devices. The transducer used to convert from electrical to acoustic energy transmits acoustic energy in two directions. One direction is usually unwanted, as it travels away from the receiving transducer, and so contributes to a loss. With a transmit and receive transducer, this represents a 6 dB loss immediately. In addition, because the acoustic wave travels on the surface of a crystal, diffraction occurs and the wave spreads, which also results in a loss of energy, since it is not received at the output transducer. Both these loss mechanisms can be reduced to a certain extent by using unidirectional transducers and/or acoustic waveguides. However, it is difficult to produce a device with all the above factors optimised.

Acousto-optic devices exploit the interaction between an acoustic wave and an optical signal. For example, a simple spectrum analyser can be made by diffracting an optical beam from refractive index perturbations caused by the passage of a SAW on $LiNbO_3$. The angle of diffraction depends on the frequency of the SAW, and photodiodes placed at appropriate positions pick up the diffracted beam, hence indicating the frequency of the SAW. Clearly, the limitations of this technology are similar to the acoustic technology described above.

Charge coupled devices produce a signal delay by the process of charge transfer. Small packets of charge are transferred between capacitors by the application of control signals. Transversal filters can be constructed using this technology and can be fully programmable. However, the bandwidths are generally low in comparison with SAW and superconductive technologies.

A comparison with digital technology is also shown in Figure 6.7.1. This is accomplished by considering the fast Fourier transform (FFT) and the number of operations required to perform it using a computer.[19,38] The radix II FFT requires $(N/2)\log_2 N$ butterfly operations for an N-point transform and each butterfly takes ten arithmetic operations. With these assumptions, the equivalent digital rate for an analogue device is at least

$$R = 5B\log_2(TB) \tag{6.7.1}$$

Direct comparisons can now be made with digital systems and analogue systems have substantial speed advantages. An example given in reference [19] is for a superconductive pulse compressor with a bandwidth of 4 GHz and a delay of 250 ns; the equivalent data rate is 10^{12} operations per second, orders of magnitude beyond digital capabilities. Digital electronics systems are also physically large and have a high power consumption compared with dedicated analogue systems.

Two technologies not shown in Figure 6.7.1 are magnetostatic devices and optical devices. Both are less well developed and not proven for microwave delay line filtering applications to the extent of the other technologies. Magnetostatic waves can potentially provide delays up to the order of a microsecond with bandwidths around 10 GHz. However, as far as potential is concerned, optical signal processing is superior to many of the technologies discussed here. For example, the optical fibre offers a very-low-loss transmission medium and it is superior to all the technologies in the higher-frequency regime, as seen in Figure 6.1.1. It also has an enormous bandwidth potential. Fibre is small in volume and weight, resulting in compact devices with large delays. Microwave delay lines can be formed from fibre or other optical transmission media by modulating a microwave carrier on to an optical signal. Such delay lines operating at microwave frequencies have been demonstrated in the form of tapped delay lines,[33,39,40] as a number of parallel delay lines forming a filter,[41,42] and recirculating delay lines;[43–47] some examples are shown in Figure 6.7.2.

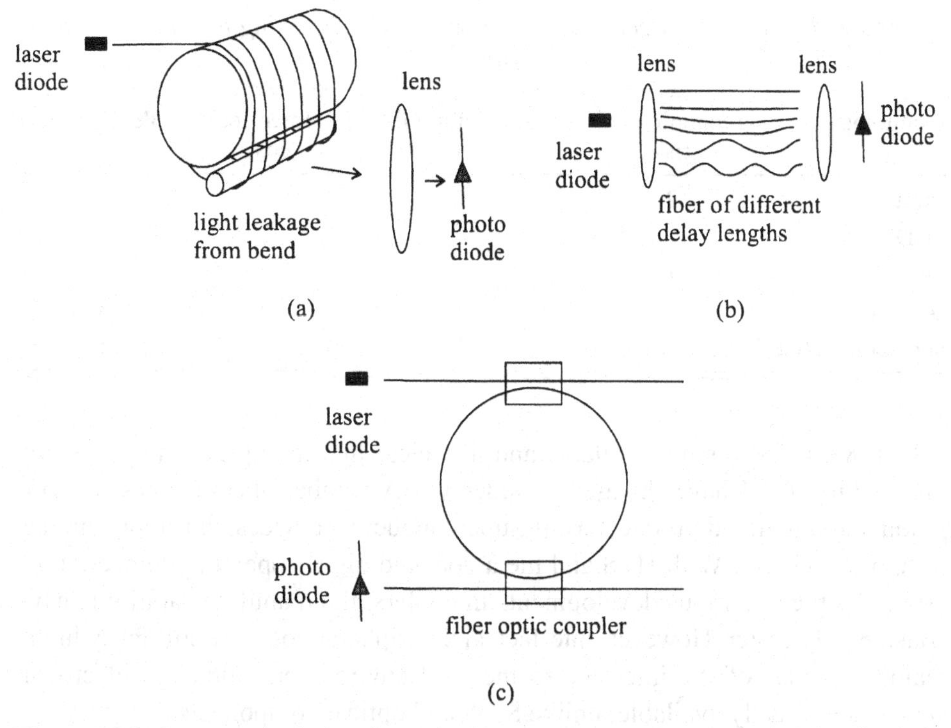

Figure 6.7.2. Some optical signal processing components: (*a*) tapped delay line, (*b*) parallel fibres with different delays, (*c*) recirculating delay line.

The limitation on the bandwidth of the system is due to the method of converting the microwave signal to an optical signal. Direct modulation of a laser diode can be done at several tens of gigahertz and external modulation goes up to slightly higher frequencies. The dynamic range of these filters is, however, severely limited by the photodiode and presents one of the main limiting factors of the technology.

The application of a new technology in the product market place is not related solely to its increase in performance over other technologies. Many technologies producing marginal improvement in performance are not used. The new technology must be demonstrated by the experts in that field right up to the level of development of systems based on the technology, and also services to engineering companies need to be developed in order to make the use of the technology as straightforward as possible. In general, system designers have little experience of these technologies and are unprepared to risk their inclusion into products which often have to be produced on short time scales at minimum cost. Table 6.7.1 shows the relative acceptance of some of the technologies discussed above, as seen by Stern in 1985.[48]

Table 6.7.1. *Relative acceptance of new signal processing technologies in 1985*[48]

Technology	Technological risk	Manufacturability	Interface costs	Peripheral costs
Digital	2	1	1	1
CCD	2	2	2	1
SAW	1	2	3	2
Optical	2	4	4	3
Superconductive	4	3	4	5

In 1985, HTS was not available and it is clear that the figures in the bottom line of Table 6.7.1 have changed considerably. A number of companies are now producing integrated products using superconductive devices, thus reducing the technological risk. With HTS and the associated higher operating temperature, as well as the enormous developments in coolers, the manufacturability is now considerably easier. However, interface and peripheral costs are still fairly high, mainly because of the interface to the cold environment, although electrical signals are directly available, unlike SAW and optical components.

6.8 References

1 Lyons W. G., Withers R. S., Hamm J. M., Anderson A. C., Mankiewich P. M., O'Malley M. L. and Howard R. E. High-T_c superconducting delay line structures and signal conditioning networks, *IEEE Trans. on Magnetics*, **MAG-27**, 2932–5, 1991

2 Jackson K. P., Newton S. A., Moslehi B., Tur M., Cutler C. C., Goodman J. W. and Shaw J. H. Optical fibre delay line signal processor, *IEEE Trans. on Microwave Theory and Techniques*, **MTT-33**(3), 193, 1985

3 Oliner A. A. Acoustic Surface Waves, *Topics in Applied Physics*, **24**, Springer-Verlag, Heidelberg, New York, 1978.

4 Anderson A. C., Withers R. S., Reible S. A. and Ralston R. W. Substrates for analog signal processing devices, *IEEE Trans. on Magnetics*, **MAG-19**(3), 485–9, 1983

5 Wong S. C., Anderson A. C. and Rudman D. A. Processing of thick-film dielectrics compatible with thin film superconductors for analog signal processing devices, *IEEE Trans. on Magnetics*, **25**(2), 1255–7, 1989

6 Liang G.-C., Withers R. S., Cole B. F., Garrison S. M., Johansson M. E., Ruby W. and Lyons W. G. High-temperature superconducting delay line filters on sapphire and thinned $LaAlO_3$ substrates, *IEEE Trans. on Applied Superconductivity*, **3**(3), 3037–41, 1993

7 Liang G.-C., Withers R. S., Cole B. F. and Newman N. High-temperature superconductive devices on sapphire, *IEEE Trans. on Microwave Theory and Techniques*, **42**(1), 24–40, 1994

8 Bourne L. C., Hammond R. B., Robinson McD., Eddy M. M. and Olson W. L. Low loss microstrip delay line in $Tl_2Ba_2CaCu_2O_8$, *Appl. Phys. Lett.*, **56**(23), 2333–5, 1990

9 Höfer G. J., Kratz H. A., Schultz G., Söllner J. and Windte V. High temperature superconductor delay lines, *IEEE Trans. on Applied Superconductivity*, **3**(1), 2800–3, 1993

10 Track E. K., Drake R. E. and Hohenwarter G. K. G. Optically modulated superconducting delay lines, *IEEE Trans. on Applied Superconductivity*, **3**, 2899–902, 1993

11 Shen Z.-Y., Pang P. S. W., Holstien W. L., Walker C., Dunn S., Face D. W. and Laubacher D. B. High T_c superconducting coplanar delay line with long delay and low insertion loss, *IEEE MTT-S Int. Microwave Symp. Digest*, **3**, 1235–8, 1991

12 Talisa S. H., Janocko M. A., Meier D. L., Moskowitz C., Grassel R. L. Talvacchio J., LePage P., Buck D. C., Nye R. S., Pieseski S. J. and Wagner G. R. High-temperature superconducting wide band delay lines, *IEEE Trans. on Applied Superconductivity*, **5**(2), 2291–4, 1995

13 Hohenwarter G. K. G., Track E. K., Drake R. E. and Patt R. Forty five nanosecond superconducting delay lines, *IEEE Trans. on Magnetics*, **3**(1), 2804–7, 1993

14 Fenzi N. O. and Aidnik D. L. An HTS 100 nano-second delay line, *1994 IEEE Applied Superconductivity Conf., Boston MA*, 1994

15 Lynch J. T., Anderson A. C., Withers R. S., Wright P. V. and Reible S. A. Passive superconducting microwave circuits for 2–20 GHz bandwidth analog signal processing, *Proc. 1982 International Microwave Symp. New York*, IEEE, 1982

16 Withers R. S., Anderson A. C., Wright P. V. and Reible S. A. Superconductive tapped delay lines for microwave analog signal processing, *IEEE Trans. on Magnetics*, **MAG-19**(3), 480–4, 1983

17 Lynch J. T., Withers R. S., Anderson A. C., Wright P. V. and Reible S. A. Multigigahertz-bandwidth linear-frequency-modulated filters using a superconductive stripline, *App. Phys. Lett.*, **43**(3), 319–21, 1983

18 Withers R. S., Anderson A. C., Green J. B. and Reible S. A. Superconductive delay-line technology and applications, *IEEE Trans. on Magnetics*, **MAG-21**(2), 186–92, 1985

19 Withers R. S. and Ralston R. W. Superconductive analog signal processing devices, *Proc. of the IEEE*, **77**(8), 1247–63, 1989

20 Dilorio M. S., Withers R. S. and Anderson A. A. Wideband superconductive chirp filters, *IEEE Trans. on Microwave Theory and Techniques*, **37**(4), 706–10, 1989

21 Lyons W. G., Withers R. S., Hamm J. M., Anderson A. C., Mankiewich P. M., O'Malley M. L. and Howard R. E. High-T_c superconducting delay line structures and signal conditioning networks, *IEEE Trans. on Magnetics*, **MAG-27**, 2932–5, 1991

22 Lyons W. G., Withers R. S., Hamm J. M., Anderson A. C., Mankiewich P. M., O'Malley M. L., Howard R. E., Bonetti R. R., Williams A. E. and Newman N. High temperature superconductive passive microwave devices, *IEEE MTT-S Digest*, pp. 1227–30, 1991

23 Lyons W. G., Withers R. S., Hamm J. M., Mathews R. H., Clifton B. J., Mankiewich P. M., O'Malley M. L. and Newman N. High-frequency analog signal processing with high temperature superconductors, *Picosecond Electronics and Optoelectronics Conf., Salt Lake City UT USA*, March, 1991

24 Lyons W. G., Withers R. S., Hamm J. M., Anderson A. C., Mankiewich P. M., O'Malley M. L., Bonetti R. R., Williams A. E. and Newman N. High temperature

delay line filters in superconductivity and its applications, New York: *American Institute of Physics 1992*, **251**, 639–58, 1992
25 Huang F., Cheung H. C. H., Lancaster M. J., Humphreys R. G., Chew N. G. and Goodyear S. W. A superconducting microwave linear phase delay line filter, *IEEE Trans. on Applied Superconductivity*, **3**(1), 2778–81, 1993
26 Cheung H. C. H., Huang F., Lancaster M. J., Humphreys R. G. and Chew N. G. Improvements in superconducting linear phase microwave delay line bandpass filters, *IEEE Trans. on Applied Superconductivity*, **5**(2), 2675–7, 1995
27 Huang F. Low loss quasi-transversal microwave filters with specified amplitude and phase characteristics, *IEE Proc.*, H**140**(6), 433–40, 1993
28 Huang F. Quasi-transversal synthesis of microwave chirped filters, *Electron. Lett.*, **28**(11), 1063–4, 1992
29 Wright P. V. and Haus H. A. A closed form analysis of reflective array gratings, *IEEE Ultrasonics Symposium Proceedings IEEE, New York*, pp. 282–7, 1980
30 Huang F. Frequency dependent transmission line loss in quasi-transversal microwave filters, *IEE Proc. Microw. Antennas Propag.*, **141**(5), 402–6, 1994
31 Withers R. S. A comparison of superconductive and surface acoustic wave signal processing, 1988 *Proc. Ultrasonics Symposium*, pp. 185–94, 1988
32 Oliner A. A. Acoustic surface waves, *Topics in Applied Physics*, **24**, Springer-Verlag, Heidelberg, New York, 1978
33 Jackson K. P., Newton S. A., Moslehi B., Tur M., Cutler C. C., Goodman J. W. and Shaw J. H. Optical fibre delay line signal processing, *IEEE Trans. on Microwave Theory and Techniques*, **MTT-33**(3), 193, 1985
34 Stiglitz M. R. and Sethcres J. C. Magnetostatic waves take over where SAWs left off, *Microwave Journal*, **25**, 19, 1982
35 Hoskins M. J. and Hunsinger B. J. Recent developments in acoustic charge transport devices, *Proc. IEEE Ultrasonic Symposium*, pp. 439–50, 1986
36 Munroe S. C. *et al.* Programmable four-channel, 128 sample 40-Mc/s analog–ternary correlator, *IEEE J. Sol. State Circuits*, **25**, 425–30, 1990
37 Lyons W. G. and Withers R. S. Passive microwave device applications of high temperature superconducting thin films, *Microwave Journal*, p. 85, November, 1990
38 Cafarella J. H. Wideband signal processing for communications and radar, *Proc IEEE Nat. Telesystems Conf.*, pp. 55–8, 1983
39 Newton S. A., Jackson K. P. and Shaw H. J. Optical fibre V-grove transversal filter, *Appl. Phys. Lett.*, **43**(2), 149, 1983
40 Jackson K. P., Bowers J. E., Newton S. A. and Cutler C. C. Microbend optical fibre tapped delay line for gigahertz signal processing, *Appl. Phys. Lett.*, **41**(2), 139, 1982
41 Lewis M. F. and West C. L. Novel narrow band fibre-optic microwave filter, *Electron. Lett.*, **22**(19), 1016, 1986
42 Chang C. T., Cassaboom J. A. and Taylor H. F. Fibre-optic delay line device for R. F. signal processing, *Electron. Lett.*, **19**(13), 480, 1983
43 Bowers J. E., Newton S. A., Sorin W. V. and Shaw H. J. Filter response of single mode fibre recirculating delay lines, *Electron. Lett.*, **18**(3), 110, 1982
44 Tur M., Goodman J. W., Moslehi B., Bowers J. E. and Shaw H. J. Fibre optic signal processor with applications to matrix-vector multiplication and lattice filtering, *Optics Lett.*, **7**(9), 463, 1982
45 Newton S. A., Howland R. S., Jackson K. P. and Shaw J. H. High speed pulse train generation using single mode fibre recirculating delay lines, *Electron. Lett.*, **19**(19), 756, 1983

46 Salehi J. A., Menendez R. C. and Brackett C. A. A low pass digital optical filter for optical fibre communications, *J. Lightwave Technology*, **6**(12), 1841, 1988

47 Gatenby P. V., Switzer D. L. and Green N. Broadband correlators employing fibre optic recirculating delay lines, *Electron. Lett.*, 1246, 1987

48 Stern E. Comparison of device technologies for signal processing, *Proc. of the International Seminar on Technology for High Speed Signal Processing Trondheim Norway, Aug. 1985*, Tapir Publishers, pp. 79–95, 1985

7

Superconducting antennas

7.1 Introduction

Antennas can benefit in a number of ways when superconductors are used in their fabrication. This chapter describes areas where superconductors are useful, and in which applications the greatest improvement can be obtained. The obvious application is in the improvement of the radiation efficiency of small antennas and superdirectional arrays. Antennas which are around the size of one wavelength are normally fairly efficient and superconductors cannot help. However, as the size of the antenna is reduced the efficiency reduces due to the increasing dominance of the ohmic losses. In this case superconductors reduce the losses and maintain a reasonable efficiency as the antenna shrinks in size. In principle, it is possible to obtain any directivity from any size of antenna array, and small directive antennas are called superdirective antennas. Again, superdirective antennas are very inefficient and superconductors help to make these a practical proposition, albeit with only a moderate superdirective capability. In addition to the reduction in efficiency, reducing the size of antennas and superdirective antenna arrays also makes matching to a reasonable system impedance increasingly difficult. High-Q matching networks are required, where superconductors also help considerably in performance improvement. As both these antenna types shrink, the Q also increases, restricting the bandwidth over which they can operate, and this now remains the practical limitation to their usefulness. For application purposes, the balance between efficiency or gain and Q needs to be assessed and this chapter addresses this problem in both a general way and also for some specific antennas. Superconductors have a significant role to play in the antennas of the future.

Another application of superconductors to antennas is in the feed networks of millimetre-wave antennas. The losses associated with long, narrow microstrip feed-lines can be improved considerably if superconductors are used. This is especially true for arrays with a large number of elements.

The first section of this chapter looks at some fundamental considerations of superconducting antennas. This is tackled in a very general way using spherical mode theory, and a number of specific and useful conclusions can be drawn. Following this, a section deals with small-loop and dipole antennas in detail. This emphasises the conclusion of the general section, but also looks at the design of a specific antenna. The principles can then be applied to other antenna designs. Section 7.4 discusses some small antennas and assesses their performance. Transmitting antennas are discussed in Section 7.5; here the potential problem is in the power dissipated in the superconductor as it turns normal, with possible catastrophic effects. Receiving antennas are given similar special treatment in Section 7.6. This time it is the signal-to-noise ratio that is the important parameter. The discussion of superdirective antennas begins in Section 7.7 with a development of the theory associated with one of the simplest superdirectional arrays, a two-element dipole array. This discussion complements the general discussion of Section 7.2. Expressions are developed for the gain and Q of the array as a function of its size, and agree well with experimental results. Following this a number of other superdirectional arrays are discussed in Section 7.8. The final section of this chapter looks at the future and at another application of superconducting antennas in system applications, where the signal processing components of Chapters 5 and 6 are included in the antenna array. A number of advanced system ideas and examples are discussed.

Dinger,[1] Williams[2] and Hansen[3,4] have written general review articles on superconducting antennas which complement the discussion in this chapter, and are useful sources of further information.

7.2 Fundamental considerations

It is possible to look at antennas, and hence superconducting antennas, in a very general and fundamental way. The calculations and discussion set out in this section consider a source of electromagnetic waves in terms of spherical wavefunctions. The radiation from any antenna can be decomposed into a number of spherical wavefunctions. The wavefunctions, or spherical modes, generated depend upon the geometry of the antenna. The complexity of the radiation pattern depends on the number of modes N which are present. In fact, the 3 dB beamwidth can be approximated by[5]

$$\Phi_{3\ \mathrm{dB}} = \frac{\pi}{N} \tag{7.2.1}$$

Looking at an antenna from this general point of view allows us to come to some very general conclusions about superconducting antennas. The discussion that follows draws heavily on Harrington's[6,7] work done in the late 1950s on antenna fundamentals.

Any electromagnetic field, external to a sphere containing sources of radiation, can be written as[6]

$$\mathbf{E} = -\nabla \times (\mathbf{r}\psi) + \frac{1}{j\omega\varepsilon}\nabla \times \nabla \times (\mathbf{r}\widehat{\psi}) \tag{7.2.2}$$

$$\mathbf{H} = \nabla \times (\mathbf{r}\widehat{\psi}) + \frac{1}{j\omega\mu}\nabla \times \nabla \times (\mathbf{r}\psi) \tag{7.2.3}$$

where $\mathbf{r}$ is the radius vector from the origin and

$$\psi = \sum_{m,n} A_{m,n} H_n^{(2)}(k_0 r) P_n^m(\cos(\theta)) \cos(m\phi + \alpha_{m,n}) \tag{7.2.4}$$

$$\widehat{\psi} = \sum_{m,n} B_{m,n} H_n^{(2)}(k_0 r) P_n^m(\cos(\theta)) \cos(m\phi + \widehat{\alpha}_{m,n}) \tag{7.2.5}$$

The coefficients $A_{m,n}$, $B_{m,n}$ and α depend only upon the excitation of the modes and are not dependent upon any particular antanna. $H_n^{(2)}$ is a Hankel function of the second kind and P_n^m is the Legendre function of order n and degree m. The substitution of Equations (7.2.4) and (7.2.5) into Equations (7.2.2) and (7.2.3) in order to find the E and H fields is done explicitly in Marcuvitz.[8]

7.2.1 Maximum antenna gain

Knowing the electric and magnetic fields at any point in space, the directive gain of an antenna can be calculated in terms of its spherical modes and is given by[6]

$$G(k_0 r) = \frac{4\pi r^2 S_{max}}{P} = \frac{\mathrm{Re}\left[\left(\sum_n a_n F_n + j b_n F'_n\right)\left(\sum_n b_n F_n + j a_n F'_n\right)^*\right]}{4\sum_{m,n} \frac{n(n+1)(n+m)!}{\varepsilon_m (2n+1)(n-m)!}\left(|A_{m,n}|^2 + |\eta B_{m,n}|^2\right)} \tag{7.2.6}$$

Here S_r is the real part of the radial component of the Poynting vector and P is the outward directed power over a sphere of radius r; $\varepsilon_m = 1$ for $m = 0$, and 2 for $m > 0$, and

$$F_n(k_0 r) = k_0 r H_n^{(2)}(k_0 r) \tag{7.2.7}$$

$$a_n = n(n+1)A_{1,n} \quad \text{and} \quad b_n = \eta n(n+1)B_{1,n} \tag{7.2.8}$$

What is remarkable about this expression is that there exists no limit to the directive gain of the antenna. Provided that an antenna can generate the required number of modes, the directive gain can increase arbitrarily. This applies to both small and large antennas. This can be looked at in another way. Any antenna radiation pattern, containing any number of modes, can be propagated backwards to any source distribution of any size. However, if there is a limit to the number of spherical modes generated by an antenna, then a maximum directive gain does exist and, for the far-field Equation (7.2.6) reduces to the simple expression

$$G_{max} = N^2 + 2N \tag{7.2.9}$$

where N is the maximum number of modes generated by the antenna. This equation is plotted in Figure 7.2.1, showing the increase in maximum directive gain as the number of modes excited increases.

So far there has been no indication of any dependence upon the size of the antenna. In order to consider this, the impedances of the spherical modes needs to be calculated. For positive travelling waves, these are given by[6]

$$Z_n^{TM} = \frac{E_\theta}{H_\phi} = -\frac{E_\phi}{H_\theta} = j\frac{F'_n(k_0 r)}{F_n(k_0 r)} \tag{7.2.10}$$

$$Z_n^{TE} = \frac{E_\theta}{H_\phi} = -\frac{E_\phi}{H_\theta} = j\frac{F_n(k_0 r)}{F'_n(k_0 r)} \tag{7.2.11}$$

These are the impedances seen by any of the modes travelling outwards at any particular radius $k_0 r$. There is one impedance for transverse magnetic (TM) waves and one for transverse electric (TE) waves, exactly as in a more

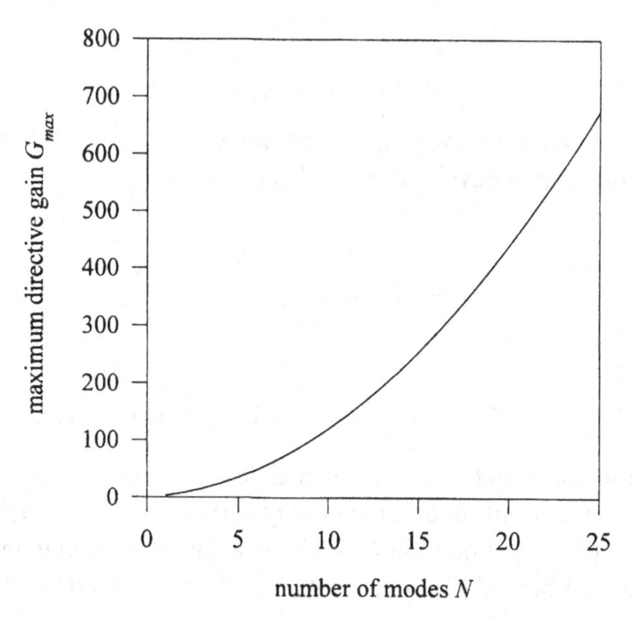

Figure 7.2.1. The maximum directive gain for an antenna containing spherical modes up to order N.

conventional waveguide. Equations (7.2.10) and (7.2.11) are predominately reactive when $k_0 r < n$, and predominately resistive when $k_0 r > n$. When $k_0 r = n$, there is a gradual changeover from reactive to resistive impedances. Also, Chu[9] has expanded the impedance of Equations (7.2.10) and (7.2.11) as a series which can be interpreted as a ladder network of series capacitances and shunt inductors; this behaves as a lowpass filter. Hence, for fixed radius it is less easy to use the high-order modes for power transfer and most normal antennas only contain about $k_0 a$ modes. Here a is the radius of the smallest sphere which contains the antenna. It is possible to define the critical radius of a normal antenna[5,10] such that $a_c = N/k_0$. If the size is below this critical size, then the antenna is dominated by reactively stored energy. This critical size represents a changeover from the near field to the far field of the antenna. Superdirective antennas can be defined as antennas which excite more modes than $k_0 a$, the higher-order modes giving the higher directive gain. Electrically small antennas are antennas which are smaller than the critical size for low numbers of spherical modes. The maximum directive gain of a normal antenna, containing $k_0 a$ modes, can now be written, using Equation (7.2.9), as

$$G_n = (k_0 a)^2 + 2k_0 a \tag{7.2.12}$$

Hence if the ordinate in Figure 7.2.1 is changed to $k_0 a$, then the abscissa will be the directive gain of the normal antenna; superdirective antennas will improve on this directive gain.

7.2.2 Antenna Q

The Q of an antenna containing spherical modes up to order N can also be calculated in this general development and is given by[6]

$$Q = \frac{\sum_{n=1}^{N}(2n+1)Q_n(k_0 a)}{N^2 + 2N} \tag{7.2.13}$$

where

$$Q_n = \frac{k_0 a}{2}|F_n(k_0 a)|^2 X'_n(k_0 a) \quad \text{and} \quad X_n(k_0 a) = \mathrm{Im}\,(Z_n^{TM}(k_0 a)) \tag{7.2.14}$$

Chaloupka[5] points out that if the antenna is less then the critical size, and it contains a large number of modes, then the Q varies as $Q \sim (a_c/a)^{2n+1}$. Figure 7.2.2 shows the Q using Equation (7.2.13) as a function of antenna size. The diagram has a number of different plots which correspond to the different number of modes excited by the antenna.

For a specific number of modes contained in an antenna the Q increases

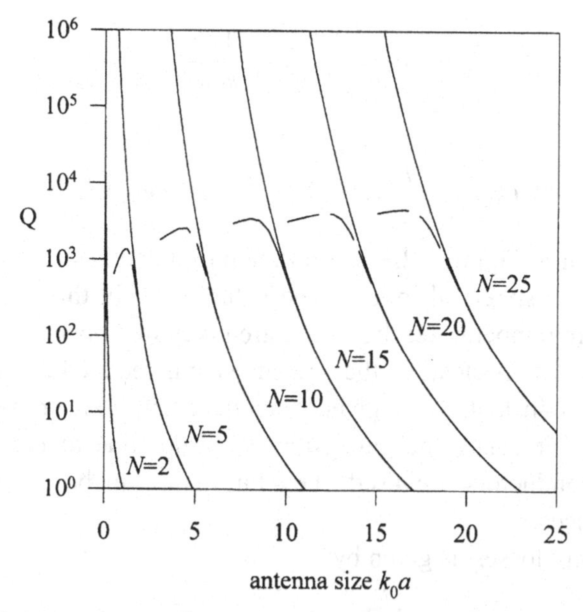

Figure 7.2.2. The antenna quality factor as a function of antenna size. Each line represents a different maximum number of spherical modes excited by the antenna. The dotted lines show the effect of losses, and are discussed later.

rapidly as the antenna size is reduced. This is expected as there is more energy stored close to the antenna. This occurs in particular if the size is less than the critical size (when $N = k_0a$ in Figure 7.2.2) and the energy is stored between its actual radius and the critical radius.[10–13] It can be seen from Figure 7.2.2 that if miniaturisation is to occur such that there is a size reduction below the size of a normal antenna, then the more directive antennas, containing a large number of modes, will have a higher Q for the same reduction in size. For example, for an antenna containing five modes and of size $k_0a = 5$, then for a Q of 10^4 a reduction of about a factor of 4 is allowed. However, for an antenna containing 20 modes with a normal size of $k_0a = 20$ and an allowed Q of 10^4, then the reduction allowed is only a factor of about 1.5. Clearly, it is important to consider the specific application before attempts are made to reduce antenna sizes whilst still maintaining Q. For most practical cases, large directivities are not a practical possibility because of the large associated Q. Efficiency and gain are the other factors of practical importance and are discussed next.

7.2.3 Consideration of losses

So far losses have not been considered. Harrington[6,7] deduces an expression for the gain of an antenna which generates N spherical modes which is given by

$$G = \sum_{n=1}^{N} \frac{2n+1}{1 + D_n(k_0 a)} \tag{7.2.15}$$

where

$$D_n(k_0 a) = \frac{R_s}{2\eta}\left(|F'_n(k_0 a)|^2 + |F_n(k_0 a)|^2\right) \tag{7.2.16}$$

This gain is shown in Figure 7.2.3 as a function of antenna size for a maximum number of different spherical modes. The value of R_s in this diagram is that, for copper at room temperature; the values are given in Appendix 1.

The effect of losses is clear; as the antenna size is reduced the directive gain is approximately constant for a given antenna exciting N modes. However, there comes a point where the gain suddenly drops due to the effect of the losses. If superconductors are used, then this gain can be kept high as the antenna size reduces.

The Q, including losses, is given by[6]

$$Q = \frac{\sum_{n=1}^{N}(2n+1)Q_n(k_0 a)}{\sum_{n=1}^{N}(2n+1)(1 + D_n(k_0 a))} \tag{7.2.17}$$

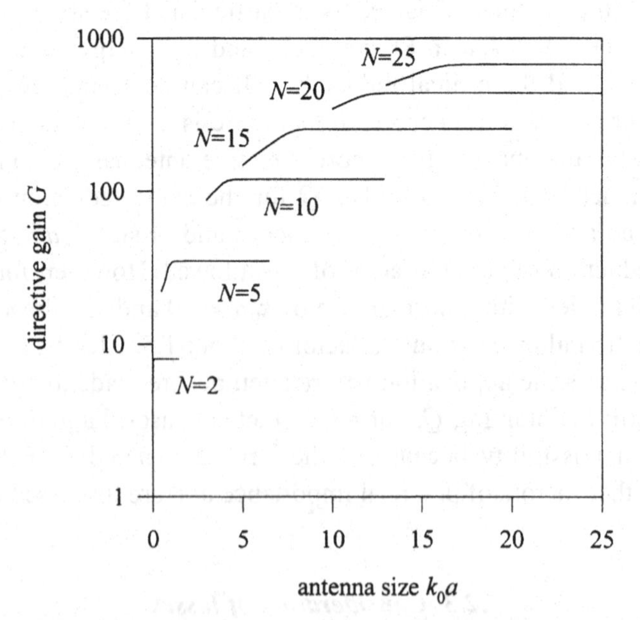

Figure 7.2.3. The directive gain of an antenna as a function of antenna size. Each line represents a different maximum number of spherical modes excited by the antenna. The graph is for copper at 300 K and 10 GHz.

Figure 7.2.2 shows this Q for copper, and it can be seen that the effect of the losses is to limit the Q. Looking at Figures 7.2.2 and 7.2.3, it is clear that any attempt to miniaturise the antenna must take into account both a possible limitation on the gain due to the finite conductivity of the metal making up the antenna and a large rise in Q. Both of these may make the antenna unsuitable for the application. Superconductors can help with the gain but will not hold the Q down. However, it is possible to reduce the Q or to increase the bandwidth to a certain extent, as described below.

7.2.4 Maximum antenna bandwidth

The question now arises: Can the bandwidth of the antenna be improved by including a matching network at its input terminals? The answer is 'yes', to a certain extent. The impedance of the antenna using a series equivalent circuit is

$$Z = R(1 + jQ\Delta(f)) \tag{7.2.18}$$

This expression is discussed in Chapter 4 with reference to cavity resonators. In Equation (7.2.18),

$$Q = \frac{2\pi f_0 L}{R} \quad \text{or} \quad Q = \frac{R}{2\pi f_0 C} \tag{7.2.19}$$

Here R is the total antenna resistance, including the loss resistance and the radiation resistance, and L and C are the inductance and capacitance of the antenna respectively. Which expression is used depends upon the antenna geometry. The quantity $\Delta(f)$ is given by

$$\Delta(f) = 1 - \frac{f_0^2}{f} \approx 2\frac{f - f_0}{f_0} \tag{7.2.20}$$

The reflection coefficient $S_{11}(f)$ can be calculated using this impedance. This is the same calculation as was done for the reflection coefficient of a cavity resonator in Chapter 4. The resultant magnitude of the reflection coefficient is given by Equation (4.4.10) as

$$|S_{11}(f)| = \left(\frac{S_{11}^2(f_0) + \dfrac{Q^2}{(1+\beta)^2}\Delta^2(f)}{1 + \dfrac{Q^2}{(1+\beta)^2}\Delta^2(f)} \right)^{1/2} \tag{7.2.21}$$

$\Delta(f)$ is just the fractional bandwidth, B, of the antenna. This can be rearranged using Equation (4.4.6) to give

$$B = \frac{1}{Q}\sqrt{\left(\frac{S_{11}^2(f)(1+\beta)^2 - (1-\beta)^2}{1 - S_{11}^2(f)} \right)} \tag{7.2.22}$$

Under normal circumstances, the antenna is critically matched, so that the magnitude of the reflection coefficient at the centre frequency is 1; this implies that $R = Z_0$ and $\beta = 1$. If the normal criteria of a 3 dB bandwidth is applied with $S_{11}^2 = 0.5$, then Equation (7.2.22) reduces to

$$B = \frac{2}{Q} \tag{7.2.23}$$

It should be noted that $\beta = 1$ is not the optimum value of β for the maximum bandwidth.[14] The maximum value of the bandwidth with respect to β can be found if Equation (7.2.22) is differentiated with respect to β and put equal to zero; this results in

$$\beta_{maxB} = \frac{1 + S_{11}^2(f)}{1 - S_{11}^2(f)} \tag{7.2.24}$$

For an antenna with a 3 dB bandwidth, this gives a value for β of 3, resulting in an increase in bandwidth by a factor of 1.4. The penalty paid by using this optimum value of β is that the insertion loss at the centre frequency is no longer zero. Using Equation (4.4.6) shows that $S_{11}(f_0)$ is 0.5 in this case.

So far, complex matching networks have not been considered and these can also improve the bandwidth. This problem was first studied by Fano in 1950,[15,16] who discovered that there is a fundamental limit to the bandwidth obtainable when using matching circuits; his maximum bandwidth is given by

$$B_F = -\frac{1}{Q}\frac{\pi}{\ln(|S_{11}(f)|)} \tag{7.2.25}$$

For the 3 dB bandwidth antenna under discussion, this is reduced to

$$B_F = \frac{9.06}{Q} \tag{7.2.26}$$

Hence the bandwidth of the antenna can be improved up to a maximum factor of about 4.5 by the inclusion of matching networks.

Knowing that the bandwidth is related to the Q by Equation (7.2.23), a simple expression connecting the bandwidth of an antenna to its Q and efficiency can be derived. The efficiency of an antenna is given by

$$\eta_e = \frac{R_r}{R_l + R_r} = \frac{1}{1 + Q_r/Q_l} \tag{7.2.27}$$

where R_l and R_r are the loss resistance and radiation resistance, respectively, in the antenna equivalent circuit; Q_r is the quality factor associated with the radiation resistance and Q_l is the quality factor associated with the ohmic losses in the antenna. Knowing $1/Q = 1/Q_l + 1/Q_r$ and using Equation (7.2.23), the bandwidth of the antenna is given simply by

$$B = \frac{2}{Q_r \eta_e} \tag{7.2.28}$$

The factor of 2 changes to 9 if the optimum matching network is used (Equation (7.2.26)).

In this section, some valuable conclusions have been drawn about superconducting antennas in general. The next section goes on to look at some specific antennas, developing the discussion through antennas which can be built and tested.

7.3 Superconducting small-dipole and loop antennas

In order to illustrate some of the important principles associated with superconducting antennas, the small-dipole and small loop will now be looked at in some detail.[17] These antennas are fundamental and are sometimes referred to as electric and magnetic dipole antennas. Because of their simplicity, closed form expressions exist for most of their properties, enabling a good understanding to be obtained when they are examined in the superconducting context. This discussion follows that of reference [17] closely.

7.3.1 Radiation pattern and directivity

For an electrically small loop, the circumference is much less than the operating wavelength. Therefore the current distribution around this loop can be taken as a constant value I_0. When this approximation is assumed, the far field electric and magnetic fields are given by[18]

$$E_\phi = \frac{\eta(\pi b^2 I_0) k_0^2}{4\pi r} e^{-jk_0 r} \sin(\theta) \tag{7.3.1}$$

and

$$H_\theta = -\frac{E_\phi}{\eta} \tag{7.3.2}$$

where η is the impedance of free space, b is the radius of the loop, k_0 is the free space propagation constant, r is the distance from the antenna and θ and ϕ are the standard spherical coordinates.

For a short dipole antenna, the current distribution can be approximated by a linear increase from zero at the tip to I_0 at the feed. The far fields of the small dipole, using this approximation, are given by[19]

$$E_\theta = \frac{j\eta(lI_0) k_0}{8\pi r} e^{-jk_0 r} \sin(\theta) \tag{7.3.3}$$

and

$$H_\phi = \frac{E_\theta}{\eta} \tag{7.3.4}$$

where l is the total length of the dipole. As can be seen, both the electric and magnetic dipoles have the same doughnut-shaped radiation pattern. They also have a directivity of 1.5. As can be seen from Equations (7.3.1) to (7.3.4), the fact that they can be superconducting has no effect on their radiation pattern.

7.3.2 Efficiency and impedance

The equivalent circuit for the small dipole and loop antennas, as with all antennas, can be put in terms of a series combination of a radiation resistance, a loss resistance and a reactance. For the dipole case, the reactance is capacitive, and for the loop it is inductive. This is illustrated in Figure 7.3.1. The radiation resistance can be considered to be an equivalent resistance, such that any power dissipated in it will actually represent power radiated.

The loss resistance is due to the conductor losses in the antenna element itself and can be calculated by consideration of the power dissipated in the antenna, that is,

$$P = \frac{1}{2} I_0^2 R l = \frac{1}{2} \int_{wire} |I(l)|^2 \frac{R_s}{2\pi a} \, dl \tag{7.3.5}$$

Here the integration is along the length of the wire which forms the antenna. The radius of the wire is a. For the wire loop under discussion, evaluation of Equation (7.3.5) gives

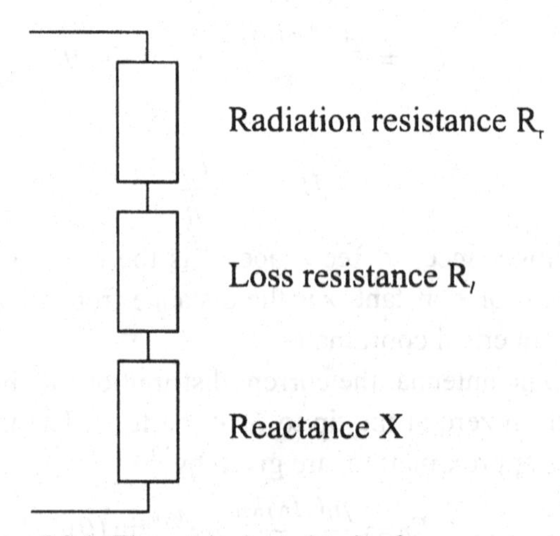

Figure 7.3.1. The equivalent circuit of an antenna.

$$R_l = R_s \frac{b}{a} \tag{7.3.6}$$

and, for the short dipole,

$$R_l = R_s \frac{l}{6\pi a} \tag{7.3.7}$$

Expressions for the radiation resistance and reactance of a small circular loop are available in the literature, and are given by[19,20]

$$R_r = \frac{\eta k_0^4 A^2}{6\pi} \tag{7.3.8}$$

$$X = \eta k_0 b(\ln(8b/a) - 2) \tag{7.3.9}$$

Here A is the area of the loop and is equal to πb^2. For the small dipole, the equivalent equations are[19]

$$R_r = \frac{\eta l^2 k_0^2}{24\pi} \tag{7.3.10}$$

$$X = -\eta \frac{2}{\pi k_0 l}(\ln(l/a) - 1) \tag{7.3.11}$$

As expected, the only contribution that a superconductor would make to the input impedance for the loop and dipole antennas is the decrease in loss resistance. There is a very small contribution to the reactance through the kinetic inductance of the superconductor, but this has a negligible effect in this case, and is not included in the equations; it may be borne in mind for other antenna geometries. It is now possible to calculate the efficiency η_e of both these antennas, that is,

$$\eta_e = \frac{\textit{radiated power}}{\textit{input power}} = \frac{R_r}{R_r + R_l} \tag{7.3.12}$$

The loss resistance and radiation resistance of both antennas decrease with decreasing antenna size. However, in both cases the reduction in radiation resistance is much more rapid. The reduction is to the fourth power of the antenna radius in the loop case and the square of the antenna length in the dipole case. This implies that at some stage the loss resistance will be more dominant than the radiation resistance in Equation (7.3.12). Consequently, producing the antenna from superconductors will improve its efficiency. However, in order to see if this is worthwhile in a practical application, we need to consider real antennas. Figures 7.3.2 and Figure 7.3.3 show the efficiencies of superconducting and copper small-loop and dipole antennas.

For the particular values chosen, the improvement in efficiency can clearly be seen, with the loop improvement being greater, as discussed above. However, this is not the whole story, as will be seen in the following sections.

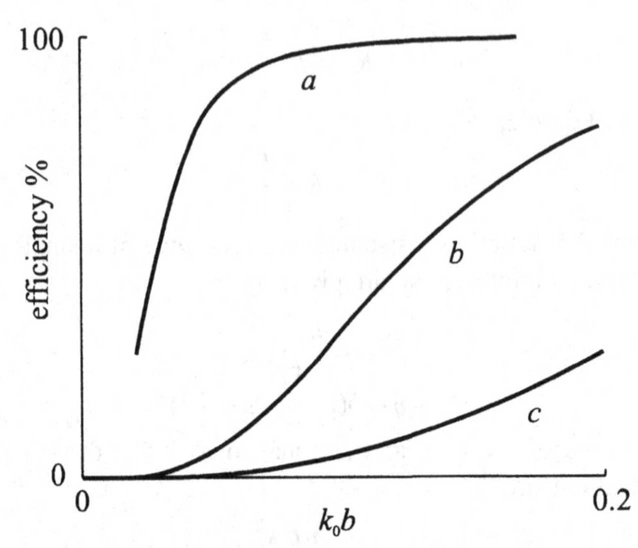

Figure 7.3.2. Efficiency of a small-loop antenna. The superconductor is assumed to be ceramic YBCO wire with a crossover frequency of 10 GHz: (*a*) HTS $a = 1$ mm 77 K, (*b*) Cu $a = 1$ mm 300 K, (*c*) Cu $a = 0.1$ mm 300 K.

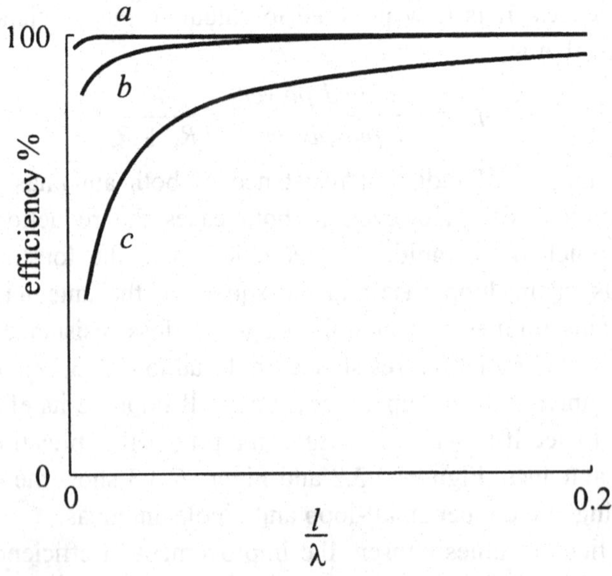

Figure 7.3.3. Efficiency of a small-dipole antenna, with the same parameters as that shown in Figure 7.3.2.

7.3.3 Matching networks

Any practical antenna must be matched to the external driving electronics. In order to do this a matching network is used to convert the antenna impedance into the source impedance, which is usually 50 Ω. Matching networks for superconducting antennas are as important as the antenna elements themselves, if not more so.[21] The dipole and loop antennas are shown in Figure 7.3.4, with simple matching networks consisting of a length of transmission line, with either a shunt capacitor in the case of the loop, or an inductive stub in the case of the dipole.

As the loop and dipole antennas become smaller, then the inductive reactance of the loop becomes small and the capacitive reactance of the dipole becomes very large, as can be seen from Equations (7.3.9) and (7.3.11). This, combined with the small resistance values, causes the antennas to have high Q values; this is discussed in the next section. However, this poor match to a

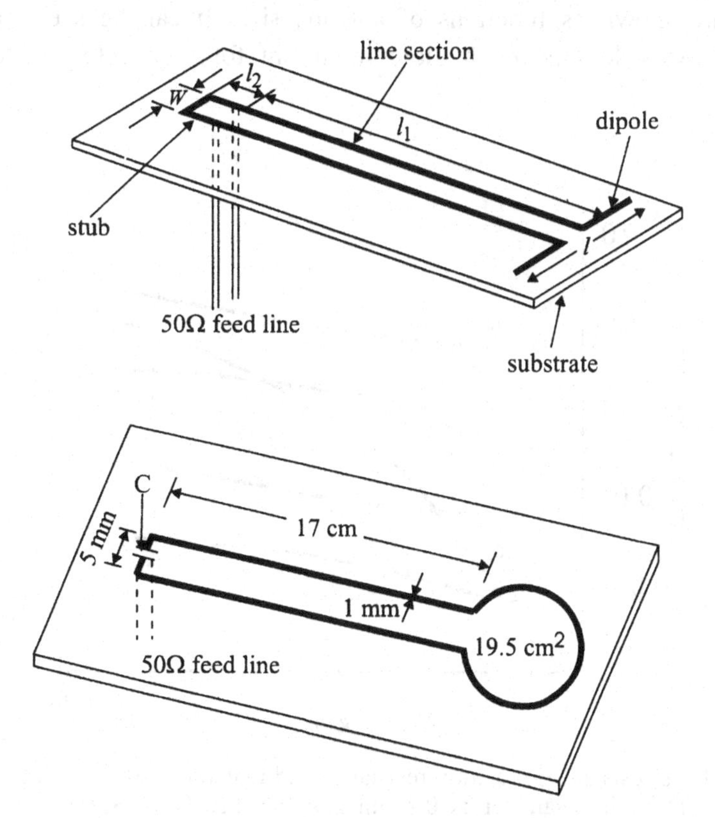

Figure 7.3.4. Small-dipole and small-loop antennas.

system impedance also causes difficulty in the fabrication of the matching network, as tolerances are tight. Considerable improvement in efficiency can be achieved if the matching networks are made from superconductors. The overall efficiency of the antenna plus the matching network can be written

$$\eta_e = \frac{radiated\ power}{input\ power} = \frac{R_r}{R_r + R_l + R_m} \qquad (7.3.13)$$

where R_m is the equivalent loss resistance of the matching network. In order to assess the importance of the value of R_m, the matching networks shown in Figure 7.3.4 will be used. The matching network consists of a transmission line section which converts the antenna admittance so that its real part has a value equal to the source admittance ($1/50\ \Omega^{-1}$). The imaginary part is cancelled by either the inductive stub or the capacitor, both forming a resonant circuit. An efficiency can be calculated assuming a lossy two-wire transmission line and Equation (7.3.13) is then used to calculate R_m, knowing R_r and R_l. The reason for doing this is shown in Figures 7.3.5 and 7.3.6; here, all the values of resistance are shown as functions of antenna size. It can be seen that the matching network loss is much more important for both HTS and copper antennas.

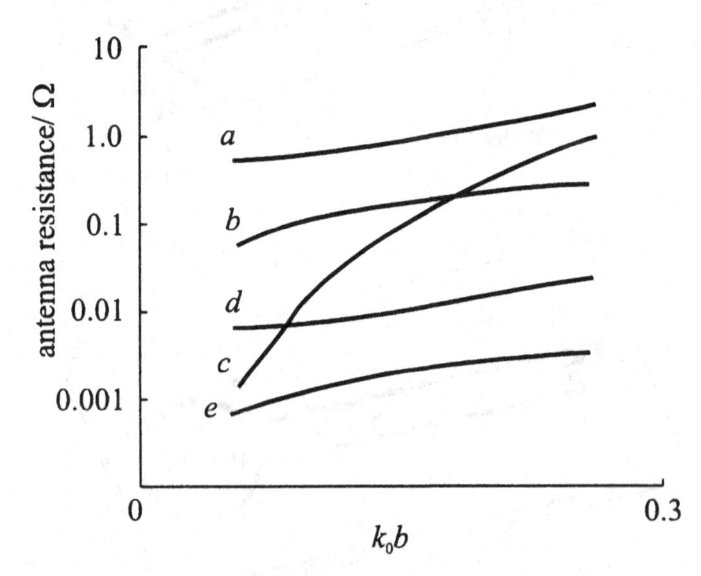

Figure 7.3.5. Loss resistance, radiation resistance and matching resistance of a small-loop antenna. The wire diameter is 0.4 mm and the YBCO HTS has a crossover frequency of 10 GHz: (*a*) copper R_m, (*b*) copper R_l, (*c*) R_r, (*d*) HTS R_m, (*e*) HTS R_l.

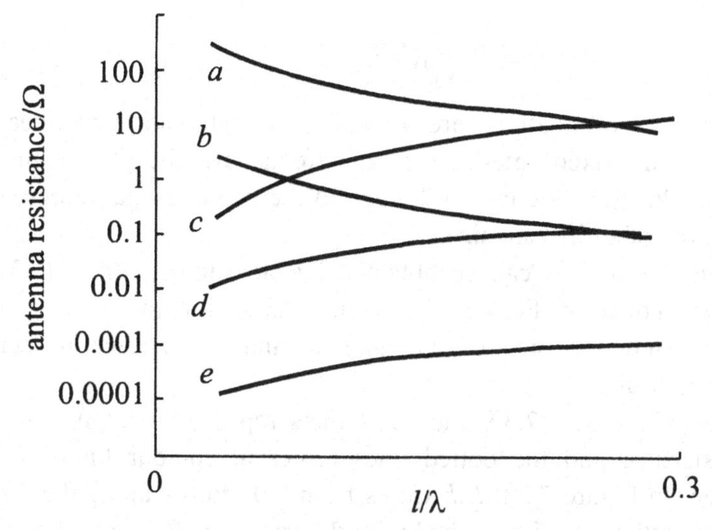

Figure 7.3.6. Loss resistance, radiation resistance and matching resistance of a small dipole antenna, with the same parameters as are given in Figure 7.3.5.

7.3.4 Bandwidth and Q

As mentioned above, as the antenna reduces in size its Q increases. In order to assess the impact that this has on practical applications, some numerical values need to be considered. The Q of the antenna can be calculated when considering it in the series circuit of Figure 7.3.1, giving $Q = X/(R_r + R_l)$. Using Equations (7.3.6), (7.3.8) and (7.3.9), the Q of the small-loop antenna is given by

$$Q = \frac{120}{20k_0^3 b^3 \pi + R_s/\pi k_0 a}(\ln(8b/a) - 2) \tag{7.3.14}$$

Similarly, the Q for the dipole using Equations (7.3.7), (7.3.10) and (7.3.11) gives

$$Q = \frac{240}{5k_0^2 l^2 + R_s k_0 l^2/6\pi a}(\ln(l/a) - 1) \tag{7.3.15}$$

Furthermore, knowing the efficiency given in Equation (7.3.12), these can be rewritten in terms of efficiency. For the loop:

$$Q = \frac{6}{k_0^3 b^3 \pi}(\ln(8b/a) - 2)\eta_e \tag{7.3.16}$$

and for the dipole:

$$Q = \frac{48}{k_0^3 l^3}(\ln(l/a) - 1)\eta_e \tag{7.3.17}$$

As expected, these equations are very similar and both vary linearly with efficiency for any fixed antenna size and frequency. The Q as a function of efficiency is plotted in Figure 7.3.7. Each of the two lines can represent a loop or a dipole, but of a different size.

An alternative picture can be obtained by plotting Equations (7.3.14) and (7.3.15), as shown in Figures 7.3.8 and 7.3.9, for the loop and dipole respectively. Here the size of the antenna and the surface resistance are included specifically.

In Figures 7.3.8 and 7.3.9 the solid lines represent contours of constant surface resistance and the dotted lines represent contour lines of constant antenna size. In Figure 7.3.8 $k_0 b$ varies from 0.02 to 0.2 along the dotted line from the low-efficiency side to the high-efficiency side for each of the different surface resistance lines. In Figure 7.3.9 the dipole size $k_0 l$ varies from 0.016 to 0.4 in the same fashion. The three solid lines are for a practical ceramic YBCO superconductor, with a crossover frequency of 10 GHz and an operating frequency of 500 MHz.

The antenna Q has a maximum (seen best for the 77 K copper line in Figure 7.3.8). This is due to the increase in loss resistance at the end of the line,

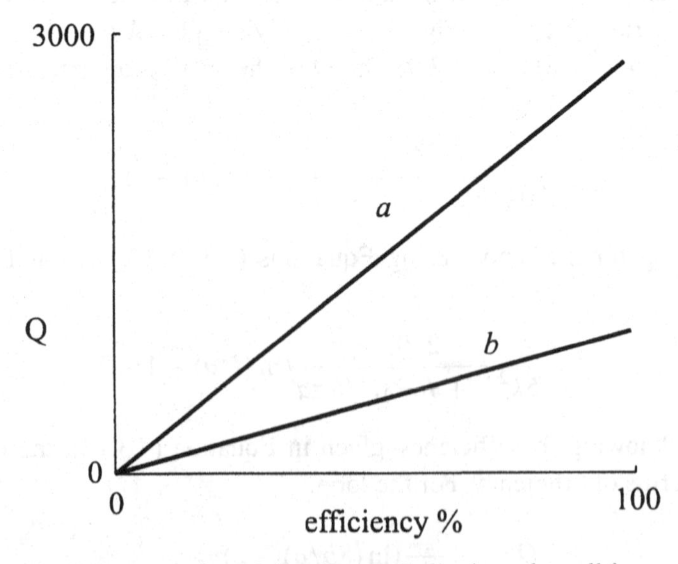

Figure 7.3.7. Efficiency against Q for the small-dipole and small loop antennas: (a) a dipole with $k_0 l = 0.2$ or a loop with $k_0 b = 0.14$, (b) a dipole with $k_0 l = 0.6$ or a loop with $k_0 b = 0.4$.

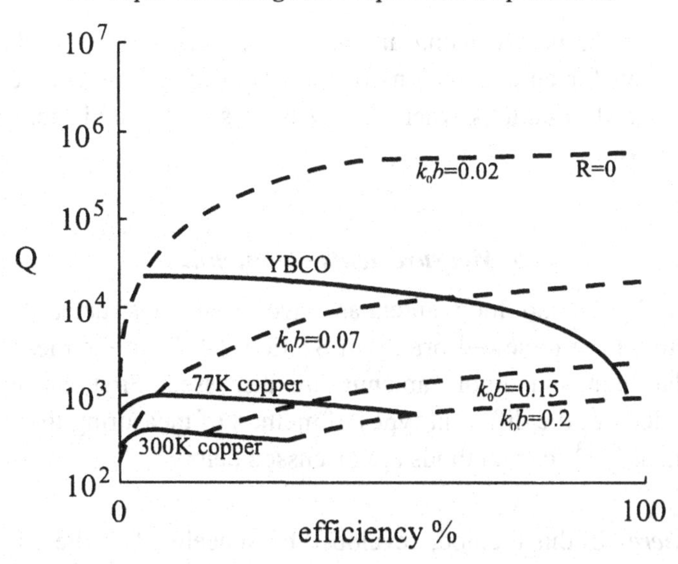

Figure 7.3.8. Q against efficiency for a loop antenna. The solid lines represent the contours of constant surface resistance and the dotted lines represent lines of constant antenna size. The wire diameter is 0.3 mm.

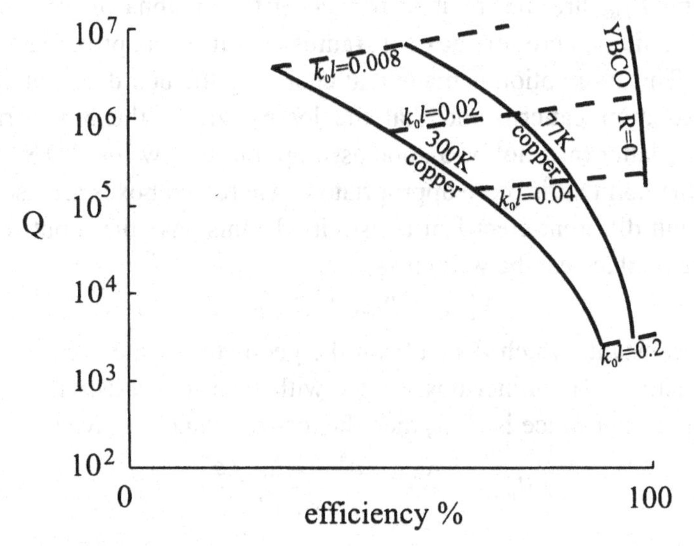

Figure 7.3.9. Q against efficiency for a small-dipole antenna.

corresponding to small antenna size and the larger radiation resistance at the other end. The shape of the antenna is clearly important in determining the particular value of Q for a given efficiency, although there is a fundamental limitation, dependent upon the volume occupied, as discussed in Section 7.2. Clearly, the antenna Qs are very large for a large part of the graphs shown in

Figures 7.3.8 and 7.3.9, and using antennas with very high Q is difficult in practice. However, for certain regions of the graphs, the YBCO antennas can be used in practical situations where the Q is reasonable. and there is still a benefit over copper.

7.3.5 Measurements on antennas

A number of dipole and loop antennas have been constructed from HTS material and these are reviewed briefly in Section 7.4. Here the measurements on a particular loop and dipole antenna are discussed. First, however, it is appropriate to look at the different types of method of measuring the efficiency of small antennas;[22,23] four methods are discussed below.

1. Radian sphere. In this method, developed by Wheeler,[24–26] the efficiency is derived from the relationship

$$\eta_e = \frac{R_{in} - R_{loss}}{R_{in}} \tag{7.3.18}$$

where R_{in} and R_{loss} are the input resistance of the antenna in free space and inside a metallic sphere or box of radius equal to approximately $\lambda/2\pi$ respectively. The assumption is made that enclosing the box does not affect the near field radiation pattern, and that the losses due to the box surface are negligible; the latter may not be a good assumption for low-loss HTS antennas. However, a method that is more appropriate is where two boxes are used, of the same shape but different electrical resistivity. In this case the input resistance with a sphere in place can be written as

$$R'_{loss} = R_{loss} + K\sqrt{\rho'} \tag{7.3.19}$$

Here K is a constant, which depends on the geometry of the box and ρ' is its electrical resistivity. If another box is used with electrical resistivity ρ'' and the measured input impedance is R''_{loss}, then the loss resistance is given by

$$R_{loss} = \frac{R'_{loss}\sqrt{\rho''} - R''_{loss}\sqrt{\rho'}}{\sqrt{\rho''} - \sqrt{\rho'}} \tag{7.3.20}$$

2. The Q method. The efficiency of an antenna can be written in terms of its radiation Q and fractional bandwidth as given in Equation (7.2.28), that is,

$$\eta_e = \frac{2}{Q_r B} \tag{7.3.21}$$

An upper estimate of the radiation efficiency of the antenna can be made using the spherical mode theory of Section 7.2. Equation (7.2.13) can be used to calculate a maximum Q_r. Also, if the fractional bandwidth B is measured for both an antenna made from a normal conductor, and one made from HTS, then it follows from Equation (7.3.21) that the ratio of the bandwidths will be the ratio of the efficiencies of the antennas. The radiation Q should be constant in this case as the antenna geometry has not changed.

3. Relative gain measurements. If the antenna to be measured is used as a transmitter antenna and the signal from it is measured at a particular distance, then this can be compared with another standard antenna whose characteristics are known. The efficiency is then given by

$$\eta_e = \frac{\eta_{es} G_s}{G} G_{rel} \tag{7.3.22}$$

Both the efficiency η_{es} and the directive gain G_s of the reference antenna must be known. This can be estimated from analytical calculations. Also, the directive gain, G, of the test antenna must be calculated from analytical calculations. Relative gain measurements are also useful in comparing antennas made from HTS and normal conductors of exactly the same shape; the improvement in gain can be measured directly.

An example of a number of printed small-loop antennas with their matching networks is shown in Figure 7.3.10.[22] These antennas are made from thick film YBCO, printed on to the surface of zirconia substrates. Antenna (*a*) does not have a matching network, but the other antennas have single-stub tuner matching networks in different geometrical configurations; antenna (*b*) has a straight transmission line in the matching network; antenna (*c*) has a meandered transmission line; and antenna (*d*) has the transmission line internal to the loop. The structures show how the size of the total antenna, including matching network, can be reduced.

Measurements on these antennas are discussed in reference [22] and summarised in Table 7.3.1. All the antennas give the expected results, with improved efficiencies over similar normal metal antennas, the smaller antenna

Table 7.3.1. *Measurements on the loop antennas in Figure 7.3.10*

	Loop (a)	Loop (b)	Loop (c)	Loop (d)
Operating frequency/MHz	260	430	410	140
Efficiency (%)	$>90\%$	42%	80%	10%
radius/mm	20	9	9	20
radius/wavelength	0.017	0.013	0.012	0.001
Bandwidth (%)	1.56	2.3	1.4	3.2
Supergain factor S_g	6.2	0.4	2.3	1.2

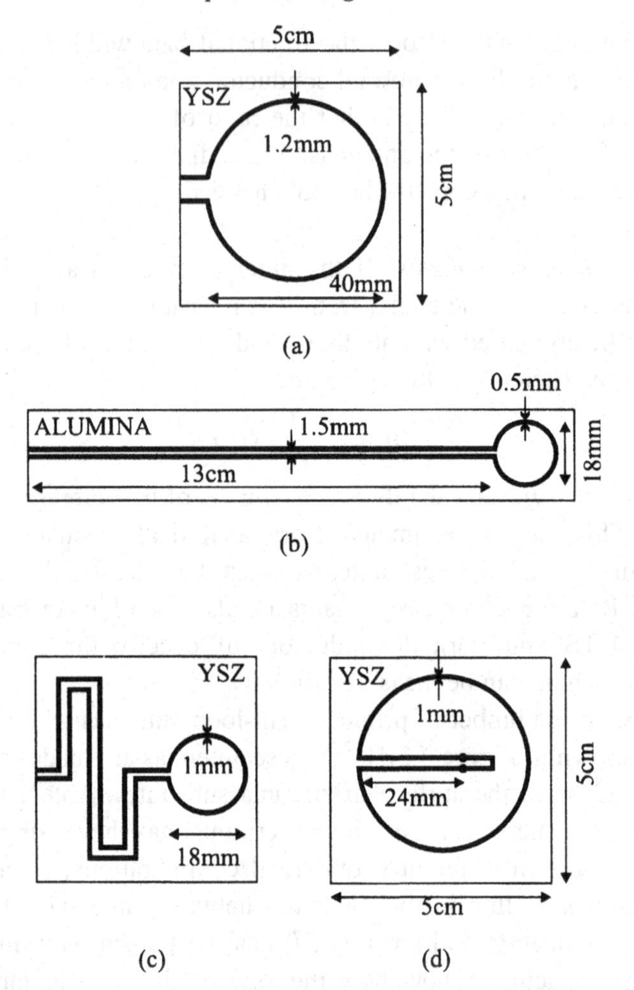

Figure 7.3.10. Printed small-loop antennas and matching networks.[22] The different antennas (*a*)–(d) show a variation in the design of matching networks.

not performing as well. The bandwidths are small, again as expected. The supergain factor is also included in Table 7.3.1; since all antennas are potentially supergain, this is an important factor. The supergain factor is defined as

$$S_g = \frac{\eta_e G}{G_n} \tag{7.3.23}$$

Here G is the directive gain of the antenna, taken as 1.5 because all the antennas are small. G_n is the directive gain of a normal antenna defined in Section 7.2 (Equation (7.2.12)) as $G_n = (k_0 a)^2 + k_0 a$, a being the radius of the antenna. Supergain rather than superdirective is used here so that the efficiency of the antenna is a factor in the assessment. It can be seen from Table 7.3.1 that antennas (*a*), (*c*) and (*d*) show supergain properties. Supergain and superdirectivity are discussed extensively in Sections 7.2, 7.7 and 7.8.

7.4 Small antennas

Having seen that there is a potential for miniaturising antennas, the next consideration is to look at the mechanisms of the size reduction. The simplest way to achieve miniaturisation is by taking a conventional antenna and then reducing it in size. For example, a dipole and a loop can easily be shrunk, and this is discussed in detail in Section 7.3. Conventional antennas need not be altered in shape. The high dielectric constant of some HTS substrates shrinks down the size of the antenna. However, more interestingly, new antennas can be designed from scratch which have the required radiation characteristics for the application. This is not an easy task, and considerable experience may be required. Some self-resonant antennas can be considered to be composed of a small inductor and capacitor as their basic elements, and altering the shape of these alters the radiation characteristics. Also, any resonant cavity not completely enclosed behaves as a self-resonant antenna. In order to make it a useful antenna, it is only a question of getting the correct balance between the radiation losses and ohmic losses in the cavity.

It is important that all antennas are matched to the system impedance (usually 50 Ω), and a matching network is needed to do this. The matching network converts the antenna impedance into the system impedance for maximum power transfer. With self-resonant antennas the matching network is simple; since it is self-resonant the input impedance is reduced to a real resistive quantity at the resonant frequency. Simple capacitive or inductive coupling can attain the required input impedance. For non-resonant antennas, the matching is to a complex impedance, and standard techniques are available to design such networks. They may be constructed from transmission line sections or lumped elements. The simplest form of matching network will provide a single resonant structure. However, bandwidth can be improved by designing multi-resonance, wide band matching networks. For self-resonant antennas, a wider bandwidth can be obtained by using multi-elements resonating at slightly different frequencies. The bandwidth can only be increased until the fundamental limit is reached, as described in Section 7.2.

Dipole and loop antennas, with their matching networks, have probably been the most studied of all HTS antennas, as they are the most fundamental type of antenna. For this reason, some theoretical consideration is given in Section 7.3; here experimental results are reviewed, as well as results for other antennas.

Demonstrations of superdirectional antennas are reviewed in Section 7.8. The first experimental superconducting antenna was reported in 1969,[27] although previous attempts at production had been made.[28] The first superconducting antenna was a small loop antenna with a radius of 0.0007 wavelengths, operating at 400 MHz and made out of lead. The loop had a matching network consisting of a coaxial resonant cavity with the coupling through a moveable probe in the side of the cavity. Upon cooling to liquid helium temperatures, the Q increased from 156 to 20 000 with a corresponding increase in gain of 27 dB. Radiation patterns of the loop were also determined. Despite being the first demonstration of a superconducting antenna, this still remains one of the most impressive. There is a later example of conventional superconducting antennas (before HTS came along) in Japan;[29,30] however, the main application has been with HTS, where the cooling is less prohibitive. The first high-temperature superconducting antenna was developed at Birmingham University in the UK in 1988.[31] This consisted of a short dipole made out of bulk polycrystalline YBCO material. The length was 0.037 wavelengths and it operated at 550 MHz. The gain over copper at room temperature was 12 dB. Since then a number of groups have demonstrated small antennas of dipole,[32–35] loop[36–38] and helical[39–41] geometries. All these antennas operate in the frequency range of several hundreds of MHz and demonstrate an improvement in gain over copper antennas (usually several decibels over a similar antenna made from copper at a similar temperature, and around 10 dB over similar copper antennas at room temperature). Actual gain values depend on the size of the antenna and the quality of the material used in their manufacture.

Modern communication systems are increasingly using patch or microstrip antennas, and one of the most widely quoted superconducting patch antennas is the 'H' antenna, first demonstrated in its superconducting form at Wuppertal University.[42] The antenna is shown in Figure 7.4.1. It consists of three sections of microstrip transmission line, two wide sections behaving essentially as capacitors and a thin narrow section forming an inductive section of transmission line. The structure can thus be thought of as forming a resonant circuit from short transmission line sections behaving as lumped impedances. This has to be compared with a conventional microstrip resonator, which is just a rectangular patch above a ground plane. The miniaturisation is considerable, not only from the viewpoint of the change in shape but also in the use of $LaAlO_3$ ($\varepsilon_r = 24$) as a substrate, reducing the size by a factor of around 5. For a 2.4 GHz antenna, the conventional rectangular patch would have a length of 42 mm on RT-duroid, with $\varepsilon_r = 2.2$, and 12.5 mm on $LaAlO_3$; the 'H' antenna is 6 mm in length. The coupling to the antenna is made by a coaxial feed from

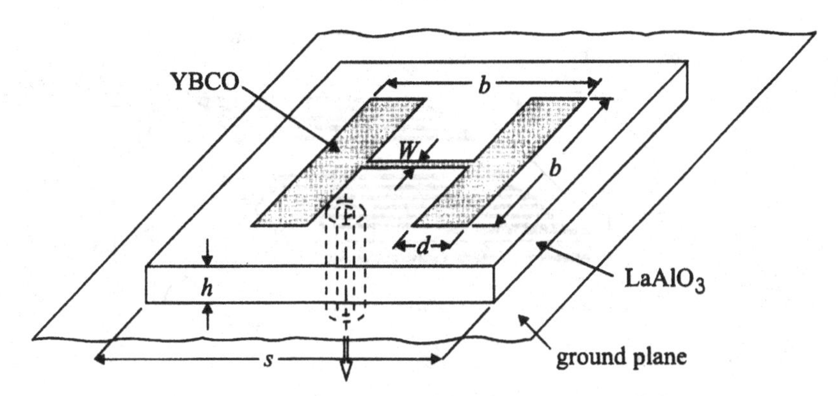

Figure 7.4.1. 'H' antenna. (Taken from reference [42].)

beneath the ground plane and can be made adjustable in order to match the 50 Ω input impedance required. Rigorous analysis of the structure has been done by a method of moments analysis in order to determine the current distribution and hence the radiation pattern.[42] The radiation is normal to the structure with a directivity of 5 dB. Experimental results indicate that the antenna has a bandwidth of about 3 MHz (0.17%) at the centre frequency of 2.45 GHz. The efficiency increases from around 2% and 5% for an all copper antenna at 300 K and 77 K, respectively, to about 60% for an all YBCO antenna. Clearly, with this type of antenna, the distance between the antenna and its ground plane is important. If the distance is small then the fields will be more tightly confined within the antenna structure, and losses in the superconductor will be more dominant. If the antenna is further away from the ground plane, then the fields are less confined and the relative amount of radiation increases. This is confirmed by numerical results in reference [42]. The bandwidth of this type of patch antenna has been improved towards the theoretical maximum by stacking an additional patch above the first. Experimental investigations were able to double the bandwidth and give an increase in efficiency by a factor of 1.5.[43] Distortion of the output signal from these antennas has been observed at high power levels.[44,45]

Another antenna for which it is possible to use this multi-resonance technique in order to increase the bandwidth is the meander-line antenna.[46] This consists simply of a microstrip transmission line in a meander shape, as shown in Figure 7.4.2.

The meander structure can be operated as an antenna with a resonant frequency corresponding to the total length of the meander line when all the bends are taken out. However, although this is a very small antenna, only one resonance is available, and this is narrow band. If the antenna is used when the

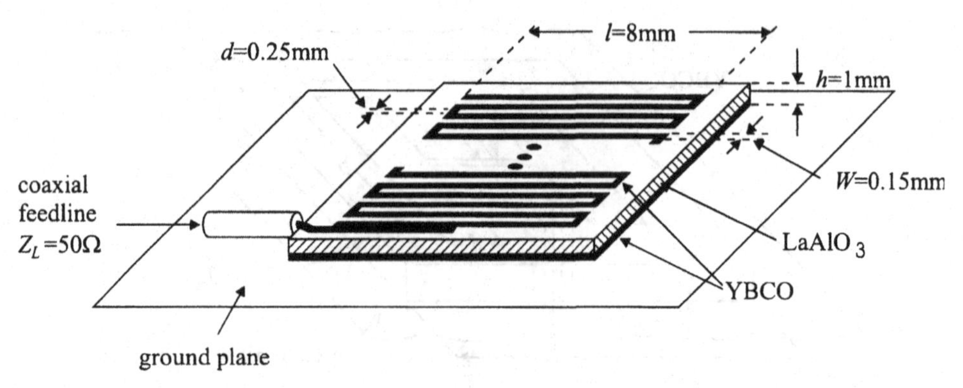

Figure 7.4.2. The meander line antenna. (Taken from reference [46].)

length of the meander-line sections are half-wavelength, then multiple resonances occur due to series sections and parallel coupling within the structure. These resonances are not quite at the same frequency, resulting in a multiple resonance structure, with the input reflection coefficient shown in Figure 7.4.3. The antenna shown in Figure 7.4.2 is $1/10$ of a wavelength long, the miniaturisation arising from the high substrate dielectric constant; it has a bandwidth of 4% and is 50% efficient. Other meander line antennas are presented in [47] although they are not discussed in the multiple resonance context.

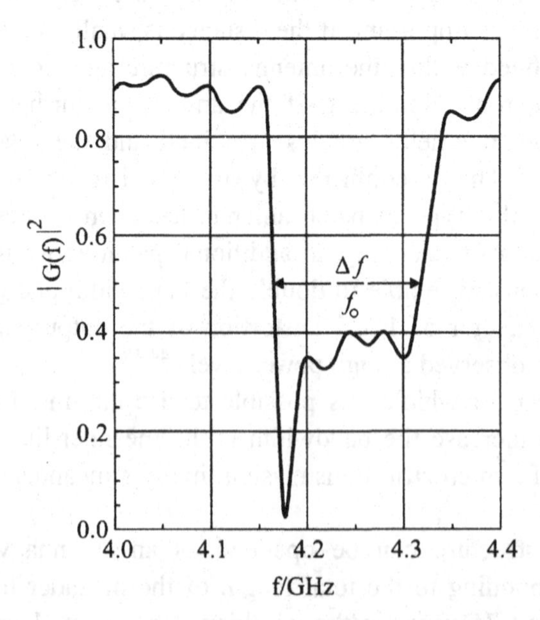

Figure 7.4.3. The measured input reflection coefficient of the meander line antenna. (Taken from reference [46].)

7.5 Transmitting antennas

Conventional antennas made from normal conductors are limited in handling high powers due to overheating of the conductors, by dielectric heating or by dielectric breakdown. However, these antennas can be designed to have a very large power handling capability. The obvious limitation of transmitting antennas when made from superconductors is the power handling capability of the superconductor itself. Increasing the power input to the antenna will eventually produce high microwave fields in the antenna or its matching circuit and will limit the output. An idea of the power handling capability can be taken directly from Equation (4.6.10), rearranged to give

$$P_{max} = \frac{\pi Z_0 l J_c^2 A^2}{2\lambda Q_l} \quad (7.5.1)$$

Here A is the cross-sectional area occupied by the current in the antenna, l is the length of the antenna and its matching network, Q_l is the loaded Q of the antenna system, Z_0 is the characteristic impedance of the antenna feed and J_c is the critical current density in the superconductor. This equation suggests that the power handling capability of the HTS thin film material is substantial. For example, a thin film with critical current of 10^7 A/cm^2 with a small antenna fitting on a 2" substrate and a line width of 1 mm, operating at 10 GHz with a loaded Q of 1000, has a maximum power handling capability of around 100 kW. This, however, is reduced to only a few watts if the line width is 10 μm. This approximation does not take current peaking on the antenna and its matching network into account, but gives only an idea of the order of magnitude of the power handling capability for antennas where the current is approximately uniformly distributed.

Figure 7.5.1 shows the power handling capability of three short dipoles as a function of critical current. Again this shows that substantial power outputs may be obtained for small antennas. As before, this calculation takes no account of any peaking in current density.

Although the non-linearity is usually associated with the problem of output power limitation, it could also be put to good effect in the antenna itself, for example using the non-linearity for mixing in a down converter or for a switching element. Such a switch element could be used to shape a beam or, alternatively, a transmit–receive switch could be designed. Cutting out the use of semiconductors has the advantage of simplifying the construction considerably and removing parasitic elements associated with the contacts. This is discussed further in Section 7.10.

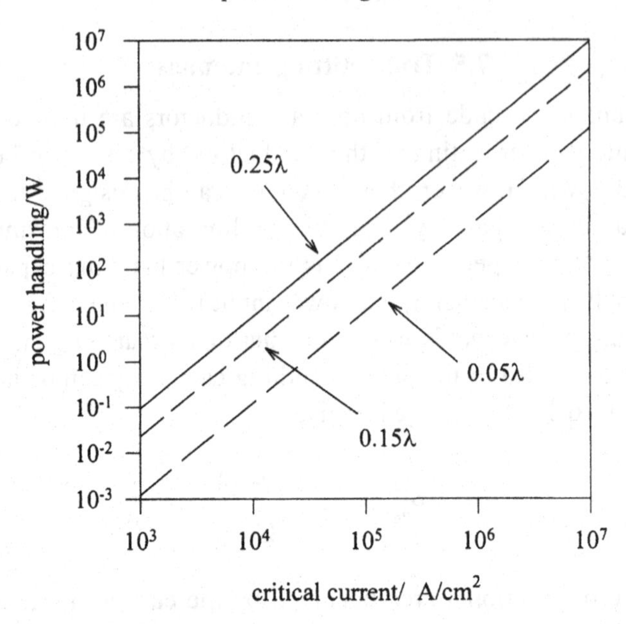

Figure 7.5.1. Power handling capability of three short-dipole antennas as a function of critical current at 1 GHz. The lengths are 0.25, 0.15 and 0.05 λ, and the frequency of operation is 1 GHz. The wire diameter a is 1 mm and the penetration depth of the bulk material is assumed to be 10 μm.

7.6 Receiving antennas

The important criterion for receiving antennas is the signal-to-noise ratio obtained at the antenna terminals (or after some pre-amplification). A highly efficient antenna may be of little or no benefit in many cases, as described below. Figure 7.6.1 shows an equivalent circuit of an antenna with equivalent noise sources. The signal received through the radiation resistance is given by[48]

$$P_s = \frac{E^2}{2\eta}\frac{G\lambda^2}{4\pi}\eta_e \tag{7.6.1}$$

Here P_s is the received signal power, E is the incident electric field, G is the directive gain and η_e is the antenna efficiency. The received noise power from external sources is

$$N_e = \eta_e k_B T_{external} B \tag{7.6.2}$$

Here k_B is Boltzmann's constant, B is the bandwidth and $T_{external}$ is the external temperature as seen by the antenna. In addition to the external noise sources, noise sources are present in the antenna structure itself. If these are assumed to

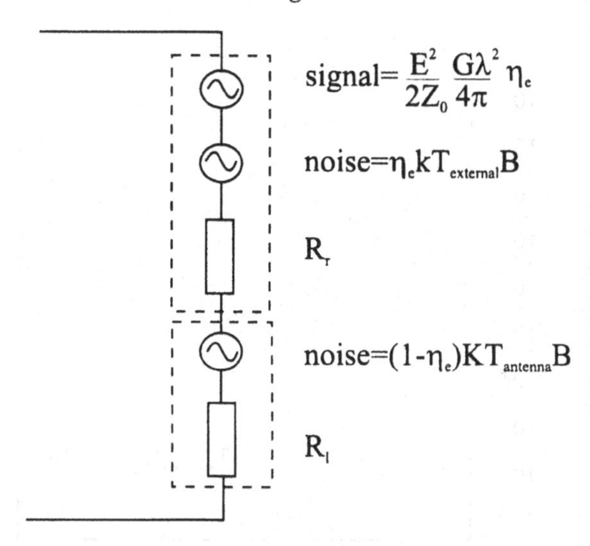

Figure 7.6.1. The antenna equivalent circuit including noise sources.

be of thermal origin, then the relative magnitude of the noise source associated with the loss resistance is

$$N_a = (1 - \eta_e)k_B T_{antenna} B \tag{7.6.3}$$

The signal-to-noise ratio of a receiving antenna can now be calculated from Equations (7.6.1), (7.6.2) and (7.6.3), and is given by

$$SNR = \frac{E^2}{2\eta}\frac{G\lambda^2}{4\pi k_B B}\left[\frac{\eta_e}{\eta_e T_{external} + (1 - \eta_e)T_{antenna}}\right] \tag{7.6.4}$$

The term in square brackets in Equation (7.6.4) is plotted in Figure 7.6.2. The terms outside the square brackets are constant with respect to signal-to-noise ratio. A number of plots are shown, each with a different value of $T_{antenna}/T_{external}$, and it is clear that for low values of $T_{antenna}/T_{external}$ the signal-to-noise is independent of efficiency unless the efficiency is very low. This is also apparent from Figure 7.6.1. The situation does, however, change if $T_{antenna}/T_{external}$ is increased and it becomes increasingly more important to have good efficiency in order to maintain the signal-to-noise ratio available at the input of the antenna.

The magnitudes of both the external and antenna temperatures depend considerably upon the environment and system in which the antenna is to be used. The mere act of cooling the antenna reduces thermal noise, but noise may not arise solely from thermal noise, especially for the case of superconductors. However, to simplify the discussion, take an antenna system being used in an outdoor communication link. Here the external temperature will be the sky temperature. This is very variable with frequency, but at frequencies less than

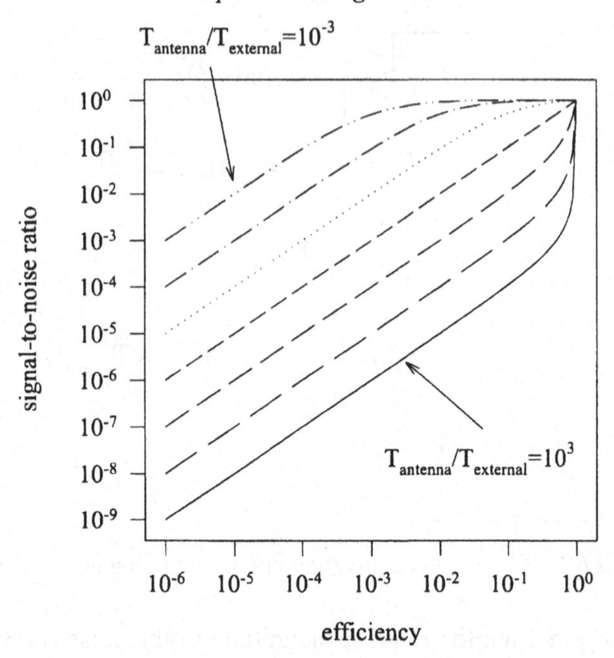

Figure 7.6.2. Signal-to-noise ratio as a function of antenna efficiency for various ratios of $T_{antenna}/T_{external}$. From 10^{-3} to 10^3 in multiples of 10 from top to bottom.

about 30 MHz it becomes very large.[49] The situation then occurs that it is not important to have an efficient antenna to increase the signal-to-noise ratio. As frequency is increased, the sky temperature is reduced and above about 500 MHz the ratio $T_{external}/T_{antenna}$ becomes of the order of unity; there will then be a significant advantage in using a HTS antenna.

The discussion given above is very general and for any particular system the properties of the receiving antenna need to be considered. For example, the antenna may not be operated where the sky temperature is dominant, or it may for certain applications be of interest to achieve a maximum signal if the signal-to-noise ratio of the system is adequate. Consideration may need to be given to the amplification stages after the antenna element and the effect of these upon the signal. However, the development discussed above does shed light on some very fundamental considerations of receiving antennas.

7.7 Two-element superdirective antenna array

The two-element superdirective array was the first demonstration of a superdirective antenna array using high-temperature superconductors.[50] In this configuration two dipoles, half a wavelength long, are placed side-by-side, as shown in Figure 7.7.1; superdirectivity can be observed as the elements draw closer together. The importance of this array is not necessarily in its practical value but, rather, in the insight it gives into the mechanisms of superdirectivity. Because of the simple nature of the array, all the calculations can be done analytically. In addition, by varying the spacing d between the dipoles the gain of the array can be changed, and this can be compared with experimental results. More complex arrays are discussed in the next section.

The two dipoles are fed with a coaxial transmission line, through a balun and a simple single stub tuner matching circuit, using a transmission line section and a discrete capacitor as shown in Figure 7.7.1. This results in the two elements having equal and opposite currents, one of the prerequisites, among others, for superdirectivity. It is possible to analyse this antenna by considering it as a radiating transmission line. However, although this is simpler, the more general analysis of considering it as an array of coupled dipoles will be considered here. This allows much more understanding of superdirectivity which can then be applied to more complex antenna systems.

Much of the analysis given below is summarised in Kraus,[51] who first studied arrays of closely spaced dipoles experimentally in 1937. He used the theoretical work of Brown.[52] At that time the use of superconductors to increase the gain of the dipoles was not considered. For a single dipole of length $\lambda/2$ the electric field in the far field can be written

$$\mathbf{E}_d = I_d \mathbf{K}(r, \theta) e^{-jk_0 r} \tag{7.7.1}$$

where I_d is the peak current on the antenna and

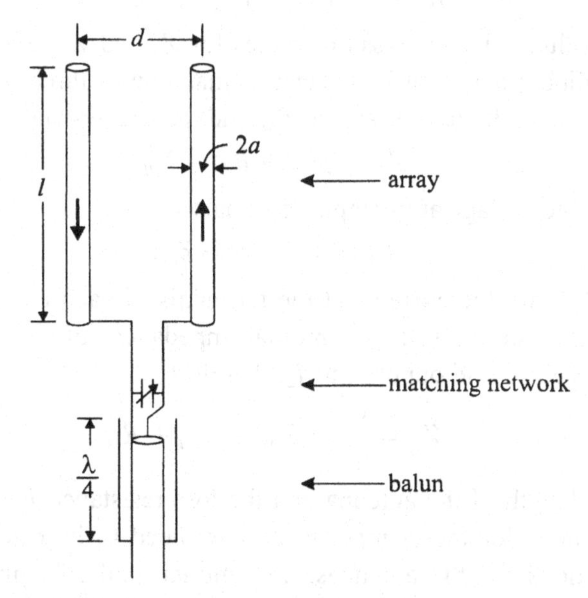

Figure 7.7.1. Two-element superdirective antenna array.

$$\mathbf{K}(r,\theta)=\frac{j\eta}{2\pi r}\frac{\cos(\pi/2\cos(\theta))}{\sin(\theta)}\mathbf{a}_\theta e^{j\omega t} \quad (7.7.2)$$

If the input impedance of the antenna is Z_d, then the power dissipated is given by

$$P_d=\frac{1}{2}\operatorname{Re}(Z_d)I_d^2 \quad (7.7.3)$$

So Equation (7.7.1) can be written

$$\mathbf{E}_d=\mathbf{K}(r,\theta)\sqrt{\left(\frac{2P_d}{\operatorname{Re}(Z_d)}\right)}e^{-jk_0 r} \quad (7.7.4)$$

Now consider the two-element array of Figure 7.7.1. The electric field in the far field is given by

$$\mathbf{E}_a=\mathbf{E}_{d1}\exp(jk_0 d\cos(\theta))+\mathbf{E}_{d2} \quad (7.7.5)$$

If the currents on the antennas are 180° out of phase, then $\mathbf{E}_{d1}=-\mathbf{E}_{d2}$. This gives a total electric field in the far field of the array of

$$\mathbf{E}_a(r,\theta,\phi)=2jI_d\mathbf{K}(r,\theta)\sqrt{\left(\frac{P_d}{\operatorname{Re}(Z_a)}\right)}\exp(-jk_0 r)\exp(-jk_0 d\cos(\phi)/2)$$
$$\times\sin(k_0 d\cos(\phi)/2) \quad (7.7.6)$$

Here Z_a is the input impedance of the array and P_a is the power dissipated in the array. The gain of the array over the single dipole is thus given by

$$G(\phi)=\frac{|\mathbf{E}_a|^2}{|\mathbf{E}_d|^2}=\frac{2\operatorname{Re}(Z_d)}{\operatorname{Re}(Z_a)}\sin(k_0 d\cos(\phi)/2) \quad (7.7.7)$$

In order to evaluate this expression, values for Z_a and Z_d need to be obtained. For a single dipole the input impedance is made up of three parts; the antenna reactance X_d, the radiation resistance R_{dr} and the loss resistance R_{dl}, where

$$Z_d=jX_d+R_{dr}+R_{dl} \quad (7.7.8)$$

For the array, the voltage at the input terminals V_1 is

$$V_1=I_1 Z_{as}+I_2 Z_{am} \quad (7.7.9)$$

where I_1 and I_2 are the currents at the terminals of each element in the array and Z_{as} and Z_{am} are the self- and mutual impedances of the antenna element. However, I_1 is equal and opposite to I_2; therefore

$$Z_a=\frac{V_1}{I_1}=Z_{as}-Z_{am}+R_{al} \quad (7.7.10)$$

Here l is the length of the antenna and the loss resistance R_{al} has been added arbitrarily. Values for these impedances now need to be calculated from the field distributions for the antennas. The mutual and self-impedances of the antenna are given by Kraus[51] as

$$Z_{s,m} = \frac{I'_{1,2}}{I_1 I_{1,2}} \int_0^l E_{(1,2)}(z) \sin(k_0 z)\, dz \qquad (7.7.11)$$

The subscripts refer either to the self- or to mutual impedances, $I'_{1,2}$ is the peak current on element 1 or element 2 and $I_{1,2}$ is the terminal current at the antennas. $E_{(1,2)}(z)$ is the electric field in the z-direction at the antenna, which is integrated over the antenna length l. Equation (7.7.11) is general for any feed point to the antenna. The feed point under consideration has a current null and a voltage peak. This results in Z_a approaching infinity. However, because we are calculating the ratio of impedances in Equation (7.7.7), the null currents cancel and the expression reduces to the same as the ratio of impedances with any feed point. Conveniently, these calculations of the fields and the integration in Equation (7.7.11) have been calculated previously. The self- and mutual impedances are a centre fed, thin lossless half-wavelength dipole and dipole array are given by[51]

$$Z_d = \frac{\eta}{4\pi}(\mathrm{Cin}(2\pi) + j\,\mathrm{Si}(2\pi)) = 73 + j42.5 \qquad (7.7.12)$$

$$\begin{aligned} Z_a = \frac{\eta}{4\pi}\{2\,\mathrm{Ci}(k_0 d) - \mathrm{Ci}(k_0(\sqrt{(d^2+l^2)} + l)) - \mathrm{Ci}(k_0(\sqrt{(d^2+l^2)} - l))\} \\ + j\frac{\eta}{4\pi}\{2\,\mathrm{Si}(k_0 d) - \mathrm{Si}(k_0(\sqrt{(d^2+l^2)} + l)) - \mathrm{Si}(k_0(\sqrt{(d^2+l^2)} - l))\} \end{aligned} \qquad (7.7.13)$$

where Si, Ci and Cin are sine and cosine integrals.[53] Using these expressions in Equation (7.7.7), for the case of a $\lambda/2$ dipole, the gain of the superdirective array can be calculated. This is shown in Figure 7.7.2.

As expected, Figure 7.7.2 shows that the lossless dipole array has a gain improvement as the spacing decreases. The improvement is several decibels over the gain of a single dipole. As the loss resistance is introduced, the gain at small spacing is rapidly reduced, but is still quite reasonable at the higher spacing. Loss resistances for a copper antenna can be of the order of several ohms, whereas loss resistance for HTS bulk material at around 1 GHz and 77 K can be of the order of a milliohm. Figure 7.7.3 shows the power gain as a function of direction ϕ. The gain can again be seen to increase as the separation reduces. Even with this simple array the advantage of increased gain is observed when using low-loss conductors in its construction and having small spacing. However, the penalty that must be paid is the increase in the Q of the antenna, given by

$$Q = \frac{\mathrm{Im}(Z_a)}{\mathrm{Re}(Z_a)} \qquad (7.7.14)$$

which is plotted in Figure 7.7.4.

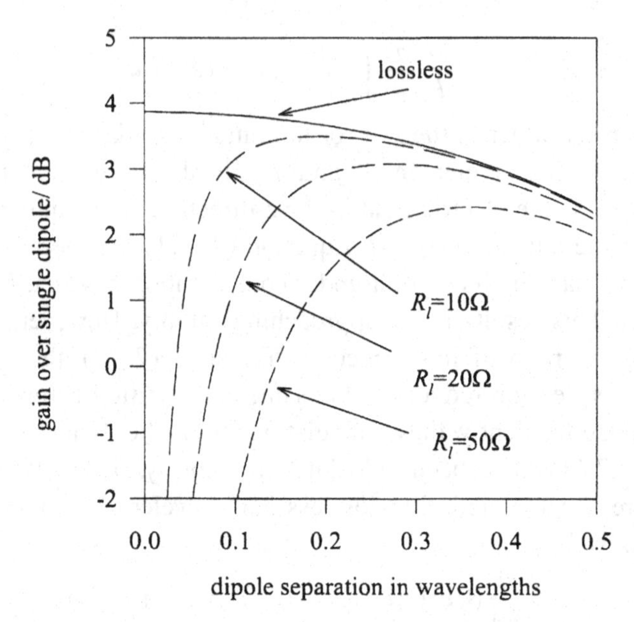

Figure 7.7.2. Gain of a two-element dipole array at angle $\phi = 0$, over that of a single dipole.

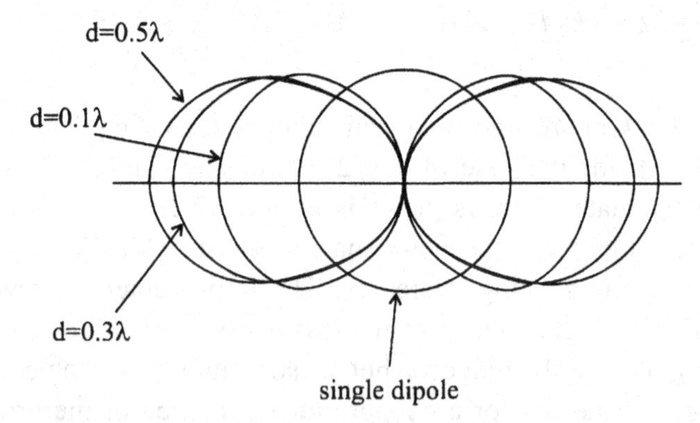

Figure 7.7.3. Gain as a function of direction for the two-element dipole array. The spacing between the elements is $d = 0.5$, 0.3 and 0.1, with increasing gain. The central circle is the corresponding gain for a single dipole.

Figure 7.7.4 demonstrates the dramatic increase in the quality factor as the dipole spacing becomes small. If the loss resistance is finite, then the Q is reduced; this is at the expense of gain. This increase in Q is due to a relative increase in the localised stored energy compared with the radiated energy and is apparent in the values of the antenna reactance and radiation resistance as the spacing becomes small.

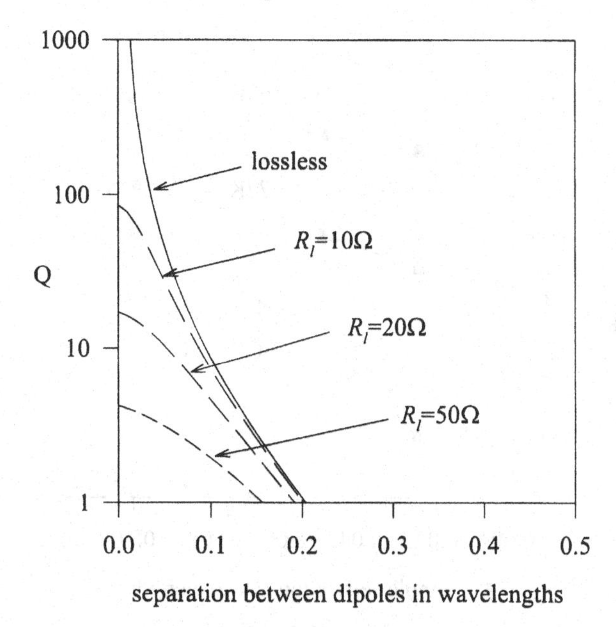

Figure 7.7.4. The Q of a two-element dipole array as a function of separation of the dipoles.

Because of the relatively simple nature of this dipole array, experimental work on the gain of both copper and superconducting antennas as a function of separation is possible. The results of such work are shown in Figure 7.7.5; further details of the experiment are given in reference [50]. The array has been constructed from both copper and thick film YBCO wire. The YBCO consisted of a zirconia rod coated with YBCO, giving a total wire diameter of 1 mm. Three arrays were made from both copper and YBCO, the YBCO tested at 77 K and the copper at 300 K and 77 K. In each case the antennas were matched to 50 Ω and a sleeve balun was used. The antennas were operated at a frequency of 1 GHz.

The results shown in Figure 7.7.5 agree qualitatively with those shown in Figure 7.7.2; as expected, more benefit is gained when small-element spacing is used. However, detailed quantitative comparisons have not been made since the matching network also provides significant improvement in gain for the superconducting antenna.

It is possible to describe qualitatively what is happening as the two dipoles become superdirective. Because the elements have currents 180° out of phase, as the dipoles come closer the radiation field tends to cancel out, leading to large reactive fields in the near zone. However, as long as the dipoles are separated to some extent there is never full cancellation of the field, and the

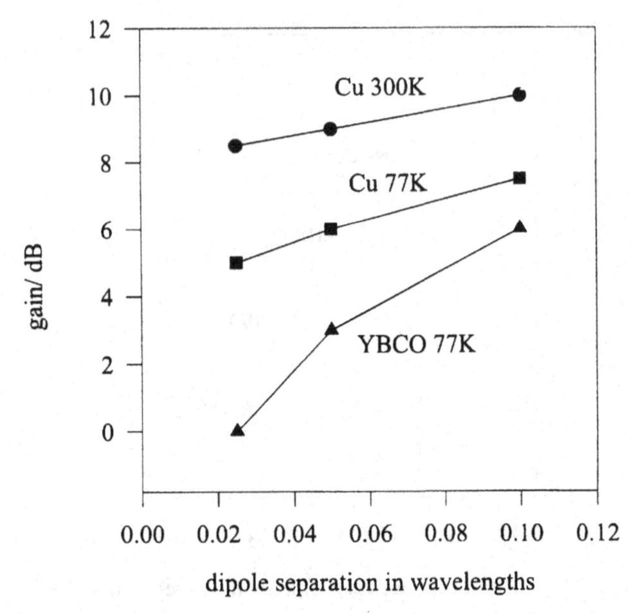

Figure 7.7.5. Measured gains of a YBCO array over a copper array.

directivity of this radiation increases slightly as the two dipoles draw closer. It is possible to maintain the same output power as the antenna elements draw closer by increasing the input power to the array. However, this leads to greater losses and reduced efficiency. The increase of Q is due to the increase in the reactively stored energy in the near field.

7.8 Superdirective antenna arrays

It is possible to design an antenna array with arbitrarily small dimensions with a directivity as high as desired, as discussed in Section 7.2. This is rather surprising, as most people imagine larger arrays to have higher directivities and, in particular, arrays which are less than a wavelength to have low directivities. This is because the conventional conception of a source of waves assumes approximately a uniform excitation across an aperture. If the current distribution is not uniform, then any directivity can be obtained from an antenna array of any size. There have been numerous definitions of super-directivity; for example:[54] superdirectivity occurs when a directivity is higher than would be obtained with the same antenna configuration which is uniformly excited (constant amplitude and phase). Although this is a workable definition, it does not give an impression of how the antenna improves over a conventional array, where the elements may be spaced by a distance of the order of $\lambda/2$. The definition used here will be that a superdirective antenna is one which has a directivity larger than that of a normal antenna. A normal antenna is one defined in Section 7.2 in terms of spherical modes, and has a maximum directive gain of

$$G_n = (k_0 a)^2 + k_0 a \tag{7.8.1}$$

where a is the radius of the smallest sphere which can enclose the antenna and k_0 is $2\pi/\lambda$. This definition allows a numerical factor to be defined for the superdirectivity as

$$S_d = \frac{G}{G_n} \tag{7.8.2}$$

where G is the directive gain of the antenna of interest. This definition also gives a numerical figure of supergain which is related to our normal impression of gain for conventional antennas and arrays. In the past, the term 'supergain' has sometimes been used in the literature to mean what is defined here as 'superdirective'. The term 'supergain' will be used only when the efficiency of the antenna is included, so that the supergain factor is

$$S_g = S_d \eta_e \tag{7.8.3}$$

where η_e is the efficiency of the antenna of interest. This definition is quite stringent as Equation (7.8.1) gives the *maximum* normal gain.

Inherent limitations with superdirectivity are (i) an increase in antenna Q, or a reduction in bandwidth, (ii) a reduction in radiation efficiency, (iii) difficult tolerances to meet in construction, and (iv) susceptibility to random variations in element excitation. This has led many authors to dismiss it as impracticable. However, as noted in Section 7.2, there exist situations where superdirectivity could be useful if antennas are operated in regimes of moderate superdirectivity, reasonable Q and high efficiency. Superconductors help enormously in widening these regimes of potential application.

Since their discovery by Oseen[55] in 1922, the attraction of very high gain antennas has led to a number of interesting theoretical investigations, as well as to some interesting demonstrations of real antenna arrays. Not knowing of the work of Oseen, some workers made a few early attempts to calculate the optimum current distribution for antennas,[56–58] when of course none exists. Other early work, prior to 1950, calculated current distributions for particular superdirective arrays, pointing out the fact that these are associated with large current amplitudes, and hence low efficiencies.[59,60] For example, Yaru[61] discusses a nine-element broadside array of dipoles with a total length of $\lambda/4$, giving a directivity of 8.5 times that of a single dipole. However, he points out that it has an efficiency of $10^{-14}\%$, with a required stability in the excitation frequency of $10^{-11}\%$! Some of the interesting observations of this early work were that for an endfire array the limit of the gain as the spacing between the elements approaches zero[62] varies as n^2, and the equivalent limit for a broadside array[63] is $(2n+1)/\pi$. Therefore, the endfire array is more attractive for superdirectivity for a fixed number of elements.

Because there is no optimum current distribution, the theoretical treatments usually put some other constraint on the antenna, and then try to obtain the optimum current distribution, or at least find the directive gains of the antenna as a function of some parameter. A number of constraints and parameters have been investigated and these are discussed briefly below.

Assuming a linear array of antenna elements, it is possible to calculate the optimum current excitation as a function of the separation between the elements. The actual elements remain general and are specified by their mutual resistance matrix in the work performed by Bloch.[64] He calculates the maximum gain for a linear antenna array with n elements and the element positions specified by d_l as

$$G_{max} = \sum_{m=1}^{n} \sum_{l=1}^{n} g_{lm} \cos (k_0 d_m - k_0 d_l) \tag{7.8.4}$$

where g_{lm} is the inverse of the mutual resistance matrix. A number of different antennas have been investigated in the course of this work.[65] Other authors have also looked at the problem of optimising linear arrays with respect to spacing[63,66,67] and also to two-dimensional arrays.[68]

Because of the large increase in Q with superdirective antenna arrays it is important to keep track of the Q in the antenna design; this is not considered by many authors. However, it is possible to investigate arrays where the Q factor is a constraint; this has been considered by Lo.[69] In addition, Lo also points out that the optimum current excitation for antennas less than $\lambda/2$ in length is in anti-phase, that is each successive element is 180° out of phase with the next. The amplitude of the current in the optimum array is also strongly tapered towards the ends of the array. The reason for superdirective arrays having optimum current distribution with this anti-phase excitation is discussed at the end of Section 7.7, together with a qualitative description of their operation. In addition to looking at the Q-factor, Lo also considers the signal-to-noise ratio of a receiving array.

Another problem with superdirective arrays is their susceptibility to random variations in the element positions or errors in the excitations. Thus, another possible constraint on the analysis is the susceptibility of the arrays to these electrical and mechanical tolerances. Attempts to find good superdirective arrays that are less sensitive to fabrication errors have been undertaken by a number of authors.[70–72]

There has been great interest in the theoretical aspects of superdirectional arrays, some of which are discussed briefly above. However, this has not led to the construction of many practical examples or experimental demonstrations. The first demonstration of a superdirectional array was by Bloch in 1953.[64] He constructed an array from four $\lambda/2$ dipoles designed for endfire operation. The dipoles were spaced equally along a length of 0.6λ. The array operated at 75 MHz and showed a gain of 8.7 dB over a single dipole, and a bandwidth of around 2%. The calculated theoretical gain was 10.1 dB. The difference between the theory and the experiment was attributed to losses in the feeder, dipoles and matching networks. Since then there have been a number of demonstrations of superdirective antennas using normal conductors,[72–79] but the first superconducting one was demonstrated by Walker in 1977.[80] This antenna array consisted of two loops of superconducting lead, each with a loop area of about 1 cm^2; the loops were matched through a cavity resonator. The experimental results showed a gain improvement of 24 dB between room temperature and the superconducting state at 4.2 K. The low-temperature Q of the arrangement was 6200. Another low-temperature superconducting antenna array was designed by Adachi in 1980.[81] The first HTS antenna was demonstrated in 1991. This two-dipole array is discussed in Section 7.1 and will not be considered further here. A number of other HTS arrays have been demonstrated showing several decibels gain over similar antennas made out of copper.[82–84]

Modern numerical techniques and high-speed computers allow optimisation of gain by feedback around a numerical algorithm. Four-[85–87] and sixteen-element[88] printed dipole arrays made from HTS have been demonstrated by this design technique. In this design the number of dipoles is first chosen, together with the feed network configuration. Given the constraint of substrates of a fixed size, the optimum position of the dipoles to achieve maximum directive gain is found using repeated iterations of the space domain method of moments.[89]

Figure 7.8.1 shows the 16-element superdirective array designed using the space domain method of moments. It consists of 16 dipoles, each of length about $\lambda/15$, printed on to the surface of a 1-mm-thick zirconia ($\varepsilon_r \sim 25 \tan\delta \sim 5 \times 10^{-3}$) substrate. One arm of the dipoles is printed on the upper surface of the substrate whilst the other is on the bottom. This enables convenient feeding using a 36 Ω parallel strip transmission line with one conductor on the upper-side of the zirconia and the other on the lower. Alternate dipoles are fed 180° out of phase, as can be seen in Figure 7.8.1. The

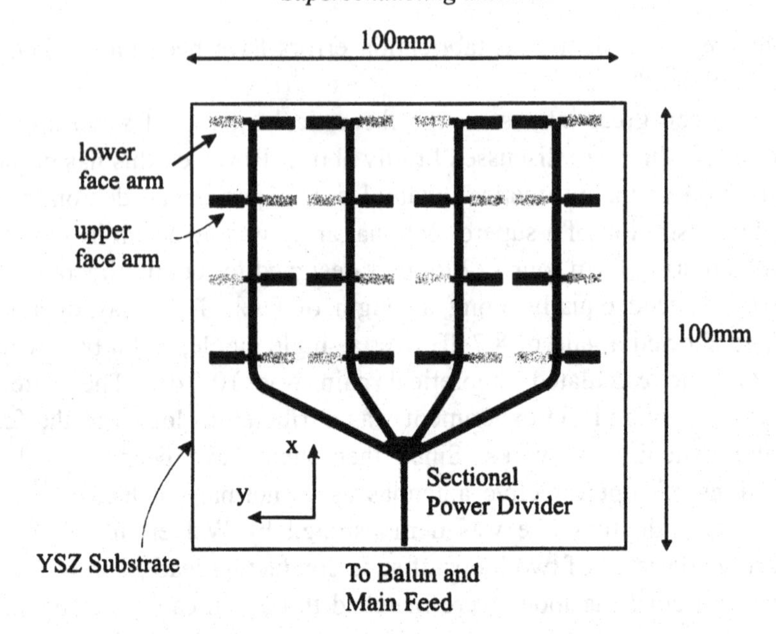

Figure 7.8.1. 16-element superdirective array. Black represents the YBCO printed on the upper side of the zirconia substrate, grey is YBCO on the lower side.

array consists of four series-fed dipoles connected in parallel with a sectional power divider[90] used to feed the configuration. The whole antenna is fed by a 45 Ω, $\lambda/4$ impedance transformer connected in parallel with a tuning capacitor. A balun is used external to the antenna during experiments to avoid spurious radiation, although in later designs a planar balun is integrated with the design.[85] The whole array is fitted on a substrate 100 mm square, representing 0.38λ at the resonant frequency of 1.14 GHz.

The measured radiation patterns of the array are shown in Figure 7.8.2. The maximum gain is at 45° to the axis of the antenna, as shown in Figure 7.8.2. Table 7.8.2 shows some of the performance parameters of the array.

Table 7.8.2. *Experimental performance parameters for the 16-element printed superdirectional array*

	Size	Directivity (experiment)	Directivity (theory)	Frequency	Bandwidth	S_g
16-element array	0.38λ	7.1 dBi (HTS) 3.9 dBi (Ag, 77 K)	11.5 dBi	1.14 GHz	0.7% (HTS)	4.2 (HTS) 1.9 (Ag, 77 K)

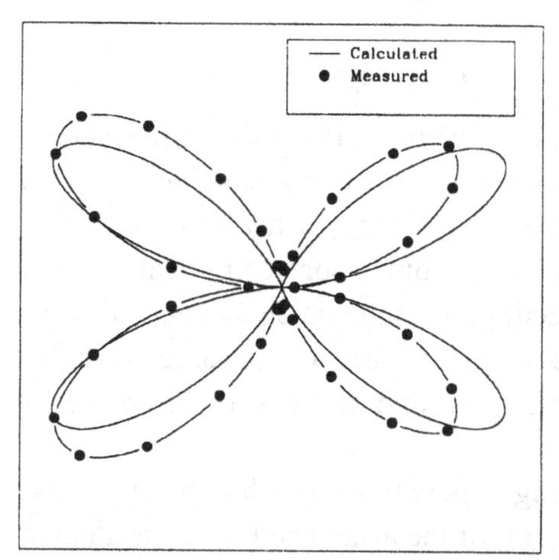

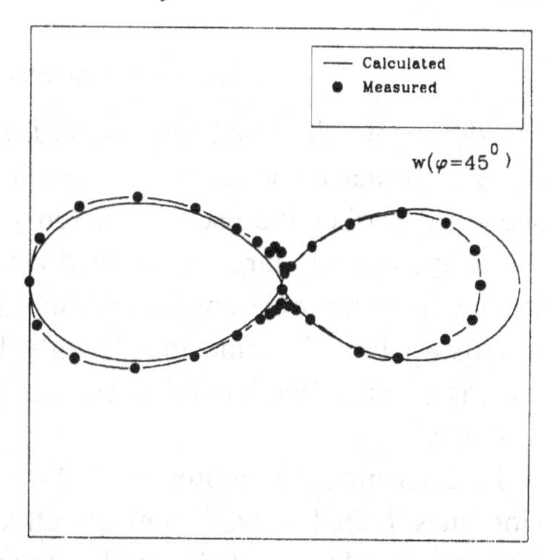

Figure 7.8.2. Radiation pattern of the 16-element superdirective array shown in Figure 7.8.1. (*a*) Radiation in the plane of the array; the main radiation is 45° to the axis of the antenna. (*b*) Radiation in the plane normal to (*a*) and around the maximum gain direction shown in (*a*). Experimental results are shown as black dots.

It can be seen that the performance is good with a directivity of 7.1 dBi,[1] and a gain of 3.2 dB over a silver antenna of the same design. This is less than the theoretical prediction due to losses in the feeding and matching system owing to the relatively high-loss dielectric used as a substrate. The supergain ratio defined by Equation (7.8.3) is larger than one, even for the supercooled silver antenna array. Other supergain antennas developed using the same design technique are discussed in references [85–87].

This section has described the wealth of work done on superdirective arrays in the past, and the many design approaches considered. A number of superdirective antennas have been constructed from normal conductors, low-temperature superconductors and high-temperature superconductors, many showing improved properties over normal arrays. The problem with the superdirective arrays is that the synthesis is difficult; for example, if an application requires a particular beamwidth, bandwidth and directive gain, how is this antenna designed? With conventional arrays, the problem is very much simpler than that of a superdirective array, and this still remains one of the major factors in the application of superdirectional arrays. Superconductors help in widening the areas where superdirectional antennas are practical, but are by no means the limiting factor in their widespread application.

To complete the bibliography on superdirective antennas, further works are listed, but not specifically referred to here, in references [91–102].

[1] dBi is the gain in decibels with reference to an isotropic antenna.

7.9 Conventional antenna arrays

In addition to the small and superdirective antenna arrays discussed above, there are other antennas which benefit from the use of superconductors. This section describes the use of superconductors in antenna array feed networks. As arrays become large, their feed networks become more complicated. This leads to long narrow transmission lines feeding the antenna elements, and these microstrip lines become increasingly lossy as the size of the antenna array increases. Superconductors overcome this loss, and make very large arrays possible.

To determine the improvement by using superconductors for the transmission lines in feed networks, the directive gain of the array needs to be evaluated and then the losses of the feed network included. Take the two-dimensional array of microstrip square patches shown in Figure 7.9.1. The approximate directive gain of such an array, with a uniform current amplitude distribution on each element of the array, is given by[103]

$$G = \frac{4\pi A_e}{\lambda^2} \tag{7.9.1}$$

where A_e is the effective aperture of the array, taken here as the area occupied by the array. The efficiency of each individual element is assumed to be high. The loss associated with the feeding network can be approximated by assuming a wide microstrip transmission line with a loss (ignoring field penetration into the superconductor), determined in Section 2.3, given by

$$\alpha = \frac{R_s}{hZ_0} \tag{7.9.2}$$

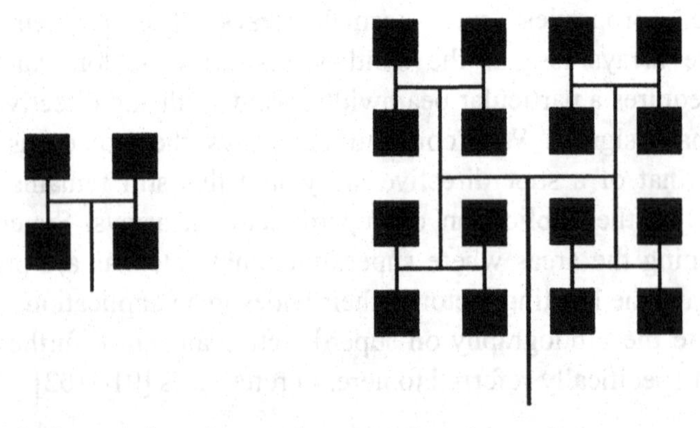

Figure 7.9.1. A four- and a sixteen-element microstrip antenna array with the feed networks.

where h is the substrate thickness. This is the loss down each of the feed lines, so the total gain referenced to the array terminals is given by

$$G = \frac{4\pi A_e}{\lambda^2} \exp\left(-\frac{2R_s L}{h Z_0}\right) \tag{7.9.3}$$

where L is the total length of a single feed line from the antenna input to the element.[104] Assuming that the patches are spaced by $\lambda/2$, then the area of the antenna is $(N\lambda/2)^2$, where N is the number of patches on each side. The length of the antenna feed can also be derived in terms of the number of elements on each side of the antenna; for the antenna shown in Figure 7.9.1, this is

$$L = \sum_{i=0}^{\log_2(N)-1} 2^{n-1}\lambda = \frac{\lambda}{2}(N-1) \tag{7.9.4}$$

Hence the gain can be written as

$$G = \pi N^2 \exp(-\alpha\lambda(N-1)) \tag{7.9.5}$$

Figure 7.9.2 shows the gain using Equation (7.9.5) at the array input port of a square antenna array at a frequency of 20 GHz. Results for copper and superconducting feed networks are shown. It can be seen that as the length of

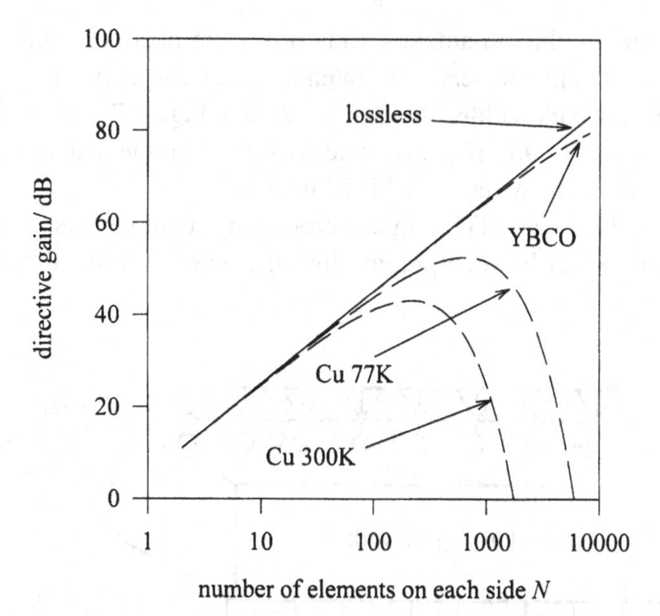

Figure 7.9.2. The directive gain of a square antenna array with the elements spaced by $\lambda/2$ as a function of the number of elements on each side. The surface resistance of the copper and HTS are given in Appendix 1. The frequency of operation is 20 GHz, the thickness of the substrate is 0.2 mm and the line impedance 30 Ω.

the feeding network increases, due to the increase in size of the antenna, then the losses eventually dominate and the gain reduces. The superconductors maintain the gain for larger-size arrays. The values of surface impedance used for copper in Figure 7.9.2 are based on ideal material given in Appendix 1; in actual practice the losses may be slightly higher, due, for example, to surface roughness.

It is obvious that the use of superconductors is only of importance if the number of elements is large. However, when a large number of elements are required, then the use of HTS will make it possible to use this type of antenna, whilst without HTS it would have been a practical impossibility.

A number of antenna arrays of this type have been constructed, mainly with a low number of elements to test the technology. Four-,[105,106] sixteen-,[107,108] and sixty-four-[104] element arrays have been demonstrated and, as expected, the increase in gain is small over a similar antenna made out of normal conductors. However, the work has highlighted the technology in the design, cooling and packaging of such antennas and paves the way for the more complex antenna arrays described below.

7.10 Arrays with signal processing capability

Signal processing within an antenna array or system may be required to shape the antenna beam and/or scan the antenna beam for each of a number of discrete-frequency bandwidths; this is illustrated in Figure 7.10.1. The potential for achieving one or all of these functions in a single antenna system is improved considerably by the use of HTS material.[109,110]

To achieve this, the HTS signal processing components described in Chapters 5 and 6 can be incorporated into the antenna system itself. Such a

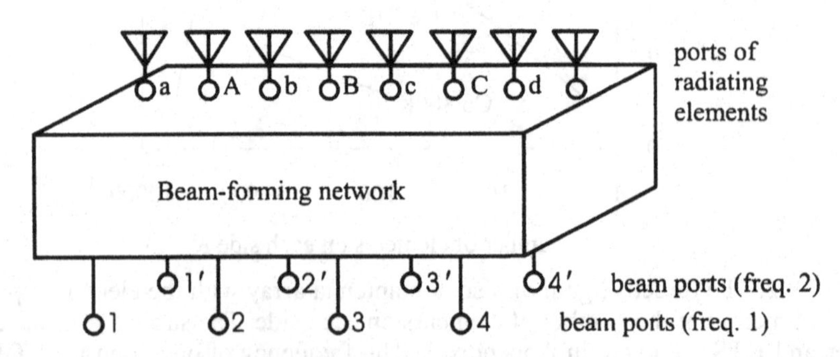

Figure 7.10.1. Generalised antenna array. (Taken from reference [109].)

system can form particular beams at particular frequencies by the linear combination of the output signals from each of the antenna elements. Such antennas may have a number of different ports, each referring to a number of multiple beams, beam patterns or frequencies. To form such a system, networks consisting of delay lines, couplers and possibly filters will have to be used, the layout complexity increasing as the number of functions is increased. It is also possible to incorporate adaptive functions. For beam scanning arrays phase shifters are required, as described in Section 5.10; if a different phase shift is applied to each antenna element, then the beam can be scanned by the application of control signals. For beam shaping, antenna elements need to be switched into or out of the array; this can be accomplished by a HTS switch as discussed in Section 5.9. Arrays tracking targets by placing the beam at an appropriate point in space, or reducing interference by placing nulls at appropriate positions, can be designed. These can be controlled by digital systems or by direct analogue feedback. Some such systems are also appropriate for normal conducting antenna systems, but the use of HTS, with its miniature components which can be integrated, allows much more sophisticated systems to be conceived with greatly reduced size.

With conventional antenna patch arrays the size of the patches is around $\lambda/3$ and they operate at a single frequency. However, if a small antenna (say the 'H' antenna) is used instead of the conventional patch, then space is available. This could be used for the signal processing components or, alternatively, for another small patch. A separate array of a different frequency can be interleaved on the first; in fact, many such arrays can be interleaved, all operating at different frequencies. Each of the small-antenna elements has an inherently narrow band, so provided that the other close elements are out of this band, coupling will be small and the arrays will operate independently. This is a channelising antenna system, with each frequency at the input emerging from a separate antenna port. Switches can be placed in any of the channels to suppress interference at one frequency. The total bandwidth of all the channels will be the total of all the antenna element bandwidths and will not be limited by the use of the small antennas.

Once the HTS signal processing components are available, then the potential for complex systems integrated with antennas is large and many applications may arise for frequencies above several gigahertz. The low loss of HTS makes the integration of the components feasible.

An example of a multi-function array is given by Chaloupka;[109] it consists of two 50-mm-diameter $LaAlO_3$ substrates on top of each other, forming a multi-frequency, multi-beam array, and has three ports. Port 1 functions as a 4.5 GHz array forming a beam perpendicular to the substrate with a half-power beamwidth of 45° and a bandwidth of 4%. Port 2 is at 9.5 GHz, with the beam pointing in the same direction as port 1, this time with a 25° beamwidth, 16 dB gain and 5% bandwidth. Port 3 is the same as port 2, except that the beam exhibits a null in the direction of the beams in ports 1 and 2. The elements included in the array are meander antennas (discussed above), quarter-wave tapered impedance transformers, and a branch line coupler with two 90° phase elements. All this is contained within an area which is less than a wavelength square at the lower operating frequency.

7.11 References

1 Dinger R. J., Bowling D. R. and Martin A. M. A survey of possible passive antenna applications of high-temperature superconductors, *IEEE Trans. on Microwave Theory and Techniques*, **39**(9), 1498–1507, 1991

2 Williams J. T. and Long S. A. High temperature superconductors and their application in passive antenna systems, *IEEE Antennas and Propagation Magazine*, pp. 7–18, Aug., 1990

3 Hansen R. C. Superconducting antennas, *IEEE Trans. on Aerospace and Electronic Systems*, **26**(2), 345–55, 1990

4 Hansen R. C. Antenna applications of superconductors, *IEEE Trans. on Microwave Theory and Techniques*, **39**(9), 1508–12, 1991

5 Chaloupka H. J. HTS antennas, *European Applied Superconductivity Conf. Göttingen, Germany, Oct. 1993*, DGM Informationsgesellschaft-verlag, ed. Freyhardt H. C., pp. 983–90, 1993

6 Harrington R. F. Effect of antenna size on gain, bandwidth and efficiency, *J. Research of the National Bureau of Standards-D. Radio Propagation*, **64**D(1), 1–12, 1960

7 Harrington R. F. *Time-harmonic Electromagnetic Fields*, McGraw-Hill, 1961

8 Marcuvitz N. *Waveguide Handbook*, Peter Peregrinus Ltd, London, UK, 1986

9 Chu L. J. Physical limitations on omni-directional antennas, *J. Appl. Phys.*, **19**, 1163–75, 1948

10 Chaloupka H. J. High-temperature superconductor antennas: utilisation of low rf losses and non-linear effects, *J. Superconductivity*, **5**(4), 403–16, 1992

11 Wheeler H. A. Small antennas, *IEEE Trans. on Antennas and Propagation*, **24**(4), 462–9, 1975

12 Wheeler H. A. Fundamental limitations on small antennas, *Proc. IRE*, **35**, 1479–84, 1947

13 Hansen R. C. Fundamental limitations on antennas, *Proc. of IEEE*, **69**(2), 170–82, 1981

14 Pues H. F. and van De Capelle A. R. An impedance-matching technique for increasing the bandwidth of microstrip antennas, *IEEE Trans. on Antennas and Propagation*, **37**(11), 1345–53, 1989

15 Fano R. M. Theoretical limitations on the broad band matching of arbitrary impedances, *J. Franklin Inst.*, **249**(1), 57–83, 1950

16 Fano R. M. Theoretical limitations on the broad band matching of arbitrary impedances, *J. Franklin Inst.*, **249**(2), 139–54, 1950

17 Lancaster M. J., Wu Z., Maclean T. S. M. and Hunag Y. Supercooled and superconducting small-loop and dipole antennas, *IEE Proc.*, H**139**(3), 264–70, 1992
18 Kraus J. D. *Antennas*, McGraw-Hill, London, 1950
19 Johnson R. C. and Jasik H. *Antenna Engineering Handbook*, McGraw-Hill, 1984
20 Collin R. E. *Antennas and Radiowave Propagation*, McGraw-Hill, 1985
21 Khamas S., Cook G. G., Kingsley S. P. and Woods R. C. Significance of matching networks in enhanced performance of small antennas when supercooled, *Electron. Lett.* **26**(10), 654–5, 1990
22 Ivrissimtzis L. P., Lancaster M. J., Maclean T. S. M. and Alford N. McN. On the design and performance of electrically small printed thick film $YBa_2Cu_3O_{7-x}$ antennas, *IEEE Trans. on Applied Superconductivity*, **4**(1), 33–40, 1994
23 Ivrissimtzis L. P., Lancaster M. J., Maclean T. S. M. and Alford N. McN. Efficiency characterisation of small printed thick film $YBa_2Cu_3O_{7-x}$ antennas, *23rd European Microwave Conf. Proc.*, pp. 906–9, 1993
24 Wheeler H. A. The radiansphere around a small antenna, *Proc. IRE*, **47**, 1325–31, 1959
25 Newman E. H., Bohley P. and Walter C. H. Two methods for the measurement of antenna efficiency, *IEEE Trans. on Antennas and Propagation*, **AP-23**(4), 457–61, 1975
26 Pozar D. M. and Kaufman B. Comparison of three methods for the measurement of printed antenna efficiency, *IEEE Trans. on Antennas and Propagation*, **AP-36**(1), 136–9, 1988
27 Walker G. B. and Haden C. R. Superconducting antennas, *J. Appl. Phys.*, **40**(1), 2035–9, 1969
28 Moore J. D. and Travers D. N. Radiation efficiency of a short cryogenic antenna, *IEEE Trans. on Antennas and Propagation*, **AT-14**, 246, 1966
29 Adachi S., Onuki S., Kimura M. and Onodera Y. An experiment on superconducting antennas, *Trans. IECE Japan*, **J59-B**(5), 299–300, 1976
30 Adachi S., Mohri Y. and Ohnuki S. Experiments on superconducting electric dipole and its array, *Int. Symp. on Antennas and Propagation Sendai, Japan*, pp. 109–12, 1978
31 Khamas S. K., Mehler M. J., Maclean T. S. M., Gough C. E., Alford N. McN. and Harmer M. A. High-T_c superconducting short dipole antenna, *Electron. Lett.*, **24**(8), 460–1, 1988
32 Fujinaka M., Ishida T., Goto T., Tabei K. and Maehara T. Trial manufacture of dipole antenna using $YBa_2Cu_3O_{7-x}$ ceramic wire, *Int. J. Electronics*, **67**(2), 245–9, 1989
33 Gough C. E., Khamas S. K., Maclean T. S. M, Mehler M. J., Alford N. McN. and Harmer M. M. Critical currents in high T_c superconducting short dipole antenna, *IEEE Trans. Magnetics*, **MAG.-25**(2), 1313–14, 1989
34 He Y. S., He A. S, Zhang X. X., Shi C. S., Lu N., Tian B., Lu D. R., Zhou M. L. and Wang W. G. Progress in high T_c superconducting ceramic antennas, *Superconductor Science and Technology*, **4**, S124–6, 1991
35 He A. S., He Y. S., Zhang X. X., Shi C. S., Lu N., Tian B., Lu D. R., Zhou M. L. and Wang W. G. Liquid nitrogen temperature superconducting antennas made from metal oxide ceramics, *Cryogenics*, **30**, suppl., pp. 946–50, 1990
36 Wu Z., Mehler M. J., Maclean T. S. M., Lancaster M. J. and Gough C. E. High T_c superconducting small loop antenna, *Physica*, **C162–4**, 385–6, 1989
37 Lancaster M. J., Maclean T. S. M., Niblett J., Alford N. McN. and Button T. W.

YBCO thick film loop antenna and matching network, *IEEE Trans. Applied Superconductivity*, **3**(1), 2903–5, 1993
38 Dinger R. J., Bowling D. R., Martin A. M. and Talvacchio J. Radiation efficiency measurements of a thin film Y–Ba–Cu–O superconducting half loop antenna at 500 MHz, *MTT-S Digest*, pp. 1243–6, 1991
39 Itoh K., Ishii O., Nagai Y., Suzuki N., Kimachi Y. and Michikami O. High T_c superconducting small antennas, *IEEE Trans. on Applied Superconductivity* **3**(1), 2836–9, 1993
40 Lancaster M. J., Wu Z., Huang Y., Maclean T. S. M., Zhou X., Gough C. E. and Alford N. McN. Superconducting antennas, *Superconductor Science and Technology*, **5**, 277–9, 1992
41 Ishii O. High T_c superconducting antenna, *MWE '92 Microwave Workshop Digest*, pp. 223–8, 1992
42 Chaloupka H., Klein N., Peiniger M., Piel H., Pischke A. and Splitt G. Miniaturised HTS Microstrip patch antenna, *IEEE Trans. on Microwave Theory and Techniques*, **39**(9), 1513–21, 1991
43 Pischke A., Chaloupka Hl, Piel H., Splitt G. Electrically small planar HTS antennas, *3rd International Superconductive Electronics Conf., University of Strathclyde, June, 1991*, pp. 340–3, 1991
44 Portis A. M., Chaloupka H., Jeck M., Piel H. and Pischke A. Critical response of an HTS microstrip patch antenna, *3rd International Superconductive Electronics Conf., University of Strathclyde, June, 1991*, pp. 320–4, 1991
45 Portis A. M., Chaloupka H. and Jeck M. Power-induced switching of an HTS microstrip patch antenna, *Superconductor Science and Technology*, **4**, 436–8, 1991
46 Chaloupka H. J. High temperature superconductor antennas: utilisation of low rf losses and of non-linear effects, *Journal of Superconductivity*, **5**(4), 403–16, 1992
47 Suzuki N., Itoh K., Nagai Y. and Michikami O. Superconductive small antennas made of Eu-Ba-CuO thin films, *Advances in Superconductivity V*, eds. Bando Y. and Yamauchi H., Springer-Verlag, pp. 1127–30, 1992
48 Kraus J. D. *Antennas*, McGraw-Hill, 1988
49 Jordan E. C. *Electromagnetic Waves and Radiating Systems*, Prentice-Hall, New Jersey, 1968
50 Huang Y., Lancaster M. J., Maclean T. S. M., Wu Z. and Alford N. McN. A high temperature superconductor superdirectional antenna array, *Physica*, C**180**, 267–71, 1991
51 Kraus J. D. *Antennas*, McGraw-Hill, 1988
52 Brown G. H. Directional antennas, *Proc. IRE*, **25**, 78–145, 1937
53 Abramowitz M. and Stegun I. A. *Handbook of Mathematical Functions*, Dover, New York, 1965
54 Hansen R. C. Fundamental limitations on antennas, *Proc. IEEE*, **69**(2), 170–82, 1981
55 Oseen C. W. Die Einsteinsche Nadelstichstrahlung und die maxwellshen gleichungen, *Annalen der Physik*, **69**, 202–4, 1922
56 LaPaz L. and Miller G. A. Optimum current distributions on vertical antennas, *Proc. of the Institute of Radio Engineers*, **31**, 214, 1943
57 Dolph C. L. A current distribution for broadside arrays which optimises the relationship between beamwidth and sidelobe level, *Proc. Institute of Radio Engineers*, **34**, 335–48, 1946

58 Reid D. G. The gain of an idealised Yagi array, *Journal IEE*, **93**, Part IIIA, pp. 564, 1946
59 Woodward P. M. and Lawson J. D. The theoretical precision with which an arbitrary radiation pattern may be obtained from a source of finite size, *Journal IEE*, **95**, Part III, p. 363, 1948
60 Jordan E. C. *Electromagnetic Waves and Radiating Systems*, Prentice-Hall, New York, 1950
61 Yaru N. A note on supergain antenna arrays, *Proc. of the Institution of Radio Engineers*, **39**, 1081, 1951
62 Uzkov A. An approach to the problem of optimum directive array design, *Comptes Rendus de l'Académie des Sciences de l'URSS*, **53**, 35–8, 1946
63 Tai C. T. The optimum directivity of uniformly spaced broadside arrays of dipoles, *IEEE Trans. on Antennas and Propagation*, **AP-12**, 447–54, 1964
64 Bloch A., Medhurst R. G. and Pool S. D. A new approach to the design of super-directive aerial arrays, *Proc. Inst. Elec. Eng.*, part III, **100**, 303–14, 1953
65 Stearns C. O. Computed performance of moderate size, supergain, end-fire antenna arrays, *IEEE Trans. on Antennas and Propagation*, **AP-14**, 241–2, 1966
66 Schelkunoff S. A. A mathematical theory of linear arrays, *Bell System Technical Journal*, **22**, 80–107, 1943
67 Ma M. T. The directivity of uniformly spaced optimum end-fire arrays with equal sidelobes, *Radio Science*, September, 1965
68 Cheng D. K. and Tseng F. I. Maximisation of directive gain for circular and elliptical arrays, *Proc. IEE*, **114**(5), 589–94, 1967
69 Lo Y. T. Optimisation of directivity and signal-to-noise ratio of an arbitrary antenna array, *Proc. IEEE*, **54**(8), 1033–45, 1966
70 Gilbert E. N. and Morgan S. P. Optimum design of directive antenna arrays subject to random variations, *Bell System Technical Journal*, **34**, 637–63, 1955
71 Uzsoky M. and Solymar L. Theory of super-directive linear arrays, *Acta Phys., Acad. Sci. Hung.*, **6**, 185, 1956
72 Newman E. H., Richmond J. H. and Walter C. H. Superdirective receiving arrays, *IEEE Trans. on Antennas and Propagation*, **AP-26**(5), 629–35, 1978
73 Seeley E. W. VLF superdirective loop arrays, *J. Research*, **67**D(5), 563–5, 1963
74 Seeley E. W. Two and three loop superdirective receiving antennas, *J. Research*, **67**D(2), 215–35, 1963
75 Newman E. H. *et al.* A wideband electrically small superdirective array, *IEEE Trans. on Antennas and Propagation*, **AP-25**, 885–7, 1977
76 Bokhari S. A., Smith H. K., Mosig J. R., Zürcher J. F. and Gardiol F. E. Super-directive antenna arrays of printed parasitic elements, *Electron. Lett.*, **28**(14), 1332–4, 1992
77 Nakamura T., Miyagawa S. and Yokokawa S. Superdirective cascaded dipole array, *Trans. IEICE Jap.* (B-II), J75-B-II, 1, pp. 62–8, 1992
78 Cummings J. A. and Harrington R. F. An experimental supergain receiving array for broadband use, presented at the *G-AP Int. Symp., Los Angeles, CA, Sept. 1971*, pp. 295–8, 1971
79 Andrasic G. and James J. R. Height reduced superdirective array with helical directors, *Electron. Lett.*, **29**(23), 2002–4, 1993
80 Walker G. W., Haden R. and Ramer O. G. Superconducting superdirectional antenna arrays, *IEEE Trans. on Antennas and Propagation*, **AP-25**(6), 885–7, 1977

81 Adachi S., Ashida K. and Ohnuki S. Superconducting dipole array antenna, *Trans. IECE Jap.* (B) J63-B9, pp. 916–23, 1980
82 Ishii O., Yamada Y., Kagoshima K., Itoh K. and Kimachi Y. Collinear supergain antennas with shaped radiation pattern, *ISAP'92 Sapporo*, pp. 1005–8, 1992
83 Ishii O. High-T_c superconducting antenna, *MWE Microwave Workshop Digest*, pp. 223–8, 1992
84 Itoh K., Ishii O., Suzuki N., Kimachi Y. and Michikami O. Superdirective array antennas composed of high-T_c superconducting small helical radiators, *ISAP' 92 Sapporo*, pp. 1025–8, 1992
85 Ivrissimtzis L. P., Lancaster M. J. and Alford N. McN. A supergain YBCO antenna array with integrated feed and balun, *IEEE Trans. on Applied Superconductivity*, **5**(2), 3199–202, 1995
86 Ivrissimtzis L. P., Lancaster M. J., Maclean T. S. M. and Alford N. McN. High gain series fed printed dipole arrays made of high-T_c superconductors, *IEEE Trans. on Antennas and Propagation*, **42**(10), 1419–29, 1994
87 Ivrissimtzis L. P., Lancaster M. J. and Alford N. McN. Supergain printed arrays of closely spaced dipoles made of thick film high-T_c superconductors, submitted to *IEE Proc. Microwaves, Antennas and Propagation*
88 Ivrissimtzis L. P., Lancaster M. J., Maclean T. S. M. and Alford N. McN. High gain printed dipole array made of thick film high-T_c superconducting material, *Electron. Lett.* **30**(1), 92–3, 1994
89 Newman E. H. and Tulyathan P. Wire antennas in the presence of a dielectric/ ferrite inhomogeneity, *IEEE Trans. on Antennas and Propagation*, **AP-26**(4), 587–93, 1978
90 Yeo S. P., Leong M. S., Kooi P. S., Yeo T. S., Zhou X. D. Contour-integral analysis of microstrip sectional power divider (with arbitrary sector angle), *IEE Proc.*, H**40**(1), 62–4, 1993
91 Cox H., Zeskind R. M. and Kooj T. Practical supergain, *IEEE Trans.*, **ASSP-34**, 393–8, 1986
92 Dawoud M. M. and Anderson A. P. Design of superdirective arrays with high radiation efficiency, *IEEE Trans.*, **AP-26**, 819–23, 1978
93 Bokhari S. A., Smith H. K. and Mosig J. R. Line source radiation in the presence of slotted cylinders, *IEEE APS Symp. Digest*, pp. 1583–6, 1992
94 Harrington R. F. Antenna excitation for maximum gain, *IEEE Trans. on Antennas and Propagation*, **AP-13**, 896–903, 1965
95 Sahalos J., Melidis K. and Lampou S. On the optimum directivity of general nonuniformly spaced broadside array, *Proc. IEEE*, pp. 1706–8, 1974
96 Shashikant S. M. and Butler J. K. Constrained optimisation of the performance indices of arbitrary array antennas, *IEEE Trans. on Antennas and Propagation*, **AP-19**, 493–8, 1971
97 Tseng F. I. and Cheng D. K. Gain optimisation for arbitrary arrays subject to random fluctuations, *IEEE Trans. on Antennas and Propagation*, **AP-15**, 356–66, 1967
98 Winkler L. P. and Schwartz M. A fast numerical method for determining the optimum SNR of an array subject to a Q factor constraint, *IEEE Trans. on Antennas and Propagation*, **AP-20**, 503–5, 1972
99 Kurth R. R. Optimisation of array performance subject to multiple power constraints, *IEEE Trans. on Antennas and Propagation*, **AP-22**, 103–6, 1974
100 Cheng D. K. Optimisation techniques for antenna arrays, *Proc. IEEE*, **59**, 1664–74, 1971

101 Bouwkamp C. J. and De Bruijn N. G. Radiation resistance of an antenna with arbitrary current distribution, *Phillips Research Reports*, **1**, 62–76, 1946
102 Riblet H. J. Note on the maximum directivity of an antenna, *Proc. of the Institute of Radio Engineers*, **35**, 620, 1947
103 Kraus J. D. *Antennas*, McGraw-Hill, 1988
104 Ngo P., Krishen K., Arndt D., Raffoul G., Karasak V., Bhasin K. and Leonard R. A high temperature superconductivity communications flight experiment, *Applied Superconductivity*, **1**(7–9), 1349–61, 1993
105 Richard M. A., Bhasin K. B., Gilbert C., Metzler S., Koepf G. and Claspy P. C. Performance of a four-element Ka-band high-temperature superconducting microstrip antenna, *IEEE Microwave and Guided Wave Lett.*, **2**(4), 143–5, 1992
106 Richard M. A., Bhasin K. B. and Claspy P. C. Superconducting microstrip antennas: an experimental comparison of two feeding methods, *IEEE Trans. on Antennas and Propagation*, **41**(7), 967–74, 1993
107 Herd J. S., Hayes D., Kenney J. P., Poles L. D., Herd K. G. and Lyons W. G. Experimental results on a scanned beam microstrip antenna array with a proximity coupled YBCO feed network, *IEEE Trans. on Applied Superconductivity*, **3**(1), 2840–3, 1993
108 Poles L. D., Herd J. S., Mittleman S. D., Kenney J. P., Champion M. H., Silva J. H., Rainville P. J., Bocchi W. J. and Derov J. S. 20 GHz High temperature superconducting phased array antenna development, *1994 Applied Superconductivity Conf. Boston MA*, paper EEE10, 1994
109 Chaloupka H. Superconducting multiport antenna arrays, *Microwave and Technology Lett.*, **6**(3), 737–44, 1993
110 Chaloupka H. Applications of high-temperature superconductivity to antenna arrays with analogue signal processing capability, *24th European Microwave Conf., Cannes, France*, pp. 23–35, 1994

8

Signal processing systems

8.1 Introduction

The basic signal processing functions applicable to high-frequency signals are shown in Table 8.1.1. Any microwave system is composed of a number of these functions in order to produce its required operation. Superconductors can be used to produce *all* these functions and in most cases exhibit superior performance compared with conventional methods. Many of these functions have been described in preceding chapters as individual components; it is the aim of this chapter to describe how these individual processing operations can be put together into systems for complex signal processing.

Although there may be advantages in using discrete superconducting components in a system, it is to greater advantage to integrate many of the functions given in Table 8.1.1, as the cooling overhead then becomes less per function. This also applies to the integration of semiconductors into cold systems, as many of them improve considerably in performance at reduced temperature.

Table 8.1.1. *Basic signal processing functions*

1. Signal generation
2. Multiplexing/demultiplexing
3. Transmission/reception (antennas)
4. Signal interconnects (transmission lines)
5. Storage
6. Switching/routing
7. Lowpass/highpass filters
8. Bandpass filters/matched filters
9. Fourier transform/spectrum analysis
10. Frequency conversion/direct detection
11. A/D conversion
12. Digital processing

This may be particularly relevant in cases where part of the system is already cooled, such as the sensors for infrared detection. The widespread use of superconducting technology will only come about if performance advantages of system level functions are superior to those using conventional methods. These performance advantages can be reduction in size, weight and volume as well as electrical performance improvements, and the criteria for selection must include the excess power consumption and increase in size associated with the use of coolers or liquid cryogen. Even a performance advantage may not be enough for a superconducting system to be adopted. Ultimately, it is a new capability which will bring superconductors into system level applications; in fact, not just new capability, but new capability which is desperately required.

Many of the uses of HTS at microwave frequencies have been highlighted in preceding chapters, including communications systems, radar systems, microwave instrumentation, radiometry and radio astronomy, medical systems and fundamental physical research. Within these groups a number of applications exist and some are discussed in detail in this chapter.

For example, a radar system can be required to identify small signals in the presence of clutter caused by atmospheric effects such as rain, or by external signals when attempts are made to jam the system. Here the signal processing requirements can be stringent in order to utilise the received information for maximum gain. Doppler radar[1] can be used to discriminate moving targets in the received signal, or pulse compression radar can be used to maximise the signal-to-noise ratio for stationary targets. Both these systems can benefit from superconducting components. The low phase noise provided by superconducting oscillators provides superior performance for Doppler radar, which is particularly important for slow moving targets where the phase noise close to the carrier is important. For pulse compression radar, the delay line chirp filter technology is important, firstly in providing a long time and wide bandwidth signals for the transmitted signal, and secondly for the delay line matched filter on reception. Here the time–bandwidth product is the important criterion for defining the performance, and although other technologies such as surface acoustic waves can provide superior time–bandwidth products, they do not have the bandwidth available to HTS. In pulse compression radar this translates to a superior range resolution.

Surveillance receivers are an important military application; here the environment is monitored for the presence of a signal and an appropriate action is taken. One such receiver is described in the next section, where spectral analysis plays the dominant role. Applications include the detection of jamming signals, which enables the frequency of the receiver to be altered in real time, thus avoiding signal suppression; front end filter banks are important in eliminating out-of-band large-amplitude jamming signals from the input electronics. Any non-linear effects introduced by such large signals can cause severe reduction in system performance. Another application where signal suppression is important is in aircraft communications and radar. Here large signals may be transmitted from one aircraft system and can be picked up by their own surveillance receiver. Obviously, such large local signals render the surveillance receiver useless while transmission occurs. However, if a bandstop filter is placed at the front end of the surveillance receiver, then this particular interfering frequency can be eliminated, and the receiver used for other frequencies. With systems with frequency agility, the bandstop filter must be of variable centre frequency, with this centre frequency controlled by the electronics within the aircraft. Such a filter is described in Section 8.3 and is being deployed in military aircraft. The same group which produced this filter also demonstrated one of the first radar systems using HTS components; this was a frequency modulated carrier wave (FMCW) radar and used a 2.2 ns delay line. Such a system was only a benchtop demonstrator but gave considerable insight into the full system integration of HTS, conventional electronics and cryocoolers.[2]

Communication systems can also benefit from HTS. One example is in mobile radio base stations, where more channels are possible with the inclusion of passband filters with very rapid roll-off. The attenuation between adjacent bands needs to be very high in order to reduce cross-channel interference. Superconductors provide this capability, as described in Chapter 5. The application is discussed further in Section 8.4. Satellite and space communications also provide an important applications area. As in the terrestrial communications case, filter banks are required with a large number of channels, each with high performance.

8.1.1 Space applications

The most ambitious demonstration of HTS technology is in the USA with the Naval Research Laboratory (NRL) high-temperature superconductivity space experiment (HTSSE).[3] The aim of the programme is to fly a number of HTS devices, subsystems and systems on satellite platforms. The goals are to focus the technology on space and to demonstrate the viability of these systems in this environment. The project launches three separate satellites; the first (HTSSE I) having HTS devices on board, the second (HTSSE II) having more complex devices and subsystems and the last (HTSSE III) demonstrating system level applications. Although expensive, this programme is realising what it set out to achieve and many devices and systems have been built for the programme. It has encouraged funding for HTS from companies in the USA, as well as the development of many new applications,[1] some of which are described later in this chapter.

On board HTSSE I were basic microwave components including a number of ring resonators, a half-wavelength linear resonator, a dielectric resonator, multipole filters, a lowpass filter, a delay line, a bulk HTS cavity, a 3 dB coupler, a patch antenna and a microwave switch. These represent a wide selection of HTS component applications, as described in the preceding chapters. The proposed HTSSE II payload included filter banks, a delay line, an analogue signal processor, a bandpass receiver, a digital subsystem, a digital multiplexer, a 60 GHz communication receiver, an A/D converter and a patch antenna array, although the inclusion of the latter three was abandoned. The progress from devices to subsystems in the few years between these experiments can clearly be seen from the list of components. These components are mainly supplied from major industry in the USA. In these satellites the components have to be cooled and microwave signals applied, and the resulting performance has to be monitored and transmitted back to ground. Although most of the goals have already been accomplished at take-off, monitoring continues for about six months in order to observe the reliability and environmental effects on the devices. HTSSE I weighed 350 lb, consumed 80 W of power and included an entire scalar network analyser to perform the measurements. The cooler for the experiment was a Stirling cooler, one of the few which had space qualification at the time of take-off; this amounted to a three-year continuous operation in the laboratory. Putting these systems together, with what is a fledgling technology, represented a significant technological achievement and required the cooperation of a large number of companies, research laboratories and universities, in addition to large financial resources from many different bodies.

In addition to HTSSE, NASA conceived an experimental superconducting communications experiment in space. The aim was to produce a communication link between the Space Shuttle bay and the Advanced Communications Technology Satellite (ACTS), again to demonstrate the viability of HTS in space systems in order to evaluate the technology.[4] The communications are via a patch antenna array consisting of nine elements, each containing 4×4 'H'-shaped elements operating at 20 GHz (gain 24 dB). NASA is also working

[1] See the June 1996 edition of *IEEE Trans. on Microwave Theory and Techniques* for a survey of the applications on board the HTSSE II space experiment.

on other components for inclusion in space systems,[5] for example those that can be used for deep space communications.

There are a number of special considerations which need to be taken into account when operating systems in space. In principle, the temperature of objects in space can be very low as the background radiation is around 3 K. However, radiative heating occurs from the sun and reflected radiation from the Earth and other bodies. Even on the side of a satellite which points into space the temperature will be as high as about 100 K due to thermal conduction through the satellite. This means that some form of cooling is almost always required for HTS devices, although it can be reduced when compared with terrestrial applications. This does not necessarily present a technological problem, but adds to the performance reduction of HTS systems in space. An example of an extreme case of cooling in space is given by the Infra-red Astronomical Observatory (IRAS), where the infra-red sensors were cooled to 1.7 K using 400 litres of liquid helium, which lasted for eight months.

Other important criteria which have to be considered for space applications include the possible thermal cycling as the orbit changes, or the position of the electronics changes relative to radiating bodies. Also, the systems must be insensitive to mechanical effects such as the vibration or the high g forces encountered during launch (≈ 5–$7g$). Radiation effects in space also have to be considered; electron energies can be larger then 1.6 MeV with fluxes of 10^5 cm^{-2} s^{-1}, proton energies can be above 700 MeV with fluxes 10^4 cm^{-2} s^{-1} and gamma rays up to 50 MeV. However, these are mostly confined to the van Allen belts, which are avoided in low Earth orbits.[6] Although the radiation in these orbits is much smaller, consideration must be given to the possibility of long time-scales of operation. All the above effects are used in space qualification programmes for all systems which fly in space, and such programmes can give a considerable insight into many areas of the practical application of HTS.

8.1.2 Systems based on spectral analysis

The heart of many communication and radar systems is some kind of spectral analysis, and it turns out that spectral analysis systems are particularly difficult to construct to the required performance from normal room-temperature components. Examples of applications include Doppler radar systems, wide band imaging radar, radar warning receivers, signal demultiplexing in communications systems and signal routing systems. One of the most demanding applications is in electronic warfare systems for the determination of the presence and origin of a source of unknown signals. Such systems require very wide bandwidth coverage, high resolution, fast response times and there must be a high probability of intercepting any signals present. Systems which can accomplish such tasks are superheterodyne receivers, filter banks or multiplexers, chirp transform spectrum analysers or instantaneous frequency meters. Superconductive versions of the latter two are described in Sections 8.5 and 8.7 of this chapter, whilst multiplexers have been described in Chapter 5. Superconductivity is not the only technology applicable to these systems; both acoustic, acousto-optic and digital devices have an important role to play with a maximum bandwidth of around 1 GHz for the acoustic-based devices. Discussion of other technologies is given in more detail in Section 6.7. An example of an application which incorporates a spectrum analyser is the cueing receiver shown in Figure 8.1.1.

The idea of such a system is that the spectrum analyser monitors a wide band frequency spectrum. When a signal (or signals) is received a local oscillator controller uses the information from the spectrum analyser to mix a delayed version of the incoming signal into the frequency range of other, more sensitive, specialised processing circuitry. Many of the components of such a system can be produced from superconductors, as indicated in Figure 8.1.1. This type of system is extremely flexible and can be used for a number of both military and civilian applications.

Such a system with some post-processing has been designed for the HTSSE II space experiment,[7,8] and is based on the chirp transform spectrum analyser discussed in Section 8.5. A considerable amount of room-temperature post-

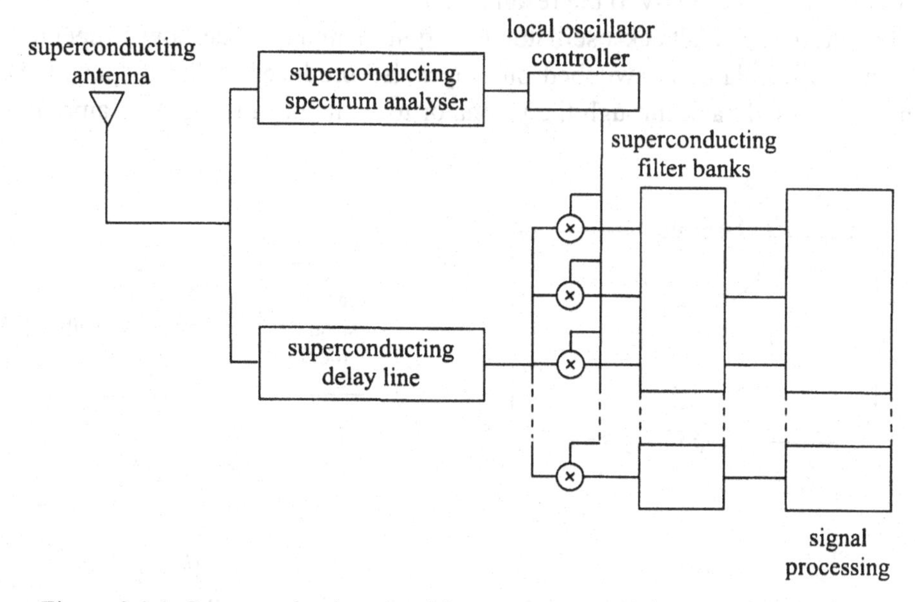

Figure 8.1.1. Superconducting signal intercept system using a cueing receiver.

processing circuitry is used in this system, resulting in a power consumption of 20 watts, although dedicated CMOS rather than ECL circuits could reduce this by an order of magnitude.

8.2 Microwave oscillators

The principal advantage of using superconducting resonators as the main stabilising element in a microwave oscillator is the reduction in phase noise over conventional oscillators. There are a number of microwave system applications where conventional oscillator circuits cannot provide the required performance. In communication systems, the continued drive for more efficient use of the spectrum brings channel spacing closer together, requiring low noise close to the carrier. In radar systems there are a number of applications of low-noise oscillators; one example is in Doppler radar, where lower phase noise translates into the identification of lower velocity targets. Applications for superconducting oscillators occur not only when very high performance is required but also when oscillators are required to be reduced in size. The same principles of size reduction apply as discussed for filters in Chapter 5.

Figure 8.2.1 shows the two basic types of microwave oscillator in schematic form. Figure 8.2.1(a) shows the basic feedback oscillator, where a resonator is located in the feedback loop of an amplifier. The second oscillator shown in Figure 8.2.1 uses a negative resistance element, which can be a transistor or diode connected directly to the resonator.

For the basic feedback oscillator it is quite simple to see how it operates. When the oscillator is switched on, any noise produced at the output of the amplifier is fed back through the resonator to its input; this signal is amplified

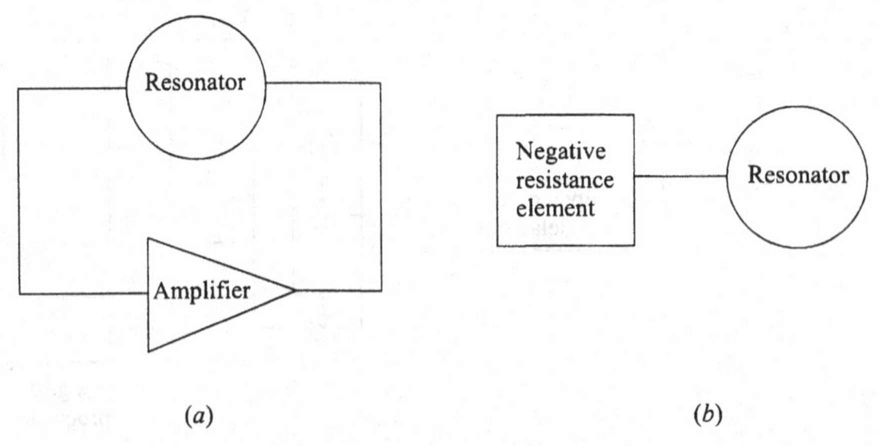

Figure 8.2.1. The two basic types of microwave oscillator.

and the process continues until the maximum output power of the amplifier is obtained and a steady state is reached. The bandpass frequency characteristic of the resonator produces the basic frequency of operation of the oscillator. For this to occur, the gain of the amplifier must be greater than the losses in the resonator and the external circuit. However, this can only occur if the total phase around the amplifier resonator loop is $2\pi n$ (n integer) at the oscillator centre frequency. This phase criterion is probably the most important in understanding the operation of an oscillator. If the loop length changes by a small amount due to some environmental effects, or noise in the amplifier, then the frequency of operation changes. For a high Q resonator the slope of the phase as a function of frequency is very high; thus the frequency change is small when this loop length changes. These small frequency changes can be measured and are referred to as phase noise. Examples are discussed below for a number of types of oscillator.

The phase noise, $L(f)$ of an oscillator in a 1 Hz bandwidth is given by[9–11]

$$L(f) = 10\log\left[N^2\left(1+\frac{f_0^2}{4Q_l^2 f^2}\right)\left(\frac{GFkT}{P}+\frac{\alpha_f}{f}\right)\right] \tag{8.2.1}$$

where f is the offset frequency from the carrier frequency f_0, Q_l is the loaded Q of the resonator, G is the gain of the amplifier, F is the noise figure of the amplifier, k is Boltzmann's constant, T is the absolute temperature, P is the power at the output of the amplifier and α_f is the flicker noise constant[12] (determined empirically). If the output signal is formed by mixing the output signal from a lower-frequency oscillator, then the frequency multiplication factor N is used. This equation can be derived quite straightforwardly by considering the voltage feedback around an amplifier with a resonator in the feedback loop. The resonator can be approximated by an LCR equivalent circuit. It should be noted that this is a linear theory, and that the noise is not correctly predicted very close to the carrier; however, for low noise oscillators the expression is a good approximation for offsets greater then about 1 Hz from the carrier.[9]

It can be observed immediately from Equation (8.2.1) that low-noise oscillators are obtained by using high unloaded Q resonators and amplifiers with a low noise figure and a high output power capability. Everard[9] has shown that there is an optimum ratio of Q_l/Q_0 to minimise the sideband phase noise. Assuming that F is independent of Q_l/Q_0, then the value is about $2/3$ and corresponds to an insertion loss of a cavity resonator of about 9.5 dB. Operating with very weakly or very strongly coupled cavities can degrade the noise performance significantly. This optimum insertion loss of the cavity resonator sets the gain of the amplifier. In addition, amplifiers with low noise figures are required, and this generally improves with cooling if the correct devices are chosen.[13] Further details of cooled active devices are given later in this chapter.

Figure 8.2.2 shows the phase noise predicted by Equation (8.2.1) for a number of different values of Q at 10 GHz. The parameters used are $P = 0.3$ W, $F = 6$ dB, $G = 15$ dB, $T = 300$ K and $\alpha_f = 4 \times 10^{-12}$ Hz. In addition, Figure 8.2.2 shows the phase noise for SAW and bulk acoustic wave (BAW) oscillators.[14–16] It can be seen that, using resonators with Qs above about 10^4, SAW and BAW oscillators can be improved upon at 10 GHz, although this must be taken as a rather approximate statement because of the number of variables involved.

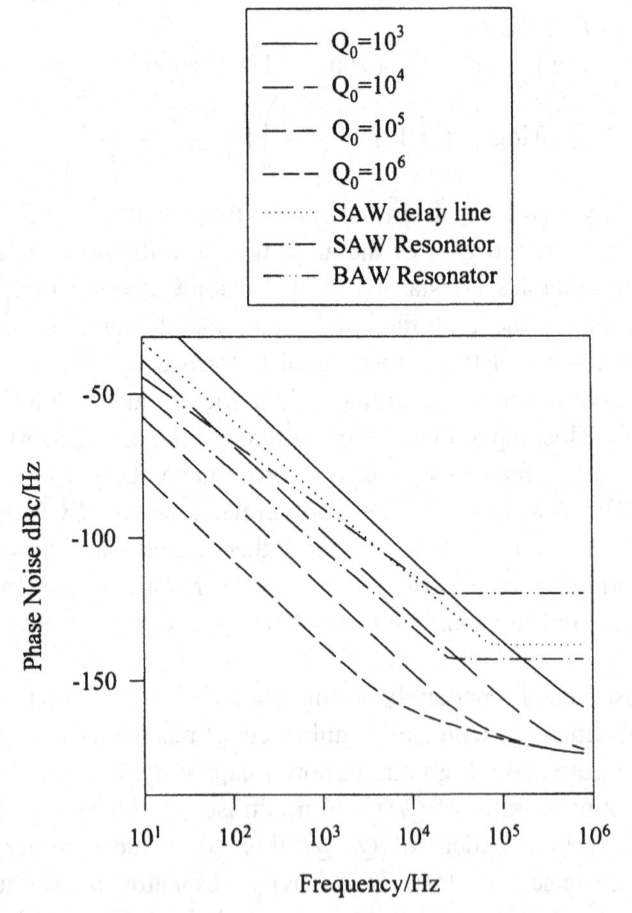

Figure 8.2.2. Values of phase noise calculated from Equation (8.2.1) for different values of unloaded Q. In addition, SAW and BAW oscillator phase noise is shown; these oscillators are multiplied up to 10 GHz.[14]

Other types of resonator used are made from normal metal, such as cylindrical cavities or microstrip; clearly, making the same cavity from a superconductor will give an improvement because of the higher Q. Probably the main competition for HTS resonators for use in oscillators is the dielectric resonator. A technology has developed so that a low-loss, high-permittivity cylindrical dielectric can be used in microstrip designs. The cylinder of the dielectric is placed close to the microstrip feed line and field coupling occurs; moving the dielectric alters the coupling coefficient. Table 8.2.1 shows the properties of some dielectrics used in these resonators. Their high dielectric constant, as well as size reduction, keeps the field inside the cylinder so that the effect of the surroundings is minimal. The inverse of the loss tangent, or dielectric Q, shown in the table can be very high. The dielectric resonators given in Table 8.2.1 compete well, in terms of Q, with microstrip or stripline HTS resonators.

It is now appropriate to look at an example of a superconducting oscillator in more detail. This particular example is of a reflection-type oscillator and has been chosen not because it gives the best performance, but because of attempts to miniaturise the entire oscillator. In fact, the oscillator described fits on to a 1-cm-square $LaAlO_3$ substrate and operates at 6.5 GHz. Similar oscillators are required in more complex microwave superconducting systems and shrinking the size down allows these systems to become more integrated. Figure 8.2.3 shows the circuit diagram of an oscillator studied at the university of Munich in conjunction with Siemens, and is in fact one of the first oscillators produced using HTS.[18] The oscillator is of the reflection type, using a coplanar half-wavelength linear resonator as the stabilising element; coupling to this resonator is by a capacitive gap, producing a 35 Ω input impedance looking across the gap to the resonator. The active element is a GaAs MESFET (NE71000) with bias fed via LC decoupling circuits. The series feedback at the source of the FET is achieved by open and shorted stubs, providing the slightly capacitive

Table 8.2.1. *Some dielectric resonator materials*[17]

Material	Dielectric constant	$1/\tan(\delta)$	Temperature coefficient/ (ppm/K)
$BaTi_4O_9$	38	10^4	+4
$Ba_2Ti_9O_{20}$	40	10^3	+2
(Zr, Sn)TiO_4	38	10^4	−4 to +10 adjustable
$Ba(Zn_{1/3}Nb_{2/3})O_2$ $Ba(Zn_{1/3}Ta_{2/3})O_2$	~ 30	2.5×10^4	0 to +10 adjustable

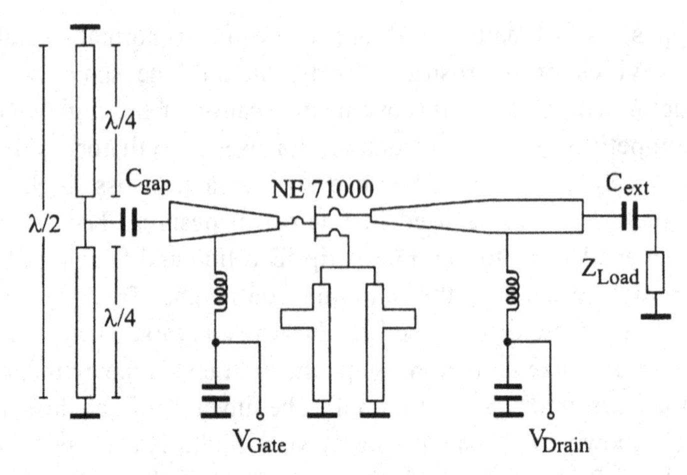

Figure 8.2.3. Circuit diagram of an integrated superconducting oscillator. The coplanar lines are shown to be wider for lower impedance. (Taken from reference [18].)

impedance required to fulfil the oscillation criteria. The output is a.c. coupled using a capacitor before the output load.

Figure 8.2.4 shows the layout of the 1-cm-square oscillator; it is formed by using a two-mask process, one for the YBCO superconducting layer and another for gold bonding wire pads. Chip capacitors are used for the bias decoupling network and, together with the FET, are glued with epoxy to the circuit. The whole structure was tested by immersion of the component in a hermetically sealed package into liquid nitrogen.

Results show that the phase noise in a 1 Hz bandwidth is 90 dB below the carrier at an offset frequency of 10 kHz (written −90 dBc/Hz@10 kHz). This

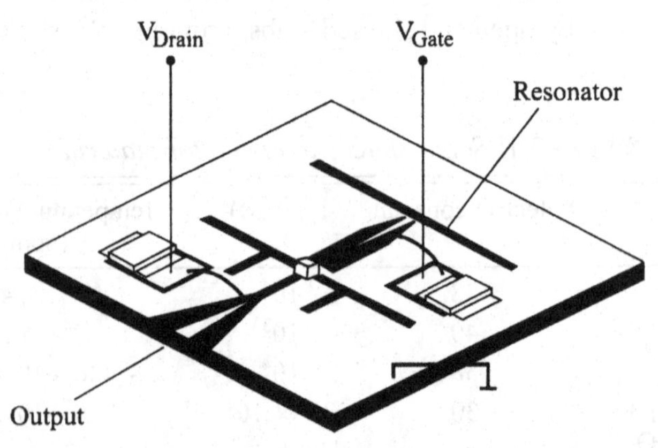

Figure 8.2.4. Layout of the superconducting oscillator. (Taken from reference [18].)

is achieved with the resonant element having an unloaded Q of 3850 at 77 K. The phase noise at other offset frequencies is shown in Figure 8.2.5. At the maximum output power of 4.9 dBm, the second and third harmonic levels are 10 dB and 30 dB below the carrier. As discussed above, this is not the ultimate in performance and can be improved considerably by using higher-Q resonators even on the 1-cm-square substrate; comparison of other types of HTS oscillator is given below.

Table 8.2.2 shows the performance of a number of superconducting and other oscillators. The non-superconductive oscillators differ mainly in their resonator types, since for many applications even a simple series inductance and capacitance suffice, but the Qs of such circuits are not high enough to produce a good phase noise performance, and are difficult to fabricate above 1 GHz. However, if these components were made from superconductors, then superior performance miniature oscillators could be produced. A transmission line resonator is a good stabilising element for an oscillator and is used with both normal conductors[19] and superconductors, as seen in the table. The example given in reference [19] is a small oscillator based on a ring resonator; some miniaturisation is achieved by placing the circuitry inside the resonator ring; the resonator is also tuneable over a 20% frequency range. The superconducting transmission line oscillators[20–25] in Table 8.2.2 show considerably improved performance in most cases. These resonators have large $1/f$ noise,[31]

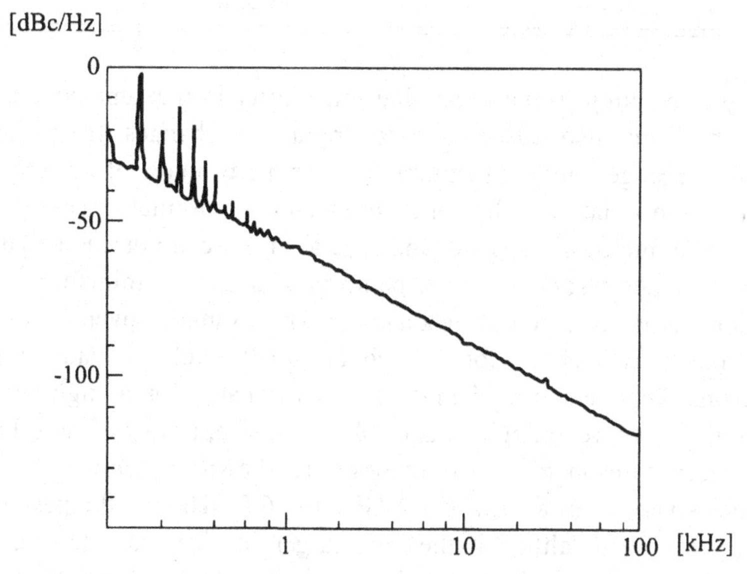

Figure 8.2.5. Oscillator single sideband phase noise performance. (Taken from reference [18].)

Table 8.2.2. *Some examples of microwave oscillators*

Reference	Type	Frequency/ GHz	Q_0 @ 77 K	Noise performance dBc/Hz @ 10 kHz offset and 77 K
Wilker[24]	HTS microstrip	5.0	< 3000	−106
Khanna[25]	HTS microstrip	2.3	1.3×10^4	−110
		10.2	6.0×10^3	−100
Khanna[23]	HTS microstrip	23	$> 7 \times 10^3$	−99 @ 70 K
Klieber[18]	HTS coplanar	6.5	3.8×10^3	−90
Rohrer[22]	HTS microstrip	10.0	2.1×10^3	−68
Phillips[33]	Thick film HTS cylinder	10	4×10^5	−120 @ 100 kHz
Button[34]	Thick film HTS cylinder	7.5	5×10^5	−135
Flory[27]	Sapphire–HTS	X-band		−160 @ 28 K
Negrete[28]	Sapphire–HTS	36.5	5.9×10^4	−103
Shen[29,30]	Sapphire–HTS	5.5	3.4×10^6	< −125 @ 70 K
Gardner[19]	Copper ring resonator	1.533–1.92		−94 @ 300 K
Everard[26]	Copper helical wire	0.9	582	−127 @ 25 kHz, 300 K
Klein[31]	Sapphire–HTS	5.6	8×10^5 @ 20 K	< −90 @ 1 kHz

probably originating in the remaining granularity in the superconducting thin films. They are also susceptible to frequency changes from the kinetic inductance change with temperature. The main advantage of transmission line resonators in oscillators is the ability to include the oscillator with other planar circuitry in a microwave system, and several of these are discussed later in this chapter. Another resonator which contains a degree of miniaturisation is an oscillator based on a helical resonator.[26] The example given in the table is made from normal conductors but could equally well be made from superconductors. The most promising type of resonator for a high-performance oscillator, but still retaining a reasonably small size, is an oscillator based on a HTS dielectric resonator. Four of these are shown in Table 8.2.2.[27–31] The frequencies vary from as low as 5.5 GHz to 36.5 GHz and the performance is generally excellent, although they are larger in size than transmission line resonators. The *Q*s are higher and the power output can be increased (also reducing the phase noise). They are also much less susceptible to any changes in kinetic inductance. Finally, two of the oscillators given in Table 8.2.2 use a thick film cylindrical cavity. These cavities have an excellent Q value at the lower frequencies but are of course much larger in size than the cavities discussed above. Excellent phase noise performance is obtained if these are integrated into an oscillator, as can be seen in Table 8.2.2.

Here, discussion has concentrated on HTS oscillators. These do not necessarily give the best performance of all oscillators since the Q of the resonators still does not compare with that of cooled sapphire resonators with superconducting shields. Such resonators are discussed briefly in Section 3.4 and can have unloaded Qs larger than 10^9, offering the best performance when used in an oscillator.[32]

8.3 Superconducting receivers

The basic receiver system shown in Figure 8.3.1 is one of the most fundamental building blocks of any high-frequency system. Its function is to take a high-frequency signal and convert it to a lower-frequency signal by mixing it with a locally generated tone. The low-frequency or IF output is much more manageable and can be processed further. Although systems can be much more complicated than the basic receiver shown in Figure 8.3.1, this contains the building blocks of any microwave system and is therefore of considerable interest. It is important that such a system be considered and demonstrated in order to see how superconductors can improve receiver performance.

In principle superconductors could be used in all the components in the receiver of Figure 8.3.1. However, in practice a mixture of semiconductors and superconductors gives the best performance. The preselect (or front end) filter is important to the system since it removes signals from the frequency band of

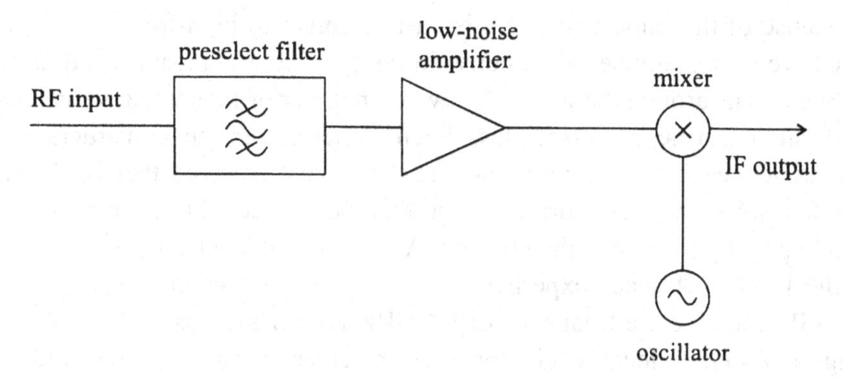

Figure 8.3.1. A basic receiver.

interest. It is important to place it at this point not only to reduce noise or to stop the amplifier limiting, but also to stop out-of-band signals mixing into the band by any non-linear effects in the amplifier. Clearly, such a filter needs to be low loss, especially for applications where the detection of small signals is of interest. The other component where superconductors can be used is in the oscillator. A superconductive resonator can provide the frequency stabilisation, giving a signal with very low phase noise. Oscillators are discussed in detail in Section 8.2. The rest of the system can be made from conventional semiconductors and GaAs-based components prove to be useful. The devices used in the demonstrations described below are principally GaAs HEMTs or FETs, both of which show improved performance at low temperatures.[35–37] This is important since the performance of such a system does not only depend on the superconducting components, but also on the semiconductor components.

The principal performance criteria of a receiver are its noise figure and gain. The noise figure represents the excess noise introduced by the components in the receiver. The gain includes both the amplification and the losses in the passive elements. Neither of these performance factors depends solely upon the superconducting components, so that it is important that any demonstration must use state-of-the-art amplifiers and mixers if improvement is to be obtained over the best room-temperature receivers. Alternatively, since the passive devices usually occupy the largest area, an additional advantage can be gained in miniaturisation of the receiver. Applications will only emerge for these devices where these performance criteria are critical to the application, or where further integration of superconducting components proves useful. An example of the former is in the reception of signals from deep space probes, where small improvements in gain, without the associated noise, may mean the difference between reception of signals or total loss of contact. The same reasoning also applies to any system where transmitted power is limited for whatever reason.

Because of the importance of this basic receiver to high-frequency systems there have been a number of demonstrations by various companies and research establishments around the world,[38,39] with a number of interesting variations on the basic construction principles. Such systems can be constructed from submodules, each containing a specific part of the receiver, that is, the filter, amplifier, mixer or oscillator; these can then be connected to form the receiver module package, which is then cooled. An example of such a receiver designed for the HTSSE II space experiment is for deep space communications[40] and uses a RF input centre frequency of 8.4 GHz, converting this to an IF of 1 GHz using a 7.4 GHz local oscillator. The receiver consists of the following submodules: (i) A four-pole edge coupled filter made from $YBa_{1.95}La_{0.5}Cu_3O_7$

(YBLCO) on a 15 × 7 × 0.51 mm $LaAlO_3$ substrate. It has an insertion loss of 0.2–0.3 dB, which is 0.5 dB better than an equivalent copper filter. (ii) A 28 dB gain, low-noise amplifier based on a Fujitsu FHX15X HEMP and associated matching circuitry. (iii) A reflection mode oscillator based on an HTS resonator with a Q of the order of 1000 and a GaAs MESFET producing an output power of 3 dBm. (iv) A single balanced mixer using a low barrier silicon Schottky diode (M/ACOM 40132). This was chosen over a double balanced mixer since it requires half the local oscillator power and the additional bandwidth of a double balanced mixer is not required. The construction is microstrip using alumina substrates. The receiver has a VSWR of 1.7:1 at the RF port and 2.4:1 at the LO port, with a 33 dB isolation between the two. These submodules were assembled into a single package and at 77 K produced a noise figure of 0.7 dB and a conversion gain of 18 dB across the 300 MHz bandwidth. This is shown in Figure 8.3.2.

For the ultimate in phase noise performance of the local oscillator, the Q of the resonator has to be as high as possible. This can be improved in a number of ways, as described in Section 8.2. A receiver based on a thick film cavity with a Q of 4×10^5 at 10 GHz and 77 K has been demonstrated;[41] although this is a high Q, the penalty is paid in the increase in size compared with planar components. This particular receiver uses an 11 GHz oscillator to give a 1 GHz IF output. The submodules are as follows: (i) A four-pole, 1% bandwidth Chebyshev filter with an insertion loss < 1 dB made from $GdBa_2Cu_3O_7$ on MgO measuring 7 × 14 mm. A similar gold filter has an insertion loss of 12 dB. (ii) An oscillator with a thick film cavity and a HEMT used as an active

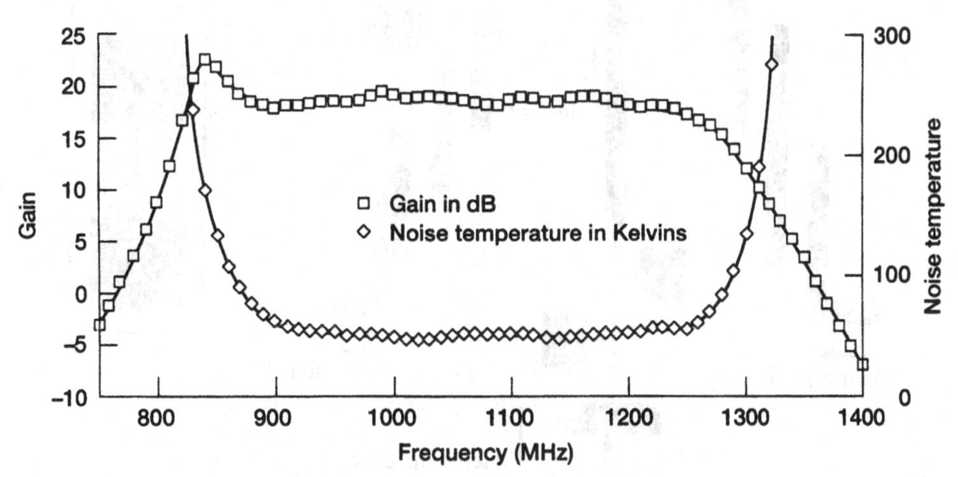

Figure 8.3.2. The gain and noise performance of a receiver with respect to the IF frequency. (Taken from reference [40].)

element producing 13 dBm of output power. The phase noise is −120 dBc at 100 kHz offset and 10 dBm output power. (iii) A silicon diode which operates at 77 K with a conversion loss of < 9 dB. The whole package is integrated into a Stirling cooler with 2 W of cooling power at 60 K and requires a total input power of 100 W with a 1 h cool down time.

In an ideal situation, all components would be integrated on to a single wafer which comprised the whole receiver. Such a construction has been demonstrated on a 3" $LaAlO_3$ wafer using double-sided deposition of YBCO.[42] This device consists of a HEMT amplifier, a five-pole diplexer, two oscillators, a balanced mixer, a filter and a video detector. This two-channel system operates at 9.29 and 9.66 GHz.

Further integration can be considered if the active elements are made from superconductors. Figure 8.3.3 shows such a device, although no amplifier or oscillator is included.[43] The device consists of a RF filter centred around 9.0 GHz and a local oscillator frequency of 5 GHz. The mixing element is just a short piece of transmission line of low impedance. The high current density introduces non-linear effects and produces the mixing. The conversion loss of the device varies from −30 to −15 dB, with temperature between 80 and 85 K and input powers at the LO of 3–15 dBm. Clearly, the conversion loss is too high at present, but further improvements could make this a practical choice of method.

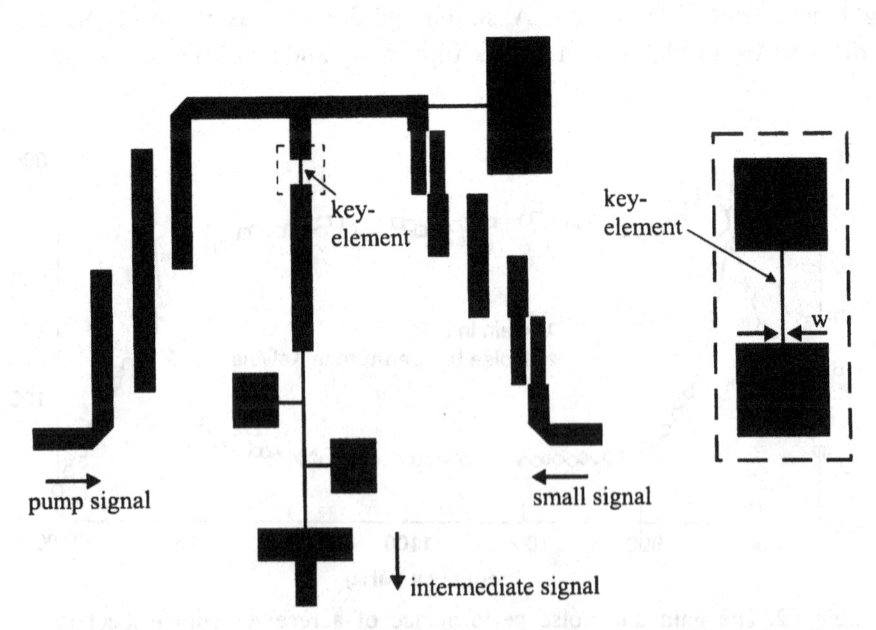

Figure 8.3.3. All-superconducting receiver. (Taken from reference [43].)

A HTS junction mixer can be investigated in order to examine greater integration. In addition to the work described in the microwave frequency regime, some demonstrations have been performed using receivers at frequencies above 200 GHz; here the junction mixers are used together with flux flow oscillators.[44,45]

8.4 Mobile communications

Filters for mobile communications base stations form a potentially huge market and have been investigated by a number of US companies. There are many different systems for mobile communications with numerous digital and analogue variants, and there are substantial differences around the world. Here, we will take as an example the system operating at about 800 MHz in the USA, as this is a large single market and has specific features which make the use of superconductors a viable proposition. There are over 15 000 base stations in the USA and the market for retrofitting these with superior filtering systems has been estimated at $750 million. The market for new sites is estimated at $150 million per year. Clearly, there is little need to cover the subject of mobile communications in detail here, and readers are referred to further comprehensive literature on the subject.[46,47]

Figure 8.4.1 shows the 800–900 MHz frequency band allocations for mobile radio in the USA. The allocations are separated into two user areas labelled A and B. This introduces competition into the market place by allowing two separate companies to use one band each, within a given area of operation. Historically, bands A and B were allocated first, then further allocation of the spectrum was given to mobile radio by the allocation by the FCC of bands A′ and B′. The bandwidth is duplicated, as shown in Figure 8.4.1, so as to have one frequency for transmission from the mobile base station and one for transmission from the mobile phone. This enables full duplex conversations. The total allocated spectrum gives 416 channels, each of 30 kHz bandwidth for each of the operators.

This complex spectrum allocation leads to potential operational problems. Cross-talk at the band edges between the two operators can lead to interference problems, with the consequence of unusable channels. Consider the situation shown in Figure 8.4.2, where a car is in communication with operator A at the edge of its geographic region. It could be close to operator B's base station and, therefore, if it is transmitting close to the allocation band edges, could cause interference. This problem can be overcome to a certain extent by the two operators coordinating their band allocation, but a better solution is that the front end filters of the receiving base station cut out the interfering signal. It

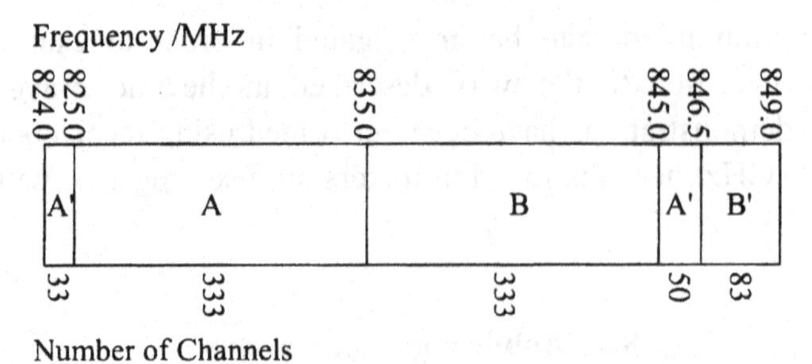

(a) Mobile Transmit Frequency Allocation

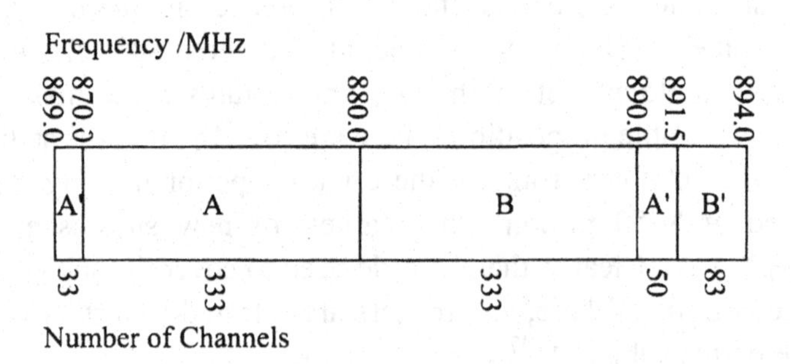

(b) Base Station Transmit Frequency Allocation

Figure 8.4.1. Band designations for mobile radio in the USA.

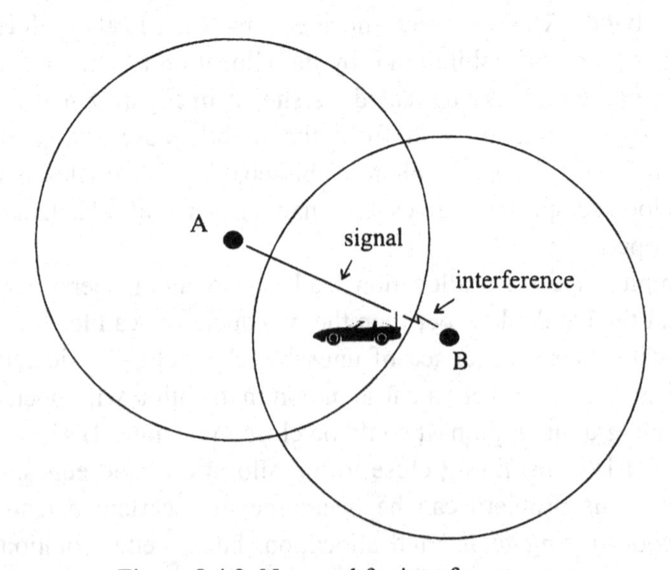

Figure 8.4.2. Near and far interference.

turns out that the specification for such filters is extremely difficult to achieve with conventional design techniques. Such a filter must have a very rapid roll off (within one 30 kHz channel) and provide high adjacent channel attenuation at the operator's band edges. In addition, because of the complex spectrum, the filters require two passbands for the regions A and A′ (or B and B′). Examples of some filters are given in Chapter 5, with a filter specifically designed for this problem being discussed in Section 5.4.

8.5 Spectrum analysers based on chirp filters

This spectrum analyser is based on the use of chirp filters to determine the spectrum of an input signal over a limited but wide bandwidth. It has the advantage of having almost 100% probability of intercepting an incoming signal, and is able to do this rapidly. It also has the advantage that it is small and lightweight, even for the coverage of a large number of frequency channels.

Consider an input to the spectrum analyser, $s_i(t)$; its Fourier transform can be represented by

$$s_0(\tau) = \int_{-\infty}^{\infty} [s_i(t) \exp(j\alpha t^2)] \exp(-j\alpha((\tau - t))^2)\, dt \qquad (8.5.1)$$

Here the term in square brackets represents multiplication of the input signal by a chirp waveform with increasing frequency with time; integration gives the convolution of the resultant signal with a similar chirp but with a reverse time characteristic. By multiplying out the terms in the exponential, Equation (8.5.1) becomes

$$s_0(t) = e^{-j\alpha\tau^2} \int_{-\infty}^{\infty} s_i(t) e^{j2\alpha\tau t}\, dt \qquad (8.5.2)$$

It can be seen that this integral is a Fourier transform with an effective frequency variable of $2\alpha\tau$. The magnitude of this gives the amplitude of the spectrum as discussed below or, alternatively, post-multiplication with a further chirp removes the quadratic phase term (the prefactor of Equation (8.5.2)). The practical implementation of this process is called a chirp transform spectrum analyser.

Figure 8.5.1 is a diagram of a chirp transform spectrum analyser. The input signal is first multiplied by a chirp waveform and then convolved with a similar chirp with inverse characteristics by passing through a chirp filter. The output of this chirp filter gives the spectrum of the input signal with the different frequency components appearing as a function of time. This arrangement is known as a multiply–convolve spectrum analyser. It does not, however,

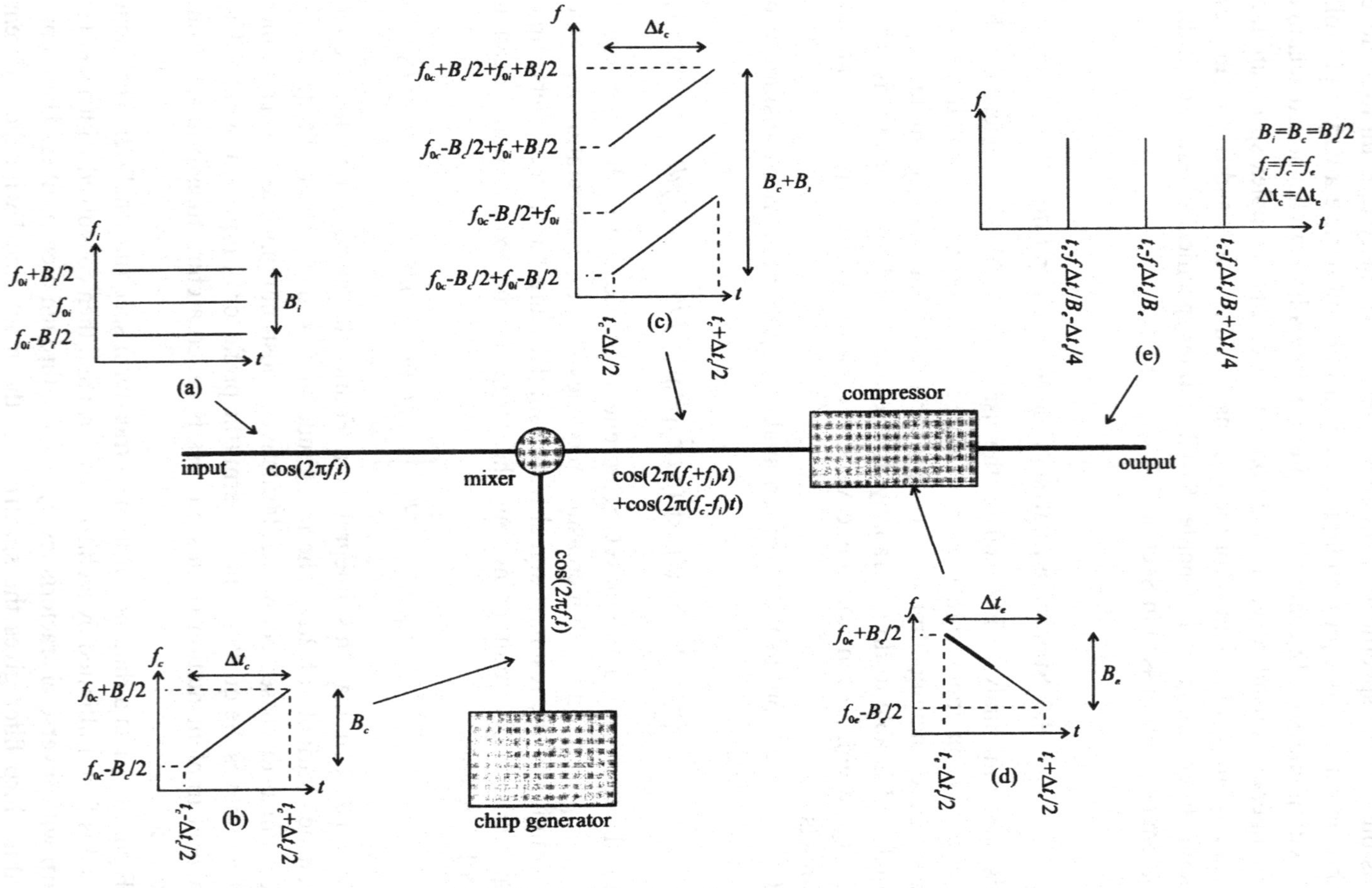

Figure 8.5.1. Chirp transform spectrum analyser.

generate the correct phase of the spectrum. This can be achieved by a further multiplication by the original chirp waveform, as seen above. This arrangement is known as the multiply–convolve–multiply spectrum analyser. It is not shown in Figure 8.5.1, since most applications only require the magnitude spectrum.

In order to explain how the spectrum analyser works, a series of frequency–time graphs is shown in Figure 8.5.1, labelled (a)–(e). Consider an input waveform which contains three frequency components, as depicted in Figure 8.5.1(a). One component is at frequency f_{0i} and two others are at frequencies $B_i/2$ above and below this frequency. This waveform is now mixed with a chirp waveform. The characteristics of the chirp waveform are shown in Figure 8.5.1(b). It has a bandwidth B_c and a time delay ΔT_c, each centred on f_c and t_c respectively. The method of generating this chirp can either be by applying an impulse to a delay line, chirp-generating filter, or by the use of a voltage controlled oscillator. At any one time the mixer shown in the diagram produces signals given by

$$2\cos(2\pi f_i t)\cos(2\pi f_c t) = \cos(2\pi(f_i + f_c)t) + \cos(2\pi(f_i - f_c)t) \quad (8.5.3)$$

The signal containing the sum frequency is the one of interest since the following filter removes the lower-frequency components. Figure 8.5.1(c) shows the frequency–time graph of this signal. At any one time the chirp frequency is shifted by the frequency of the input signal, resulting in the three-tone input becoming a signal containing three chirps, each separated in frequency by $B_i/2$, and of bandwidth B_c. It can be seen that the total bandwidth of this resulting signal is $B_c + B_i$, with the same duration as the chirp Δt_c.

The signal of Figure 8.5.1(c) is now input to a delay line chirp filter, which has the reverse characteristics of the chirp shown in Figure 8.5.1(b). This characteristic is shown in Figure 8.5.1(d). However, this needs to cover the wider bandwidth $B_e = B_i + B_c$ in order to process the full range of signals. A chirp signal with bandwidth B_e and time delay t_e input to this filter would be compressed to an impulse which occurs at time t_e. However, the signal in Figure 8.5.1(c) contains chirp waveforms which do not cover this full bandwidth and thus produce an output impulse at a different time. For example, the bold part of the line in Figure 8.5.1(d) is the highest-frequency chirp input to the filter. It can be seen that the time delay when passing this through the chirp filter will not be at the central time, but will be shifted by an amount $-\Delta t_e/4$. In practice, the multiplying chirp is increased in duration in order for the whole aperture of the compressor to be filled. The waveform depicted in Figure 8.5.1(e) is the final output, with three impulses, representing the three input tones. In this case, and in practical implementations, it is convenient to take $B_c = B_i = B_e/2$, $\Delta t_e = \Delta t_c$ and $f_i = f_c = f_e$.

The output signals in Figure 8.5.1(e) are delta functions and have no width. In practice the output signal has both a finite width and sidelobes. The sidelobes can be reduced by weighting the amplitude of the compressor response by the use of standard weighting functions. This not only has the effect of reducing the sidelobe levels in the output, but it also increases the width and reduces the resolution. The resolution of the spectrum analyser is the inverse of the compressor time delay, multiplied by a factor to take this weighting into account.

A number of spectrum analysers based on this principle have been constructed from both low-temperature and high-temperature superconductors. The first one was a niobium spectrum analyser reported in 1985.[48,49] This analyser has a total bandwidth of 2.4 GHz centred on 4.0 GHz with a designed resolution of 39.2 MHz. The system uses two niobium delay lines, each consisting of 39 μm-wide, 3.32-long striplines using 125-μm-thick silicon substrates. The delay lines are of the backward wave coupler type described in Section 6.5 and are designed for a total dispersion or delay of 37.5 ns over the 2.32 GHz bandwidth. The delay line used for the compressor has a Hamming weighting, whilst the delay line used to generate the chirp is flat weighted. In order to produce a long multiplying chirp, the flat-weighted expander device is used twice and the resultant signal is passed through a frequency doubler. Only the niobium delay lines are at 4.2 K; the rest of the electronics are at room temperature. The results show a ± 1.2 dB amplitude uniformity over the 2.8–5.2 GHz range of the analyser, with a -4 dB peak width of 0.7 ns, corresponding to a resolution of 43 MHz. The sidelobe levels are at -18 dB, slightly higher than -24 dB expected from preceding experiments using the chirp filters as the expander and compressor. The difference in sidelobe levels is attributed to the room-temperature electronics. The dynamic range of the device is 46 dB, being limited by thermal noise at -43 dBm and non-linearity at 3 dBm.

A similar design using YBCO delay lines has also been demonstrated. This spectrum analyser is included in the cueing receiver described in Section 8.1.2 and is designed for space applications.[50,51] The bandwidth of the device ranges from 7.0 to 10.0 GHz, with a 120 MHz resolution using a compressor with a 12 ns delay over the 3 GHz bandwidth.

It is interesting to note that other signal processing functions can be performed with chirp transform systems.[52] Two examples are shown in Figure 8.5.2. The first is a filter in which the frequency response can be varied by the application of a gating function. This gating function is applied when the signal is in the spectral domain. By attenuating the signal via the gate or by completely removing the frequency bands, the filter can be adjusted to the required frequency response. It is necessary to use two channels with timing interleaved in order to provide continuous filtration. Such filters have been conceived and demonstrated using surface acoustic wave technology.[53]

The other function, shown in Figure 8.5.2(*b*), is a signal processor which can add a delay to a signal. By changing the frequency of the local oscillator, the delay of the output signal can be varied given the constraints of the bandwidth and the time delay of the chirp filters. Again, SAW technology provides a demonstration.[54] In addition, it is possible to realise a variable time compression, expansion or inversion using chirp filters,[55] as well as other signal

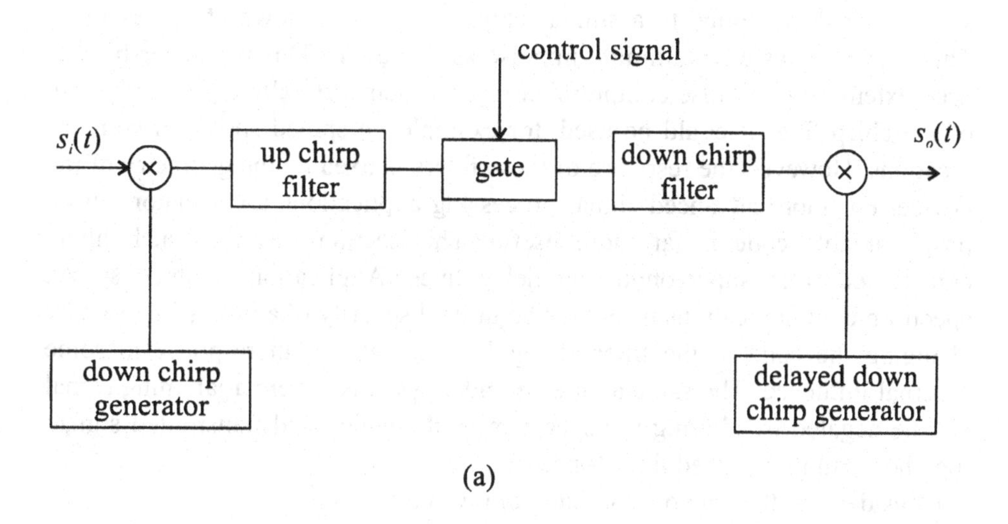

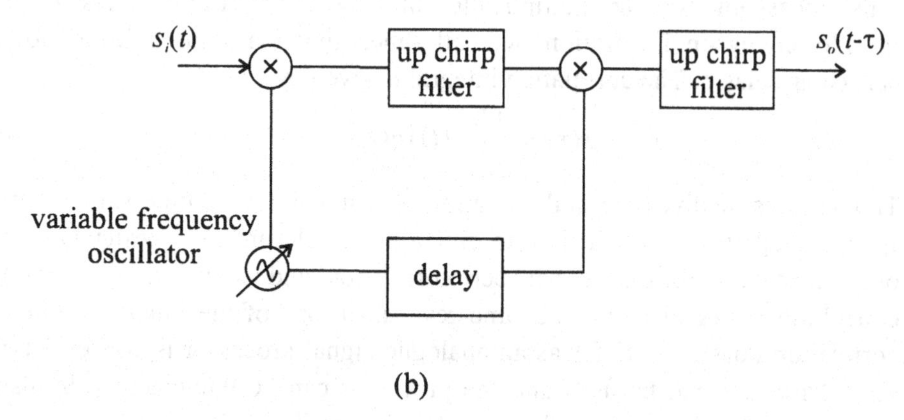

Figure 8.5.2. (*a*) Variable frequency response filter implemented by using two successive chirp transforms. (*b*) Variable time delay generation.

processing applications. All these signal processing systems require a high-performance chirp filter. Increasing both the time delay and the bandwidth of the devices is the goal for better performance, in addition to reducing spurious effects and noise.

8.6 Convolvers and correlators

To some extent, correlators and convolvers have already been looked at in the chapter on delay line filters. A delay line filter is a fixed code, matched filter or correlator. For example, correlation of an up chirp signal can be obtained if such a signal is applied to a similar chirp filter with a down chirp response. This system gives a bandlimited impulse as its output. This particular filter is used extensively in pulse compression radar systems. By altering the response of the chirp filter it could be used, for example, in spread spectrum communications. However, the response of these filters is fixed by the geometry of the device. For more advanced signal processing applications a correlator with a programmable code is far more useful. This section describes such filters constructed from superconducting delay lines. Applications include spread spectrum communications where the additional security of a link is imposed by changing the code in the filter at regular intervals, or in response to some external influence. The same applies in radar systems, where a jamming signal can be negated by altering the spectrum of the wide band transmitted signal, together with its matched filter for reception.

Consider first the convolution integral given by

$$(f * g)(\tau) = \int_{-\infty}^{\infty} f(t)g(\tau - t)\, dt \tag{8.6.1}$$

This process involves the multiplication of a signal f by a time reversed signal g followed by an integration over all time, giving a result relevant to the particular delay τ. The correlation integral is given by

$$r(\tau) = \int_{-\infty}^{\infty} f(t)g(t + \tau)\, dt \tag{8.6.2}$$

This process involves the multiplication of a signal f by a time shifted signal g, and integration over all time to give a value relevant to a particular τ. The only substantial difference between correlation and convolution is that the convolution integral requires a time reversal in one of the functions whereas correlation does not. As far as an analogue signal processor is concerned, the same hardware can be used and the processes can be interchanged by using time reversed replicas of the signals to be processed.

The process of correlation/convolution requires both an integration and a multiplication. The integration can be done by the delay line transversal filter techniques discussed in Chapter 6; however, the multiplication is more difficult. Convolvers and correlators have been studied widely using surface acoustic wave (SAW) technology. Here designs have been demonstrated using the inherent non-linearity of the piezoelectric material forming the substrate of the SAW device, although external non-linear elements have also been investigated. The disadvantage of this process is that large-amplitude signals are required in order for the non-linearity to generate an output signal of sufficient amplitude to overcome noise or other spurious effects. In principle, multiplication of signals can be achieved by the inherent non-linearity in the superconducting transmission lines; however, external mixers are more efficient.

A niobium superconductive convolver is shown in Figure 8.6.1.[56,57] This consists of a 14 ns delay line in the form of a meandered stripline line, with proximity couplers spaced at appropriate intervals. Multiplication is achieved by 25 tunnel junction ring mixers and integration is achieved by summation of the output signals on to a single transmission line. The ring mixers consist of four ports, one for the bias current, two for the input signals, which are 90° out of phase, and the fourth for the output signal. Each of the legs of the device consists of several tunnel junctions in series. These junctions consist of a niobium base electrode and a niobium oxide tunnel barrier followed by a lead counter electrode. The bandwidth of the device is 2 GHz, resulting in a time–bandwidth product of 28. The maximum output of the device is −58 dBm. The performance of this integrated device is deduced by application of both tone bursts and chirp waveform, the latter producing −7 dB sidelobes.

One of the fundamental limitations of the convolver shown in Figure 8.6.1 is that it is limited to the processing of waveforms which are restricted to the

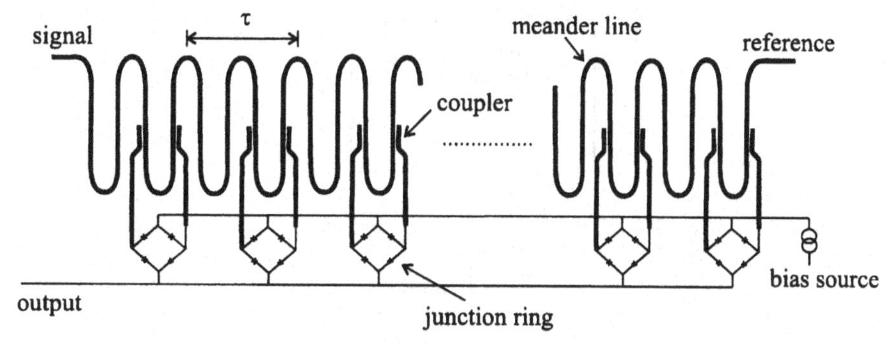

Figure 8.6.1. A superconductive convolver. (After reference [56].)

delay available on the delay lines involved. By using an integrating technique it is possible to extend this time in order to process signals of longer duration.[58] Consider the single-channel correlator in Figure 8.6.2; the input and delayed reference signals are first multiplied and then integrated over time. The integration is frequency selective and can be implemented by the use of a simple resonant circuit. For long integration periods the Q of such a circuit needs to be large. The output of the circuit represents a single point in the correlation function of the two signals s and r, but the processing time is limited solely by the time delay between readouts of the integrator. If the waveforms are repetitive, then the delay τ can be varied on each repetition in order to build up the full correlation signal. Alternatively, a tapped delay line can be used with multiple channels, each formed from the integrator in Figure 8.6.2. A single-channel device has been demonstrated, consisting of a tapped niobium delay line, junction mixers and a simple LC resonator, using similar technology to that described above. The circuit was tested with single-tone inputs.

It is interesting to note that other signal processing functions can be accomplished once the basic correlator or convolver structure is available. For example, if a signal is applied to one of the ports of the convolver and a pulse is applied to the reference port delayed by a certain time τ, the output will be a delayed version of the input signal. The delay will depend upon both τ and the length of the transmission line.

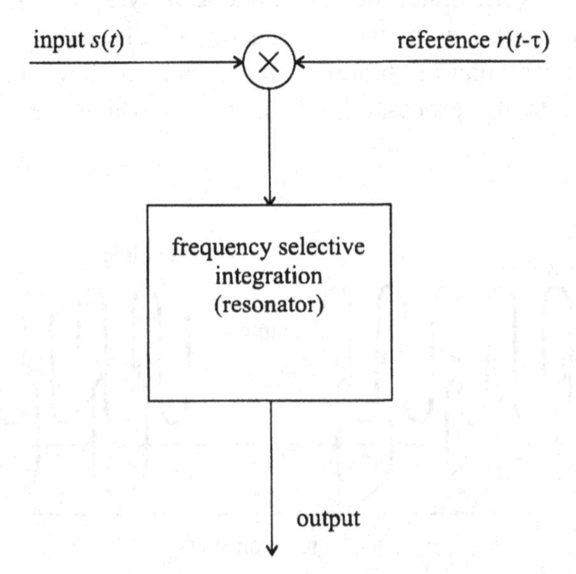

Figure 8.6.2. Time integration analogue signal correlator. (After reference [58].)

8.7 Instantaneous frequency meter

A generalised instantaneous frequency meter (IFM) is shown in Figure 8.7.1. The function of this device is to measure the frequency of an input, the output consisting of a digital code specifying the frequency. It could be used, for example, to replace the spectrum analyser in the receiver described in the preceding section. The input signal, which is usually bandlimited to the bandwidth of the IFM, is split into a number of channels by a power divider. Each of the resulting signals is further split into two and fed into two delay lines. After delays of τ_1 and τ_2, the signals are then recombined. It is the interference signal from this differential delay which forms the digital signal after detection and hard limiting. By correct choice of the delays on each of the channels, it is possible to form a digital code which uniquely identifies the input signal tone over the bandwidth of the device. Because the device is parallel in nature, any signal at the input is detected, even if it is rapidly changing in frequency; hence, the probability of intercept is 100%. The processing time before the signal is identified is about the same as the longest

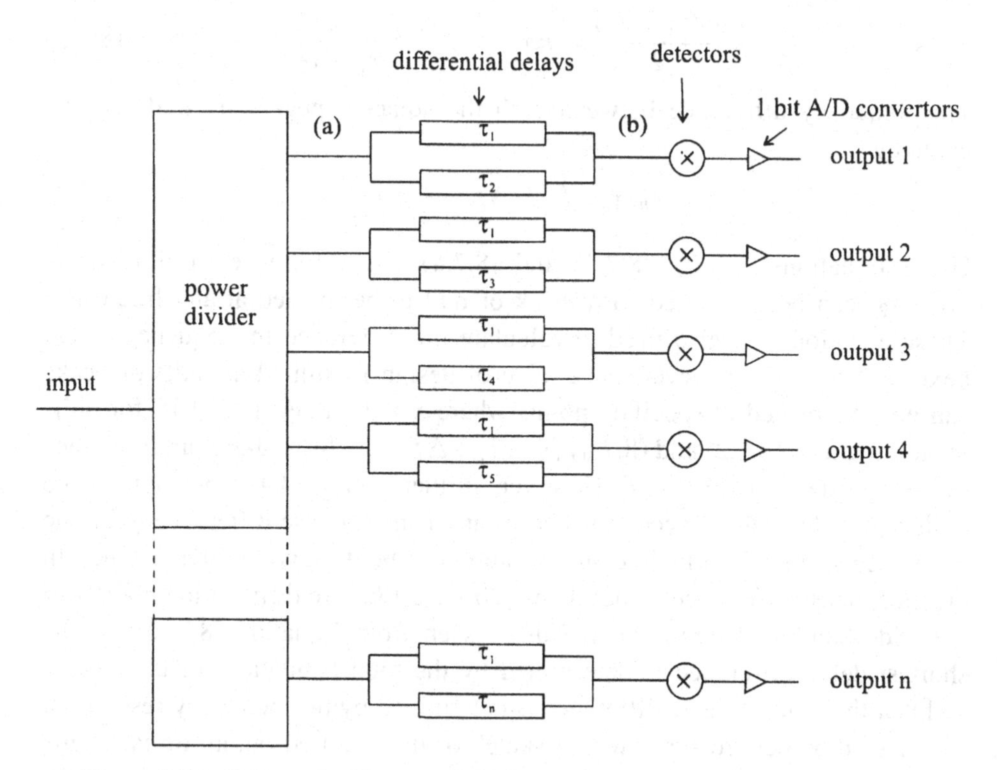

Figure 8.7.1. Schematic diagram of the instantaneous frequency meter system.

delay in the system. Clearly, there is a large amount of additional circuitry in the device besides the delay lines, but these delay lines form the central processing component and are the superconducting components. It is the ability of superconductors to produce a number of long delays in a suitable small volume which makes this device possible.

If the input to a single stage of the IFM (at point (*a*) in Figure 8.7.1) is $\sin(\omega t)$, then the output after the delay and recombination (at point (*b*)) is given by

$$s(t) = \frac{1}{2}\sin(\omega t + \omega\tau_1) + \frac{1}{2}\sin(\omega t + \omega\tau_2 + \phi)$$
$$= \sin\left(\frac{2\omega t + \omega(\tau_1 + \tau_2)}{2}\right)\cos\left(\frac{\omega(\tau_1 - \tau_2)}{2} - \frac{\phi}{2}\right) \quad (8.7.1)$$

where ϕ is a frequency independent phase shift. Setting this to zero for the moment, a zero or null occurs in this output waveform when

$$\frac{\omega(\tau_1 - \tau_2)}{2} = n\pi - \frac{\pi}{2} \quad \text{or} \quad f = \frac{n - 1/2}{\tau_1 - \tau_2} \quad (8.7.2)$$

Similarly, a peak occurs when

$$\frac{\omega(\tau_1 - \tau_2)}{2} = n\pi \quad \text{or} \quad f = \frac{n}{\tau_1 - \tau_2} \quad (8.7.3)$$

The frequency difference between both the adjacent peaks and nulls is thus given by

$$f_{n+1} - f_n = \Delta f = \frac{1}{\tau_1 - \tau_2} \quad (8.7.4)$$

Using Equations (8.7.2), (8.7.3) and (8.7.4), the difference in time delay $(\tau_1 - \tau_2)$ can be calculated for a peak or null to be located at any frequency. These equations are also used to calculate the difference in frequency to the next peak or null. However, the choice of design is limited as nulls or peaks can only be placed at specified points where n is an integer; that is, for nulls from Equations (8.7.2) and (8.7.4) $(n - 1/2)\Delta f = f$, for peaks from Equations (8.7.3) and (8.7.4) $n\Delta f = f$. However, further tuning of the position of the nulls and peaks can be accomplished by inserting a phase difference ϕ in one of the delay lines. In principle, any position can be designed in this manner. In practice, phase differences other than $\pm 90°$ or $\pm 180°$ are difficult to obtain over the wide bandwidth required. It can be seen from Equation (8.7.4) that the shortest delay difference is determined by the total bandwidth of the system and that the longest delay difference is determined by the frequency resolution. The design procedure starts with a sketch of the required output of each time delay section, as shown in the upper diagram of Figure 8.7.2. This may be modified in order to alter the required digital code. The above equations are then used to calculate the difference in time for each section, including any phase differences if required. Once this is known, τ_1 can be chosen arbitrarily and then τ_2 can be calculated for each of the sections.

A simple example is shown in Figure 8.7.2. This is a two-bit device with the first section containing delay lines, each with a delay of 2 ns and a phase difference of 90°. The second section has a delay difference of 2 ns and no phase difference. This gives a system with a centre frequency of 4 GHz and a bandwidth of 0.25 GHz. The output from each of the sections after the recombination of the delayed signals is shown in the upper diagram of Figure 8.7.2 and the signals after hard limiting are shown below this. A table of the resultant digital output is also given.

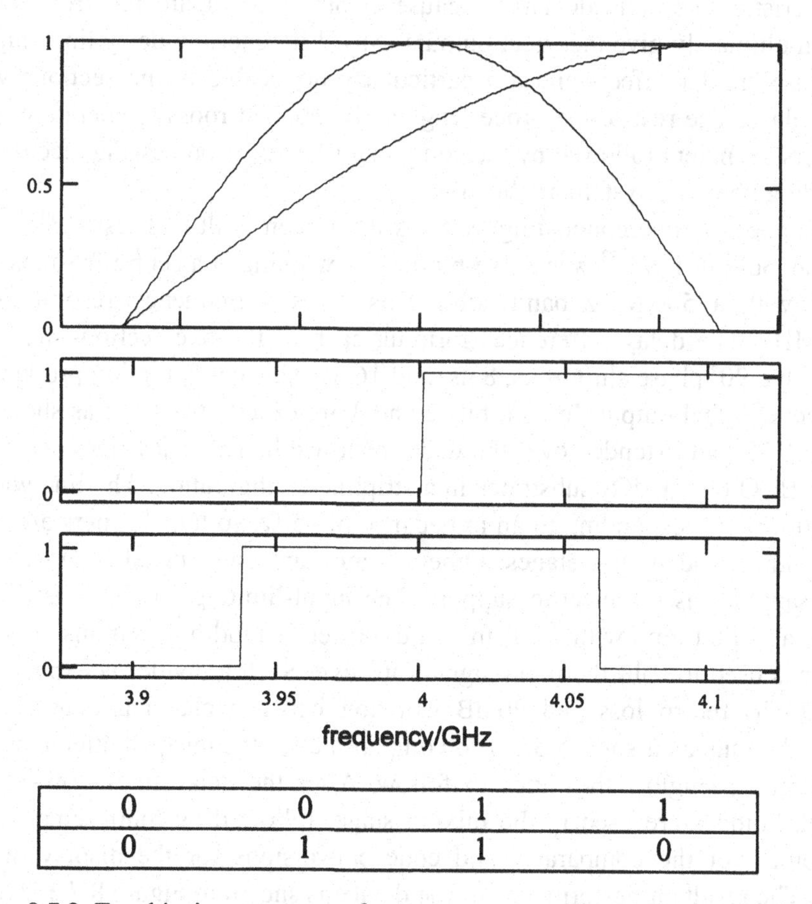

0	0	1	1
0	1	1	0

Figure 8.7.2. Two-bit instantaneous frequency meter. The upper trace shows the frequency response of each of the channels; the lower traces show digitised versions of each trace and the binary code of the output is shown at the bottom.

The delay lines in the IFM form the major signal processing element, but a large number of additional components form the whole system. On the input side, a limiting amplifier is useful in order to reduce the sensitivity to variations in input signal level and to reduce the effects of small second tones which cause interference. This is followed by a bandpass filter in order to remove the effects of out-of-band signals. Such a filter can be superconducting in order to provide ultimate sensitivity; but this will only make a small difference because of its relatively wide bandwidth. Conventional GaAs amplifiers can be used for the limiter, which may be cooled or uncooled depending upon the system configuration. A number of power dividers are required in the system, including an n-way divider and two two-way dividers in each of the delay line sections. Ideally, these can be miniaturised and placed on the substrate with the superconducting delay lines. The delay lines must have a linear phase characteristic since any deviations cause errors in the switching frequencies and can ultimately give incorrect outputs. Any differences in delay line lengths can cause similar effects and are particularly noticeable in the sections with short delays. The rest of the processing can be done at room temperature with detectors or mixers followed by the comparator and post-processing electronics depending upon the system application.

A number of superconducting systems have been built and tested.[59,60] The first one, built in 1993,[53] was a five-bit device, working at a centre frequency of 4 GHz with a 500 MHz bandwidth. This gives a frequency resolution of ±7.8 MHz. The delay differences forming each of the five sections are 2 ns, 2 ns with a 90° phase shift, 4 ns, 8 ns, and 16 ns. This configuration produces a Gray code digital output. In fact, bits 2 and 3 are exactly the same as shown in Figure 8.7.2, but extended over the wider bandwidth. The delay lines are made from YBCO on $LaAlO_3$ substrates in a stripline configuration. The line widths are 100 μm, corresponding to an impedance of 45 Ω, so transformers are used on the input and output stages. Other circuits are constructed in microstrip using sapphire as a dielectric support. The input-limiting amplifier is a five-stage GaAs FET amp with 16 dBm output power. A modified Wilkinson power divider is used to split the input signal into five. Such a divider is expected to have 20 dB return loss and 20 dB isolation between channels. The design included an unequal split in order to compensate for the unequal attenuation in the differing length delay lines to follow. After the delay lines, low-barrier Schottky diodes are used in the mixing stage, followed by room-temperature electronics for the comparator and code conversions for the display of the result. The resultant performance of the device is shown in Figure 8.7.3. This is the response of the IFM to an input tone which is increased in frequency. It is important that no channels are dropped and that the frequency transition takes

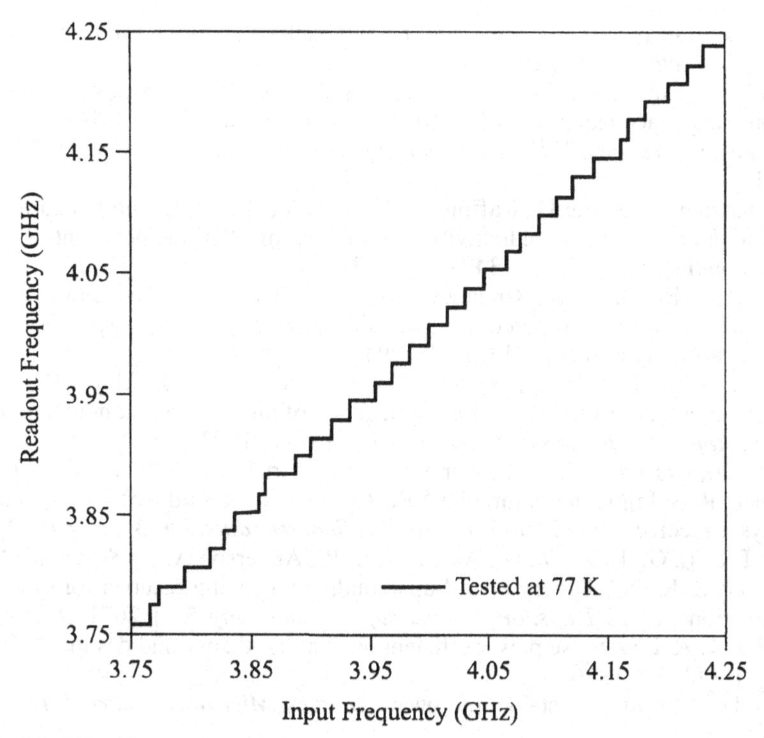

Figure 8.7.3. IFM frequency readout versus input frequency. (Taken from reference [59].)

place as close as possible to the design; in this case the average magnitude of the error is 3.1 MHz. The measured input dynamic range is from −40 dBm to 10 dBm, with dual-tone suppression, provided that the second tone is 10 dB lower than the dominant tone. This is 6 dB better with the limiting amplifier included.

A similar four-bit system operating at a centre frequency of 10 GHz with a bandwidth of 1 GHz has been investigated using niobium as well as YBCO.[60] In this case, coplanar delay lines are used. All the superconducting delay lines are fabricated on a 40 mm × 40 mm $LaAlO_3$ substrate, with the other components on separate substrates.

8.8 References

1 Fitelson M. M. Cryogenic electronics in advanced sensor systems, *IEEE Trans. on Applied Superconductivity*, **5**(2), 3208–13, 1995
2 Kapolnek D. J., Aidnik D. L., Hey-Shipton G., James T. W., Fenzi N. O., Skoglund D. L. and Nilsson B. J. L. Integral FMCW radar incorporating an HTSC delay line

with user transparent cryogenic cooling and packaging, *IEEE Trans. on Applied Superconductivity*, **3**(1), 2820–3, 1993
3 Nissenoff M., Ritter J. C., Price G. and Wolf S. A. The high temperature superconducting space experiment HTSSE-I. Components and HTSSE-II subsystems and advance devices, *IEEE Trans. on Applied Superconductivity*, **3**, 2885–90, 1993
4 Ngo P., Krishen K., Arndt D., Raffoul G., Karasak V., Bhasin K. and Leonard R. A high temperature superconductivity communications flight experiment, *Applied Superconductivity*, **1**(7–9), 1349–61, 1993
5 Leonard R. F., Bhasin K. B., Romanofsky R. R., Cubbage C. D. and Chorey C. Z. Space applications of superconducting microwave electronics, *Applied Superconductivity*, **1**(7–9), 1341–7, 1993
6 Carlberg I. A., Kelliher W. C., Wise S. A., Hooker M. W. and Buckley J. D. Environmental considerations for applications of high T_c superconductors in space, *Applied Superconductivity*, **1**(7–9), 1251–8, 1993
7 Lyons W. G., Arsenault D. R., Seaver M. M., Boisvert R. R., Sollner T. C. L. G. and Withers R. S. Implementation of a $YBa_2Cu_3O_{7-x}$ wideband real-time spectrum-analyser receiver, *IEEE Trans. on Applied Superconductivity*, **3**(1), 2891–4, 1993
8 Sollner T. C. L. G., Lyons W. G., Arsenault D. R., Anderson A. C., Seaver M. M., Boisvert R. R. and Slattery R. L. Superconducting cueing receiver for space applications, *IEEE Trans. on Applied Superconductivity*, **5**(2), 2071–4, 1995
9 Everard J. K. A. Low noise power efficient oscillators: theory and design, *IEE Proc.*, **133**(4), 172–80, 1986
10 Leeson D. B. Short term stable microwave sources, *Microwave Journal*, **13**, 59–69, 1970
11 Everard J. K. A. Low noise radio frequency oscillators, *IEE International Conf. on Frequency Control and Synthesis*, 1987
12 Parker T. Characteristics and sources of phase noise in stable oscillators, *Proc. 41st Ann. Frequency Control Symp.*, pp. 99–110, 1987
13 Llopis O., Plana R., Amine H., Escotte L. and Graffeuil J. Phase noise in cryogenic microwave HEMT and MESFET oscillators, *IEEE Trans. on Microwave Theory and Techniques*, **MTT41**(3), 369–74, 1993
14 Gerber E. A., Lukaszek T. and Ballato M. Advances in microwave acoustic frequency sources, *IEEE Trans. Microwave Theory Techniques*, **MTT34**, 1002–16, 1986
15 Montress G. K., Parker T. E. and Loboda M. J. Extremely low phase noise SAW resonator oscillator design and performance, *Proc. IEEE Ultrasonics Symp.*, pp. 47–52, 1987
16 Withers R. S. and Ralston R. W. Superconductive analogue signal processing devices, *Proc. IEEE*, **77**(8), 1247–63, 1980
17 Fiedziuszko S. J. Microwave dielectric resonators, *Microwave Journal*, pp. 1989–2000, Sept., 1986
18 Klieber R., Ramisch R., Valenzuela A. A., Weigel R. and Russer P. A coplanar transmission line superconductive oscillator at 6.5 GHz on a single substrate, *IEEE Microwave and Guided Wave Lett.*, **2**(1), 22–4, 1992
19 Gardner P., Paul D. K. and Tan K. P. Microwave voltage tuned microstrip ring resonator oscillator, *Electron. Lett.*, **30**(21), 1770–1, 1994
20 Oates D. E., Lyon W. G. and Anderson A. C. Superconducting thin film $YBa_2Cu_3O_7$ resonators and filters, *Proc. 45th Annual Symp. on Frequency Control*, p. 460, 1991

21 Oates D. E., Anderson A. E. and Steinbeck J. Superconducting resonators and high-T_c materials, *Proc. 42nd Ann. Frequency Control Symp.*, pp. 545–9, 1988
22 Rohrer N. J., Valco G. J. and Bhasin K. B. Hybrid high temperature superconductor/GaAs 10 GHz microwave oscillator: Temperature and bias effects, *IEEE Trans. on Microwave Theory and Techniques*, **MTT41**(11), 1865–71, 1993
23 Khanna A. P. S., Schmidt M. and Hammond R. B. A superconducting resonator stabilised low phase noise oscillator, *Microwave Journal*, p. 127, Feb., 1991
24 Wilker C., Shen Z. Y., Pang P., Face D., Holstein W. L., Matthews A. L. and Laubacher D. B. 5 GHz high temperature superconductor resonator with high Q and low power dependence at 90 K, *IEEE Trans. on Microwave Theory and Techniques*, **MTT39**, 1462–7, 1991
25 Khanna A. P. S. and Schmidt M. Low phase noise superconducting oscillators, *IEEE MTT-S Int. Microwave Symp. Digest*, **3**, 1239–42, 1991
26 Everard J. K. A., Cheng K. K. M. and Dallas P. A. High-Q helical resonators for oscillators and filters in mobile communication systems, *Electron. Lett.*, **25**(24), 1648–50, 1989
27 Flory C. A. and Taber R. C. Microwave oscillators incorporating cryogenic sapphire resonators, *IEEE Proc. Int. Frequency Control Symp.*, 763–73, 1993
28 Negrete G. V. An ultra low noise millimetre wave oscillator using a sapphire disk resonator and high temperature superconductor ground planes, *Microwave and Optical Tech. Lett.*, **6**, 758–62, 1993
29 Shen Z-Y., Wilker C., Pang P. S. W., Carter C. S., Nguyen V. X. and Laubacher D. B. High temperature superconductor/III–V hybrid microwave circuits, *Microwave and Optical Tech. Lett.*, **6**, 728–32, 1993
30 Shen Z-Y., Pang P. S. W., Wilker C., Laubacher D. B., Holstien W. L. and Carter C. S. High T_c superconductor and III–V solid state hybrid microwave circuits, *IEEE Trans. Appl. Superconductivity*, **3**, 2832–5, 1993
31 Klein N., Tellmann N., Dähne U., Scholen A. and Schulz H. YBCO shielded dielectric resonators for stable oscillators, *IEEE Trans. on Applied Superconductivity*, **5**(2), 2663–6, 1995
32 Luiten A. N., Mann A. G. and Blair D. G. Cryogenic sapphire microwave resonator-oscillator with exceptional stability, *Electron. Lett.*, **30**(5), 417–19, 1994
33 Phillips W. A., Jedazik D., Lamacraft K., Zammattio S., Greed R. B., Hedges S. J., Whitehead P. R., Nicholson B. F., Button T. W., Smith P. A., Alford N. McN., Peters N. and Grier J. An integrated 11 GHz cryogenic down converted, *IEEE Trans. on Applied Superconductivity*, **5**(2), 2283–6, 1995
34 Button T. W., Smith P. A., Alford N. McN., Greed R. B., Adams M. J. and Nicholson B. F. Low phase noise oscillators incorporating YBCO thick film cavity resonators, *European Applied Superconductivity Conf.*, Edinburgh, 1995
35 Wilker C., Pang P., Carter C. and Shen Z.-Y. S-parameter, I–V curves and noise figure measurements of III–V devices at cryogenic temperatures, *1992 Automatic RF Techniques Conf. Digest*, 1992
36 Smuk J. W., Stubbs M. G. and Wright J. S. S-Parameter characterisation and modelling of three terminal semiconductive devices at cryogenic temperatures, *IEEE Microwave and Guide Wave Lett.*, **2**, 111–13, 1992
37 Shen Z.-Y., Pang P., Wilker C., Laubacher D. B., Holstein W. L., Carter C. F. and Adlerstein M. High T_c superconductor and III–V solid state microwave hybrid circuits, *IEEE Trans. on Applied Superconductivity*, **3**(1), 2832–5, 1993
38 Suzuki K. *et al. ISTEC Journal*, **6**(1), paper AD–7, 1993
39 Madjar A., Biran A., Breitbard A. and Philosof A. A superconducting compact

C-Band low noise front end with an integral cooler, *Proc. 1994 European Microwave Conf., Cannes, France*, pp. 1513–17, 1994
40 Barner J. B., Bautista J. J., Bowen J. G., Chew W., Foote M. C., Fujiwara B. H., Guern A. J., Hunt B. J., Javadi H. H. S., Ortiz G. G., Rascoe D. L., Vasquez R. P., Wamhof P. D., Bhasin K. B., Leonard R. F., Romanofsky R. R. and Chorey C. M. Design and performance of low-noise hybrid superconductor/semiconductor 7.4 GHz receiver downconverter, *IEEE Trans. on Applied Superconductivity*, **5**(2), 2075–8, 1995
41 Phillips W. A., Jedamzik D., Lamacraft K., Zammattio S., Greed R. B., Hedges S. J., Whitehead P. R., Nicholson B. F., Button T. W., Smith P. A., Alford N. McN., Peters N. and Grier J. An integrated 11 GHz cryogenic downconverter, *IEEE Trans. on Applied Superconductivity*, **5**(2), 2283–6, 1995
42 Shen Z.-Y., Wilker C., Pang P. S. W., Carter C. F., Nguyen V. X. and Laubacher D. B. High temperature superconductor/III–V hybrid microwave circuits, *Microwave and Optical Tech. Lett.*, **6**, 732–6, 1993
43 Kolesov S. G., Keis V., Vendik O. G., Jeck M. and Chaloupka H. HTS—microwave mixers based on non-linear microstrip transmission lines, *European Applied Superconductivity Conf., Göttingen, Germany*, October 4–9, 1993, DGM Informationsgesellschaft-Verlag, Oberursel, Germany, pp. 1485–8, 1993
44 Koshelets V. P., Shitov S. V., Baryshev A. M., Lapitskaya I. L., Filippenko L. V., van de Stadt H., Mees J., Schaeffer H. and de Graauw T. Integrated sub-mm wave receivers, *IEEE Trans. on Applied Superconductivity*, **5**(2), 3057–60, 1995
45 Zhang Y. M., Winkler D. and Claeson T. The fabrication of an integrated superconducting sub-millimetre wave receiver, *Superconductor Science and Technology*, **7**, 235–8, 1994
46 Lee W. C. Y. *Mobile Cellular Telecommunications Systems*, McGraw-Hill, 1989
47 Calhoun G. *Digital Cellular Radio*, Artech House, 1988
48 Withers R. S. and Reible S. A. Superconductive chirp-transform spectrum analyser, *IEEE Electron. Device Lett.*, **EDL-6**(6), 261–3, 1985
49 Withers R. S., Anderson A. C., Green J. B. and Reible S. A. Superconductive delay-line technology and applications, *IEEE Trans. on Magnetics*, **MAG-21**(2), 186–92, 1985
50 Lyons W. G., Arenault D. R., Seaver M. M., Boisvert R. R., Sollner T. C. L. G. and Withers R. S. Implementation of a $YBa_2Cu_3O_{7-x}$ wideband real-time spectrum-analyser receiver, *IEEE Trans. on Applied Superconductivity*, **3**(1), 2891–4, 1993
51 Sollner T. C. L. G., Lyons W. G., Arenault D. R., Anderson A. C., Seaver M. M., Boisvert R. R. and Slattery R. L. Superconducting cueing receiver for space applications, *IEEE Trans. on Applied Superconductivity*, **5**(2), 2071–4, 1995
52 Oliner A. A. *Acoustic Surface Waves*, Springer-Verlag, 1978
53 Maines J. D., Moule G. L., Newton C. O., Paige E. G. S. A novel SAW variable frequency filter, *Proc. IEEE Ultrasonics Symp.*, pp. 355–8, 1975
54 Dolat V. S. and Williamson R. C. A continuously variable delay line system, *Proc. IEEE Ultrasonics Symp.*, pp. 419–23, 1976
55 Paige E. G. S. Dispersive filters: their design and application to pulse compression and temporal transform *Component Performance and Systems Applications of Surface Acoustic Wave Devices, IEEE Conf. Pub. No. 109*, pp. 167–80, Sept, 1973
56 Reible S. A. Superconductive convolver with junction ring mixers, *IEEE Trans. on Magnetics*, **MAG-21**(2), 193–6, 1985
57 Reible S. A., Anderson A. C., Wright P. V., Withers R. S. and Ralston R. W.

Superconductive convolver, *IEEE Trans. on Magnetics*, **MAG-19**(3), 475–80, 1983

58 Green J. B., Smith L. N., Anderson A. A., Reible S. A. and Withers R. S. Analog signal correlator using superconductive integrated components, *IEEE Trans. on Magnetics*, **MAG-23**(2), 895–8, 1987

59 Liang G. C., Shih C. F., Withers R. S., Cole B. F., Johansson M. E. and Suppan L. P. Superconducting digital instantaneous frequency measurement subsystem, *IEEE Trans. on Microwave Theory and Techniques*, **41**(12), 2368–75, 1993

60 Biehl M., Vogy A., Herwig R., Neuhaus M., Crocoll E., Lochschmied R., Scherer T., Jutzi W., Kratz H., Berberich P. and Kinder H. A 4-bit instantaneous frequency meter at 10 GHz with coplanar YBCO delay lines, *IEEE Trans. on Applied Superconductivity*, **5**(2), 2279–82, 1995

Appendix 1

The surface impedance of HTS materials

A1.1 Introduction

This appendix looks briefly at the microwave properties of superconducting materials and how these properties vary with external influences. It looks briefly at some of the issues affecting these properties and comments on reasons for specific trends. Its primary motivation is to provide typical values of these properties for the microwave design engineer. Although there are many hundreds[1] of high-temperature superconductors with varying transition temperatures, $YBa_2Cu_3O_{7-x}$ (YBCO), with a transition temperature of about 92 K, is by far the most popular. The reason for this is historical, but the surface impedance of any of the HTS materials does not improve significantly over YBCO, even though the transition temperature may be higher. As more materials are investigated in detail, with greater emphasis being placed on the optimisation of the surface impedance, the dominance of YBCO may recede. This appendix thus mainly discusses the microwave properties of YBCO, although a brief mention is made of some of the thallium-based superconductors, which have also been studied in the context of microwave engineering. The majority of measurements presented have been made using the coplanar resonator, described in Section 3.8, at The University of Birmingham. The films tested are state-of-the-art and come from a variety of sources. Since the coplanar resonator is used, the results given are therefore taken on a structure which is close to the final application, and includes any effects due to film patterning.

In order to form thin film microwave devices it is important that the patterning process does not produce deterioration of the superconducting film. A great deal of work has gone into the assessment of different types of etching method. These include dry methods such as ion beam milling, reactive ion etching and focused techniques such as ion beam lithography or laser patterning.[2,3] However, wet etching is preferable because of its convenience and reduced initial capital cost. Low concentrations of acid such as phosphoric, nitric, and hydrochloric produce etching but at a rather large etching rate.[4] Acids such as dicarboxylic adipic or tricarboxylic citric[5,6] and others are also useful. The etch used in the films discussed below is ethylenediaminetetraacetic acid (EDTA),[7,8] which has been shown to give reasonable, controllable etch rates with no significant deterioration of the surface resistance of films, provided good films are used. It also has no, or a negligible, effect upon the substrates. For the particular films described below, a diluted EDTA etch with a pH of 3.8 was used, giving an etch rate of 0.2 nm/s at room temperature.

A1.2 Surface resistance

The value of the surface resistance for HTS thin films for many of the examples given in this book is 100 $\mu\Omega$ at 77 K and 10 GHz. This represents a film on the limit of what is possible, but is a convenient round figure for the examples. Table A1.2.1 shows how this value compares with that of copper at the same temperature, as well as at 300 K. Using these values, the crossover frequency between copper and HTS thin films is 200 GHz. Table A1.2.1 also shows surface resistance values for a good and a typical thick film material. Here the crossover frequencies are 23 GHz and 10 GHz respectively. The typical thick film is used for example calculations in this text.

Another comparison of the surface resistance of YBCO with copper, niobium and niobium–tin superconducting materials is given in Figures A1.2.1 and A1.2.2. Figure A1.2.1 shows the comparison at 4.2 K, whilst Figure A1.2.2 shows the comparison at 77 K. The YBCO is depicted as a shaded band in this diagram because of the spread of experimental measurements reported. The position of a particular sample on the graph depends mainly upon the quality of the material. It is not necessarily an easy matter to produce good, consistent, low surface resistance thin (or thick) films.

Other high-temperature superconducting materials with low surface resistance also fit into the bands in the diagrams. For example, commercially available $Tl_2Ba_2CaCu_2O_8$ (2212), with a T_c of 118 K, is quoted as having a typical surface resistance of 0.25 mΩ at 77 K and 10 GHz.[14]

There have been some reports of the variation of surface resistance not being as the expected ω^2. This is mainly true for the case of thick films and bulk materials.[15] For these polycrystalline materials the most likely explanation is the inclusion of material which behaves as a normal conductor. For the thin films it is less obvious why this should occur. In coplanar resonator measurements, if the effect of the loss tangent of the substrate is not taken into account, then a deviation from the squared frequency behaviour is observed. An alternative explanation is that the finite relaxation time of the normal excitations is more important at higher frequencies, thereby reducing the surface resistance.[13,16]

Table A1.2.1. *Conductivity and surface resistance of copper compared with a good thin film. Values for a good and typical thick film are also shown*

Copper conductivity[9]	77 K	$5.22 \times 10^8\ \Omega^{-1}\,m^{-1}$
	300 K	$5.77 \times 10^7\ \Omega^{-1}\,m^{-1}$
Copper surface resistance (10 GHz)	77 K	8.70 mΩ
	300 K	26.1 mΩ
YBCO thin film surface resistance (10 GHz)	77 K	0.10 mΩ
YBCO thick film surface resistance (10 GHz)	77 K	2.5 mΩ[10,11]
YBCO thick film surface resistance (10 GHz)	77 K	8.7 mΩ

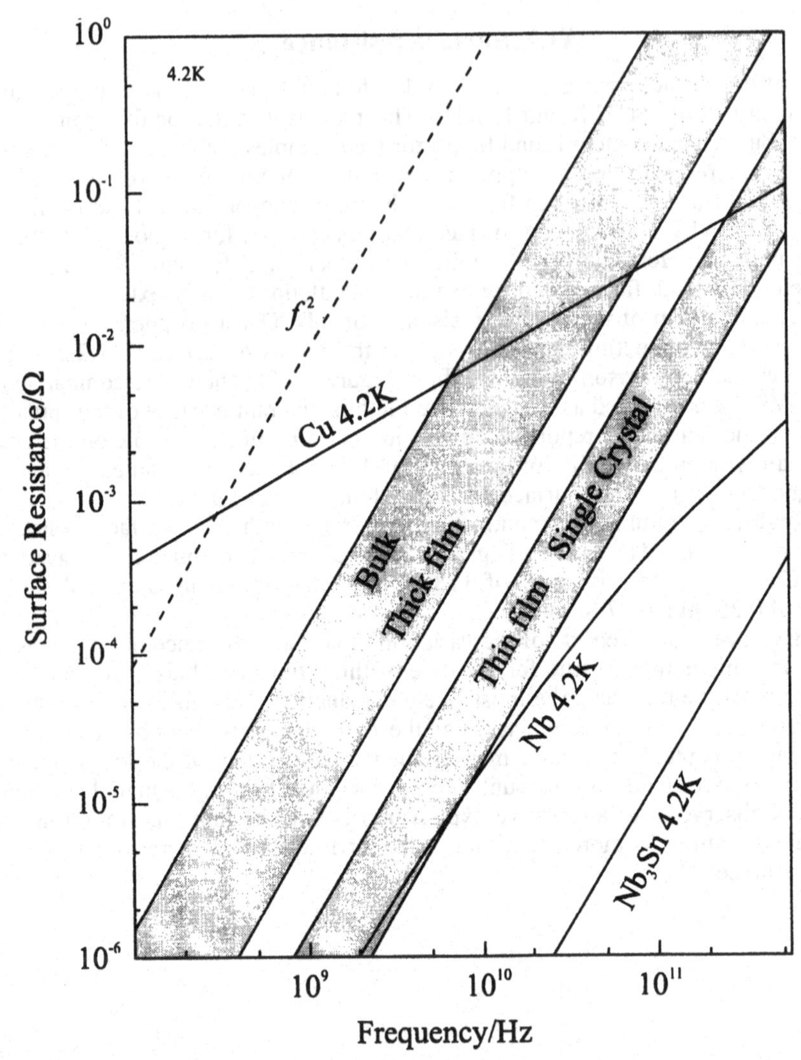

Figure A1.2.1. Comparison of the surface resistance of YBCO at 4.2 K with other superconducting materials. The grey bands represent YBCO. (After references [12, 13].)

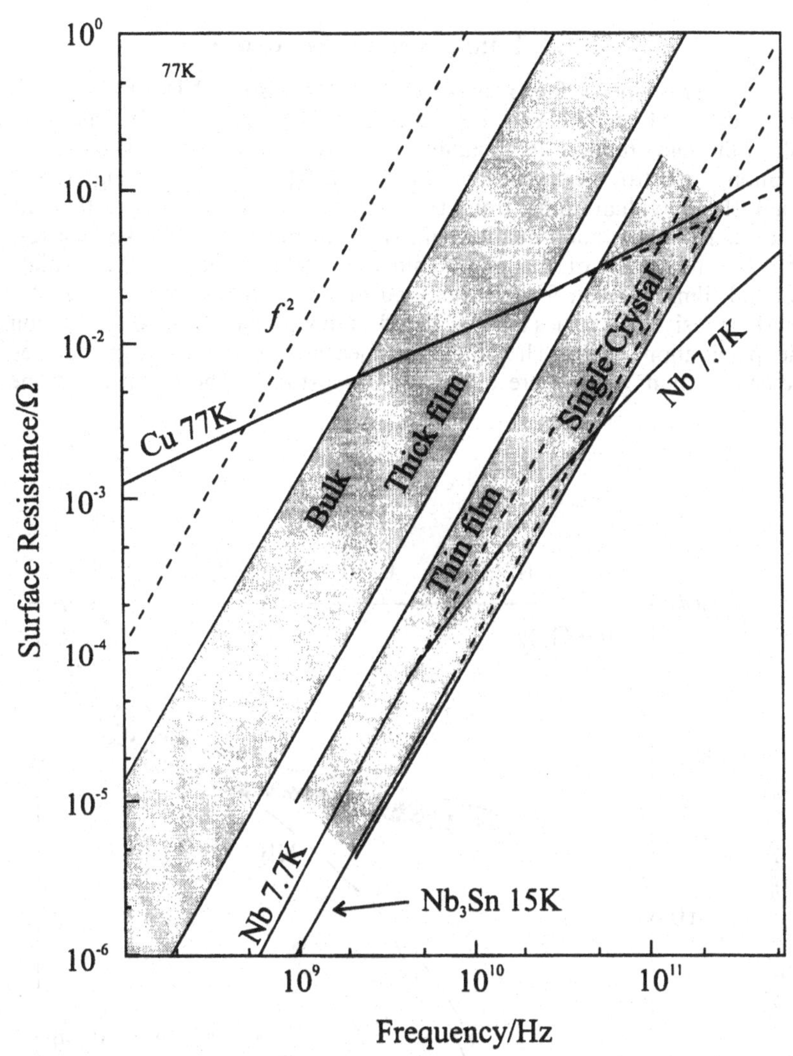

Figure A1.2.2. Comparison of the surface resistance of YBCO at 77 K with other superconducting materials. The grey bands represent YBCO. (After references [12, 13].)

A1.3 Temperature dependence

Figure A1.3.1 shows the surface resistance at 8 GHz of some YBCO films grown by a number of different techniques[17] compared with low-temperature superconducting materials at the same reduced temperature. The YBCO films are made by sputtering,[18] laser ablation,[19] electron beam coevaporation[20] and MOCVD.[21] Although it is difficult to distinguish individual films from this graph, the main point is that all these techniques can produce good quality epitaxial thin films with low surface resistance values.[22] The thickness of the above films is 350 nm for the laser ablated and coevaporated films, 400 nm for the sputtered film and 180 nm for the MOCVD film; all the surface resistance values are calculated as though the films were thick compared with the penetration depth. The films are measured using the coplanar resonator described in Section 3.8 and are all on MgO substrates. The comparative measure-

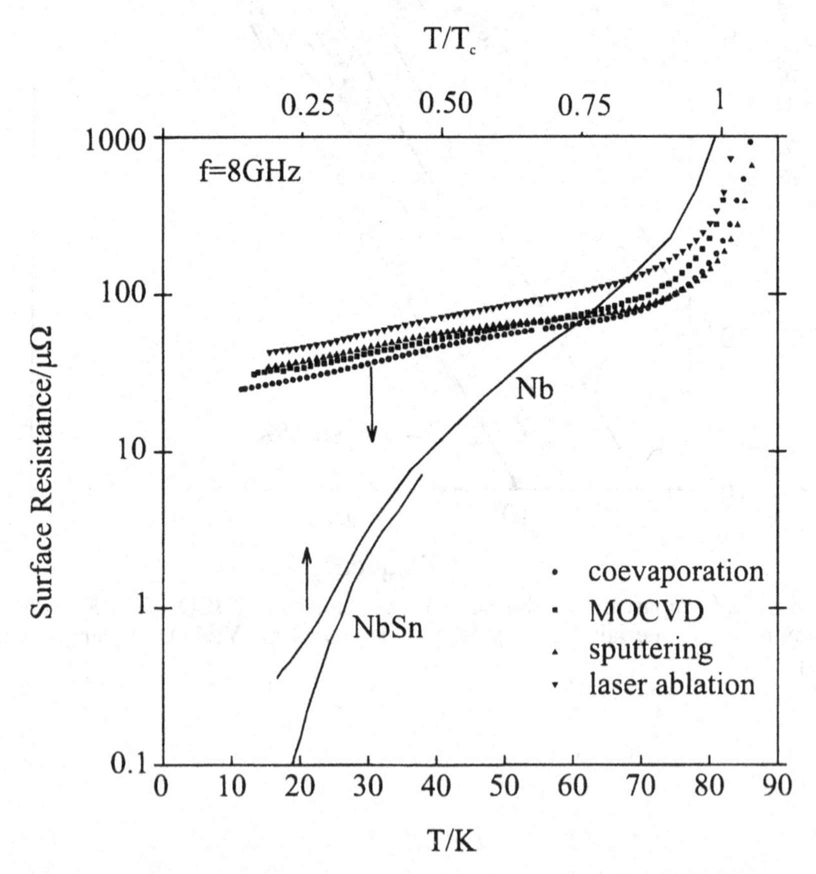

Figure A1.3.1. Temperature dependence of the surface resistance of YBCO grown by a number of techniques compared with niobium and niobium–tin superconductors at the same reduced temperature. The scale on the upper axis refers to the reduced temperature for the Nb_3Sn and Nb, whilst the lower scale refers to the actual temperature of the YBCO films. The T_c of the three materials coincides.

ments for Nn and Nb_3Sn are taken from reference [23] and scaled to 8 GHz using a quadratic frequency variation.

A1.4 Penetration depth and complex conductivity

Figures A1.4.1 and A1.4.2 show the real and imaginary parts of the complex conductivity of YBCO coevaporated films on MgO measured at 8 GHz using the coplanar resonator. The real part of the complex conductivity is dominated by the large bump between 40 and 50 K which corresponds to the region of slowly varying surface resistance given in Figure A1.3.3 in the same temperature regime. For high-purity single crystals, there is actually an increase in the surface resistance in this region as the temperature reduces.[24] This increase in σ_1 is not the BCS coherence peak as described in Section 1.9. It is an effect relating to a rapidly increasing mean free path as the temperature is reduced, competing against a decreasing number of normal electrons.[16,24] This can be seen from the Drude model (Equation (1.2.13)) where $\sigma_1 \propto n_n l$. This increase in mean free path corresponds to a reduction in quasi-particle scattering rate and is much more pronounced in good single crystals than in thin films. Measurements with impurity doping increase the scattering time and reduce the level of the peak in σ_1, with a corresponding reduction in R_s.[24] The conclusion here is that lower surface resistance values are possible with the inclusion of defects. Such defects are more apparent in thin films than in single-crystal samples and are related to grain boundaries, dislocations and/or the oxygenation state of the sample. Oxygenation studies of thin films show that the lowest surface resistance films are obtained if samples are over-doped, with the consequence of a slightly reduced T_c.[16,25] The oxygenation is altered using post-annealing processes, and the control of the oxygena-

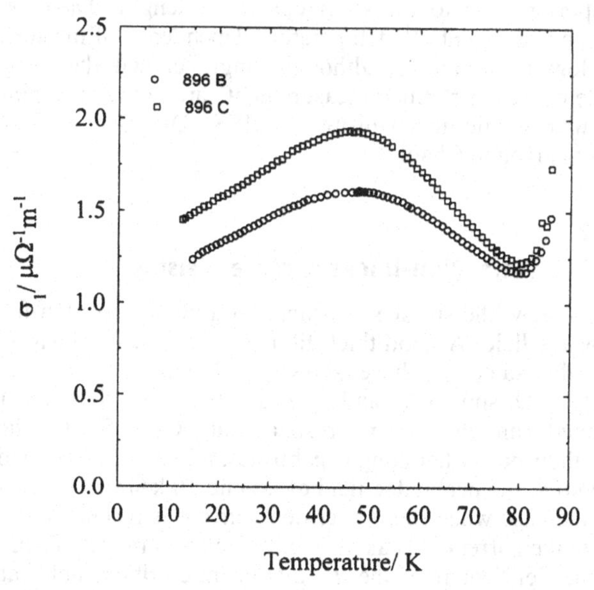

Figure A1.4.1. The real part of the complex conductivity for two patterned films at a frequency of 8 GHz.[13]

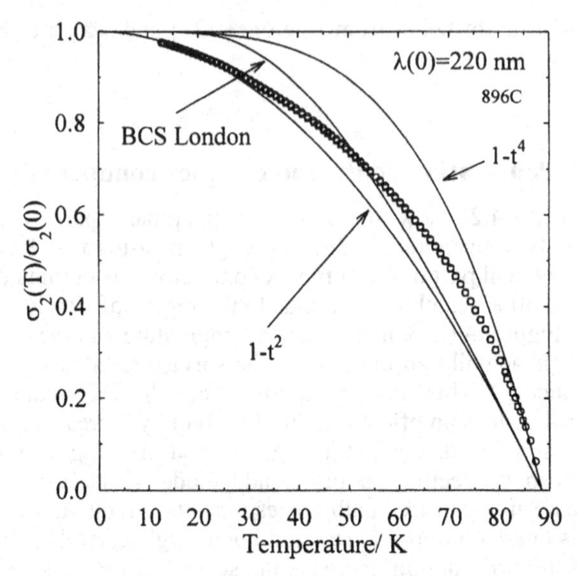

Figure A1.4.2. The imaginary part of the complex conductivity for one of the films shown in Figure A1.4.1 compared with a number of theoretical models.[13]

tion is subtle enough for it to be possible to move between different oxygen states, and return to the original if required.

The low-temperature change in surface resistance of thin films is also affected by impurities and grain boundaries, resulting in the tendency towards a residual loss; this is particularly important due to the short coherence length. There is evidence for exponential[26] and non-exponential[13] temperature dependence of the surface resistance of thin films at low temperatures, although single crystal data suggest a linear dependence with temperature.[24] Such measurements are not able to pinpoint, without reservation, the microscopic mechanisms of HTS. Discussion of some of these theoretical models is given in Chapter 1.

A1.5 Non-linear surface resistance

Figure A1.5.1 shows how the surface resistance of a number of thin films varies as a function of microwave field. A good thick film sample is also shown for comparison. The samples are the same as those shown in Figure A1.3.1 and produced by coevaporation, MOCVD, sputtering and laser ablation. The thin film measurements were again performed using the coplanar resonator at 8 GHz. The thin films show very little difference in their power handling capabilities, all having only a slowly changing surface resistance up to the peak edge field of around 10 kA/m. This field corresponds approximately to the case when the d.c. critical current is reached on the edge of the coplanar line where the current is peaked. This correlation may be fortuitous since one may expect that the field enters at the thermodynamic critical field rather than H_{c1}. Experiments on niobium and other conventional superconductors allow microwave fields greater than H_{c1} to be maintained without dissipation.[27] Poorer quality films

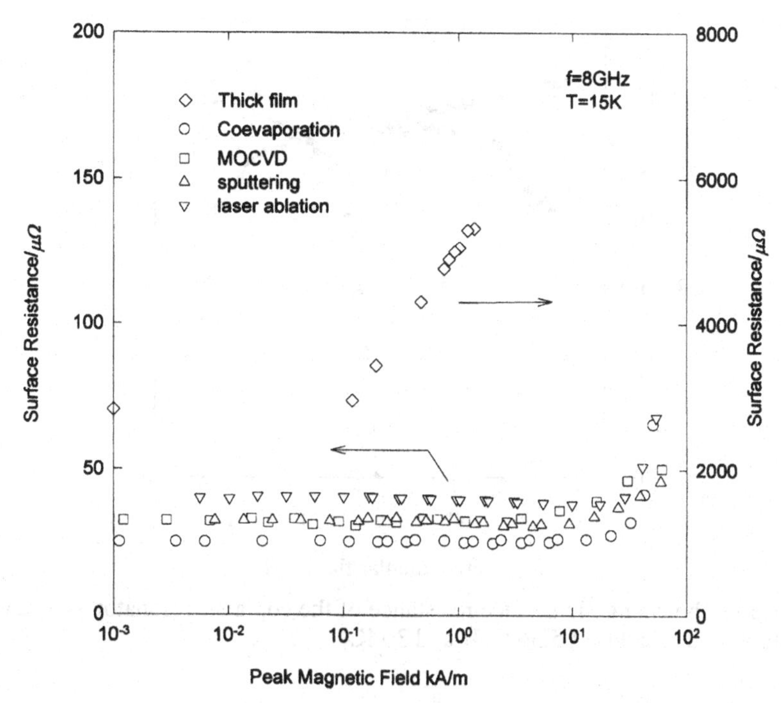

Figure A1.5.1. The surface resistance of the YBCO thin films shown in Figure A1.3.1 as a function of the peak microwave field. A good thick film sample is also shown.[10]

show a significant difference in this non-linear behaviour, although their low-power surface impedance may not be too much larger than those of the films shown in Figure A1.5.1. In this case of poor quality films there is a marked rise in the surface impedance as the power is increased, indicating the presence of serious defects such as large-angle grain boundaries.[28]

Similar experimental results have been obtained using a microstrip resonator and have been discussed with reference to the coupled grain model described in Section 1.10.[29–31] The thick film results shown in Figure A1.5.1 show that good power dependencies can be obtained. Even though the surface resistance is rising with increasing field, this is not rapid and the material is still of use above 1000 A/m.[10]

A1.6 D.c. field dependence

The change in surface resistance as a function of small d.c. applied magnetic fields is shown in Figure A1.6.1. The field is applied perpendicular to the sample. The corresponding shift in centre frequency of the coplanar resonator has the same general shape. A hysteric behaviour is observed, and increasing the field from zero produces an increase in the surface resistance. However, on reversal of the field there is a sharp reduction in surface resistance followed by a slow increase. Similar effects are obtained as the field becomes reversed and the butterfly curve can be traced out as shown in the diagram. Pinning effects alone in the sample cannot explain this shape

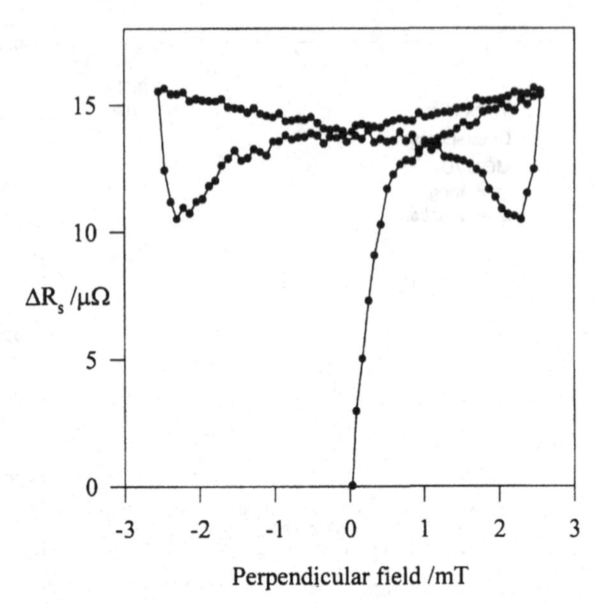

Figure A1.6.1. The change in surface resistance of the coplanar resonator as a function of small applied d.c. fields at 5.36 GHz and 30 K.

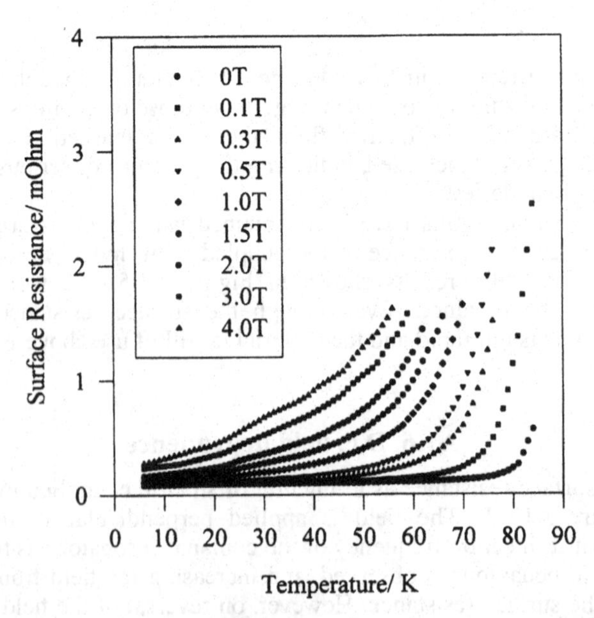

Figure A1.6.2. Surface resistance as a function of temperature at a number of different applied magnetic fields. The field is applied normal to the film surface and the frequency of the measurement is 8 GHz.

because when the field changes polarity pinning would tend to keep the surface resistance high, making the hysteric curve go in the opposite direction. At present the explanation for this effect is unclear.

The effect of large magnetic fields is shown in Figure A1.6.2. The particular measurements shown are obtained with the magnetic field applied when the HTS film is in its normal state, then cooled without altering the field. It can be seen in Figure A1.6.2, that the application of such large fields has the effect of substantially increasing the surface resistance over a wide temperature range. This is also reflected in large changes in surface reactance (or penetration depth). One of the aims of experiments which measure the surface impedance as a function of magnetic field is to validate the model described in Section 1.11 and extract the main parameters of the model. In the Coffey–Clem model these are η the viscous drag coefficient, κ_p the restoring force constant, and ω_c the depinning frequency. For the particular film whose surface resistance is shown in Figure A1.6.2 these extracted values are $\eta = 6.26\ \mathrm{Nsm}^{-2}$, $\kappa_p = 2.64\ \mathrm{Nm}^{-2}$ and $\omega_c = 4.2 \times 10^{11}\ \mathrm{s}^{-1}$ (at 40 K), with a zero-field low-temperature penetration depth of $\lambda_0 = 0.2\ \mu\mathrm{m}$ (0.214 μm at 40 K).[32]

Measurement of the basic data and extraction of these parameters is no easy matter and results can be complicated by film quality and the effects of weak links. However, the extracted parameters from a number of different groups are of the same order of magnitude as the values given above. For further discussion of the properties of thin films in high magnetic fields the reader is referred to the increasing amount of literature on the subject given in reference [33] and the references therein.

A1.7 References

1 Shaked H., Keane P. M., Rodriguez J. C., Owen F. F., Hitterman R. L. and Jorgensen J. D. *Crystal Structure of the High T_c Superconducting Copper-oxides*, Elsevier Science, Amsterdam, 1994

2 Sobolewski R., Xiong W. and Kula W. Patterning of thin film high T_c circuits by the laser writing method, *IEEE Trans. on Applied Superconductivity*, **3**(1), 2986–9, 1993

3 Ballentine P. H., Kadin A. M., Fisher M. A. and Mallory D. S. Microlithography of high temperature superconducting films: laser ablation vs. wet etching, *IEEE Trans. on Magnetics*, **25**(2), 950–3, 1989

4 Shih I. and Qiu C. X. *Appl. Phys. Lett.*, **52**, 1523, 1988

5 Ginley D. S., Ashby C. I. H., Plut T. A., Urea D. and Siegal M. P. Di and tri-carboxylic acid based etches for processing high temperature superconducting thin films, submitted to *Appl. Phys. Lett.*

6 Ginley D. S., Barr L., Ashby C. I. H., Plut T. A., Urea D., Siegal M. P. and Martens J. S. Di and tri-carboxylic acid based etches for processing high temperature superconducting thin films and related materials, submitted to *J. Materials Res.*

7 Shokoohi F. K., Schiavone L. M., Rogers C. T., Inam A., Wu X. D., Nazar L. and Venkatesan T. Wet chemical etching of high temperature superconducting Y–Ba–Cu–O films in ethylenediaminetetraacetic acid, *Appl. Phys. Lett.*, **55**(25), 2661–3, 1989

8 Ashby C. I. H., Martens J., Plut T. A., Ginley D. S. and Phillips, J. M. *Appl. Phys. Lett.*, **60**, 2147, 1992

9 Kaye G. W. C. and Laby T. H. *Tables of Physical and Chemical Constants*, Longman, 1973
10 Smith P. Private communication, University of Birmingham, 1995
11 Alford N. McN., Button T. W., Adams M. J., Hedges S., Nicholson B. and Phillips W. A. Low surface resistance in $YBa_2Cu_3O_x$ melt processed thick films, *Nature*, **349**(6311), 680–3, 1991
12 Müller G. Microwave properties of high T_c superconductors, *Presented at the 4th Workshop on RF Superconductivity at KEK Tsukuba-Shi*, Aug., 1989
13 Porch A. Microwave Surface Impedance of Y Ba Cu O, PhD thesis, University of Cambridge, UK, 1991
14 Superconductor Technologies, Microloss superconducting thin films, *Application Note*, 1995
15 Smith P. A., Alford N McN. and Button T. W. Frequency dependence of surface resistance of bulk high temperature superconductor, *Electron. Lett.*, **26**(18), 1486–7, 1990
16 Porch A., Powell J. R., Lancaster M. J., Edwards J. A. and Humphreys R. G. Microwave conductivity of patterned $YBa_2Cu_3O_7$ thin films, *IEEE Trans. on Applied Superconductivity*, **5**(2), 1987–90, 1995
17 Avenhaus B., Porch A., Lancaster M. J., Hensen S., Lenkens M., Orbach-Werbig S., Müller G., Dähne U., Tellman N., Klein N., Dubourdieu C., Senateur J. P., Thomas O., Karl H., Stritzker B., Edwards J. A. and Humphreys R. Microwave properties of YBCO thin films, *IEEE Trans. on Applied Superconductivity*, **5**(2), 1737–40, 1995
18 Poppe U., Klein N., Dahne U., Solther H., Jia C. L., Kabius B. and Urban K. Low-resistivity epitaxial $YBa_2Cu_3O_7$ thin films with improved microstructure and reduced microwave losses, *J. Appl. Phys.*, **71**(11), 5572–8, 1992
19 Stritzker B., Schubert J., Poppe U., Zander W., Kruger U., Lubig A. and Buchal C. Comparison of YBCO films prepared by laser ablation and sputtering, *J. Less Com. Met.*, **64**, 279–91, 1990
20 Humphreys R. G., Satchell J. S., Chew N. G., Edwards J. A., Goodyear S. W., Blekinsop S. E., Dosser O. D. and Cullis A. G. Physical vapour deposition techniques for growth of $YBa_2Cu_3O_7$ thin films, *Supercond. Science and Tech.*, **3**, 38–52, 1990
21 Hunder J., Thomas O., Mossang E., Ostling M., Chaudouet P., Gaskov A., Weiss F., Boursier D. and Senateur P. Metalorganic chemical-vapour-deposition, *J. Appl. Phys.*, **74**, 4631–2, 1993
22 Avenhaus B., Porch A., Lancaster M. J., Hensen S., Lenkens M., Orbach-Werbig S., Müller G., Dähne U., Tellmann N., Klein N., Dubourdieu C., Senateur J. P., Thomas O., Karl H., Stritzker B., Edwards J. A. and Humphreys R. Microwave properties of YBCO thin films, *IEEE Trans. on Applied Superconductivity*, **5**(2), 1737–40, 1995
23 Piel H. and Müller G. The microwave surface impedance of high-T_c superconductors, *IEEE Trans. on Magnetics*, **27**(2), 854–62, 1991
24 Bonn D. A., Kamal S., Zhang K., Liang R., Baar D. J., Klein E. and Hardy W. N. Comparison of the influence of Ni and Zn impurities on the electromagnetic properties of $YBa_2Cu_3O_{6.95}$, *Phys. Rev.*, **B50**(6), 4051–63, 1994
25 Chew N. G., Edwards J. A., Humphreys R. G., Satchell J. S., Goodyear S. W., Dew B., Exon N. J., Hensen S., Lenkens M., Müller G. and Orbach-Werbig S. Effect of composition and oxygen content on the microwave properties of evaporated Y–

Ba–Cu–O thin films, *IEEE Trans. on Applied Superconductivity*, **5**(2), 1167–72, 1995

26 Klein N., Poppe U., Tellmann N., Schulz H., Evers W., Dähne U. and Urban K. Microwave surface resistance of epitaxial $YBa_2Cu_3O_{7-x}$ films: studies on oxygen deficiency and ordering, *IEEE Trans. on Applied Superconductivity*, **3**(1), 1102–9, 1993

27 Piel H. and Müller G. The microwave surface impedance of high T_c superconductors, *IEEE Trans. on Magnetics*, **27**(2), 854–62, 1991

28 Porch A., Lancaster M. J., Humphreys R. G. and Chew N. G. Non-linear microwave surface impedance of patterned $YBa_2Cu_3O_7$ thin films, *J. Alloys and Compounds*, **95**, 563–6, 1993

29 Oates D. E., Nguyen P. P., Dresselhaus G., Dresselhaus M. S., Lam C. W. and Ali S. M. Non-linear surface resistance in $YBa_2Cu_3O_{7-\delta}$ thin films, *IEEE Trans. on Applied. Superconductivity*, **3**(1), 1114–18, 1993

30 Oates D. E., Nguyen P. P., Dresselhaus G., Dresselhaus M. S. and Chin C. C. Measurements and modelling of linear and non-linear effects in striplines, *J. of Superconductivity*, **5**(4), 363–71, 1992

31 Oates D. E., Anderson A. C., Sheen D. M. and Ali S. M. Stripline resonator measurements of Z_s verses H_{rf} in $YBa_2Cu_3O_{7-x}$ thin films, *IEEE Trans. on Microwave Theory and Techniques*, **39**(9), 1522–9, 1991

32 Powell J. R., Porch A., Wellhöfer F., Woodall P., Humphreys R. G. and Gough C. E. Microwave surface impedance of YBCO thin films in DC applied magnetic fields, *1995 European Conf. on Applied Superconductivity*, Edinburgh, UK, 1995

33 Gaidukov M. M., Karmanenko S. F., Klimenko V. L., Kozyrev A. B. and Soldatenkov O. I. Identification of radio-frequency and microwave residual-loss mechanisms in $YBa_2Cu_3O_{7-\delta}$ films in a magnetic field, *Superconductor Science and Technology*, **7**, 721–6, 1994

Appendix 2

Substrates for superconductors

A2.1 Substrate requirements

Superconducting films have to be grown on some sort of substrate which must be inert, compatible with both the growth of a good quality film and also have appropriate microwave properties for application purposes. From the microwave point of view, a high dielectric constant of well-known value is good for miniaturising components, as discussed in chapter 5. However, for higher frequencies this miniaturisation may be a problem in that the size of the circuit may be reduced considerably and the size reduction may make it difficult to produce the device. The design of devices is simplified if the dielectric constant is isotropic in the plane of the film and has a low dispersion for wide band devices, although compensation can be built into the device design if the parameters are sufficiently well characterised. It is also more convenient if the dielectric constant does not change much with temperature, improving the temperature stability of the final application. Whatever the dielectric constant, it must be reproducible and not change appreciably from batch to batch. For many of the devices discussed in this text a low dielectric loss tangent is of fundamental importance, especially on high-Q filters or resonators or long delay lines. If the loss tangent is not low enough, then the advantage of using a superconductor can be negated.

In addition to these microwave requirements, the substrate should be compatible with good film growth. In order to achieve good epitaxial growth the dimensions of the crystalline lattice at the surface of the substrate should match the dimensions of the lattice of the superconductor. If this is not the case, strains can be set up in the films, producing dislocations and defects. In some cases the substrates can react chemically with the substrates or atomic migrations can take place out of the substrate; this again causes impurity levels to rise and the performance of the film to reduce. Cracks can be caused in the film if the thermal expansions of the substrate and film are not appropriately matched. Some of the above problems can be overcome by the application of a buffer layer between the films and the substrates.

In addition, there are a number of mechanical requirements on substrates. The surface should be smooth and free from defects and twinning if possible. These cause unwanted growth and mechanisms which can lead to non-optimal films. The substrates should be strong and capable of being thinned to a certain extent, depending upon the application. Substrates should have good environmental stability and be available in large sizes, preferably above 3" in diameter.

A2.2 Substrate materials

With all the above requirements it is not surprising that an ideal substrate for high-temperature superconducting films has not been found. Nevertheless, a number of excellent substrates, producing good films with good microwave properties, are in common use. Table A2.2.1 shows a number of substrates which have been investigated for their microwave properties and as candidates for HTS growth. The dielectric constant and the loss tangent are given, together with the frequency and temperature they have been measured at. The properties are usually measured using a dielectric resonator similar to that discussed in Section 3.4. Much data is given in this table because of the variability of the values in the literature. The method of growth and purity can influence the microwave properties greatly, so that properties can vary between different manufacturers.

All the materials given in Table A2.2.1 have dielectric constants ranging from about 9 to 30, and therefore all are adequate for miniaturising components to a certain extent. For most of these materials there is a decrease in permittivity with temperature of a few percent from room temperature to about 10 K. The loss tangents of the materials vary widely and it is this property which mainly determines the usefulness of the substrate for high-performance applications. In addition, the loss tangent usually decreases with temperature and can be as much as one or two orders of magnitude. The most widely used substrates are $LaAlO_3$ and MgO, both of which have low loss tangents. $LaAlO_3$ has a higher dielectric constant than MgO but is generally twinned. The dielectric constant is not as well defined and can vary slightly. $SrTiO_2$, $NdGaO_3$[1] and ZrO_2 are particularly unsuitable because of their high loss tangents. A number of review articles have been published on substrates for superconducting films and the reader is referred to these for further information.[5,25–27]

Table A2.2.1. *Substrate for high-temperature superconducting films*

	Ref.	ε_r	tan (δ)	f/GHz	T/K
Amorphous quartz	2	3.8	10^{-3}	1.22	–
Crystalline quartz		4.5	7×10^{-4}	1.8	–
$AlO_2 \perp c$	3	9.4	–	10	25
$AlO_2 \parallel c$	3	11.6	–	10	25
Sapphire	4,5	–	1.5×10^{-8}	9	77
	4,5	–	4.3×10^{-7}	72	77
	6,5	–	2×10^{-9}	3	4.2
$LaAlO_3$	7	–	10^{-6}	11.6	4.3
	5,8	26	$< 5 \times 10^{-4}$	500	4.2–90
	5,9	24.5	2×10^{-6}	0.5–15	4.2

Table A2.2.1. (*cont.*)

	Ref.	ε_r	$\tan(\delta)$	f/GHz	T/K
$LaAlO_3$ (*cont*)	25	24.2	3.1×10^{-5}	10	300
	25	–	7.6×10^{-6}	10	77
	25	–	4.9×10^{-6}	10	20
$LaGaO_3$	5,10	25	1.8×10^{-3}	1 MHz	300
	5,9	25	1.5×10^{-6}	0.5–15	4.2
$SrTiO_3$	11,5	~ 230	3×10^{-2}	9.5	300
	12,5	215	1.4×10^{-1}	10 kHz	300
	13,5	310	3×10^{-2}	$10–10^3$	300
	13,5	1900	6×10^{-2}	$10–10^3$	80
	14,5	–	$1–2 \times 10^{-3}$	22	77–500
	15,5	–	$2–24 \times 10^{-3}$	22	37–600
	16	300	3×10^{-4}	3.2	300
	16	1800	1×10^{-4}	1.3	90
ZrO_2 (YSZ)	10,5	27	5.4×10^{-3}	1 MHz	300
	17,5	28	4×10^{-3}	10	300
	13,5	26	1.6×10^{-2}	$10–10^3$	300
	13,5	25.4	7.5×10^{-3}	$10–10^3$	80
	18,5	25	$2–6 \times 10^{-4}$	2–20	4.2
	19	32	3×10^{-4}	5.5	77
MgO	20,5	9.6	2×10^{-3}	1 kHz	300
	13,5	9.87	9×10^{-4}	$10–10^3$	300
	13,5	9.6	4×10^{-5}	$10–10^3$	80
	5	–	1×10^{-6}	0.5	4.2
	25	10	1.6×10^{-5}	10	300
	25	–	6.1×10^{-6}	10	20
	21	–	2×10^{-6}	8	77
$NdCaAlO_4$	22	20	$4–25 \times 10^{-4}$	200–600	5–300
$LaSrAlO_4$	23	27	1×10^{-4}	8	5

Table A2.2.1. (*cont.*)

	Ref.	ε_r	$\tan(\delta)$	f/GHz	T/K
$CaYAlO_4$	24	20	4×10^{-5}	5	77
$YAlO_3$	25	16	8.2×10^{-5}	10	300
	25	–	1.2×10^{-5}	10	77
	25	–	4.9×10^{-6}	10	20
$NdGaO_3$	25	22.8	1.1×10^{-4}	10	300
	25	–	3.2×10^{-4}	10	77
	25	–	2.3×10^{-4}	10	20
	26	23	8×10^{-3}	10	300
$SrLaAlO_4 \parallel c$	26	20	1×10^{-3}	10	300
$SrLaAlO_4 \perp c$	26	16.9	7×10^{-4}	10	300
$CaNdAlO_4 \parallel c$	26	17.6	6×10^{-6}	10	300
$CaNdAlO_4 \perp c$	26	19.6	6×10^{-4}	10	300
$LaSrGaO_4$	25	22	5.7×10^{-5}	10	300
	25	–	1.5×10^{-5}	10	77
	25	–	5.7×10^{-6}	10	20
$YbFeO_3$	27,28	7.3	2.0×10^{-3}	10	300
Liquid nitrogen	29	1.4	5.2×10^{-5}	15	77

A2.3 References

1 To H. Y., Valco G. J. and Bhasin K. B. 10 GHz $YBa_2Cu_3O_{7-\delta}$ superconducting ring resonators on $NdGaO_3$ substrates, *Superconductor Science and Technology*, **5**, 421–6, 1992

2 Anderson A. C., Withers R. S., Reible S. A. and Ralston R. W. Substrates for analogue signal processing devices, *IEEE Trans. on Magnetics*, **MAG-19**(3), 485–9, 1983

3 McCarroll C. P., Zhou B. L., Han S. C. and Luo L. Measurements of anisotropic dielectric constant of R-plane sapphire substrates for HTS thin film deposition and microwave applications, *1992 IEEE Applied Superconductivity Conf., Chicago*, paper, ETB-7, 1992

4 Braginsky V. B., Ilchenko V. S. and Bagdassarov Kh. S. Experimental observation of fundamental microwave absorption in high quality dielectric crystals, *Phys. Lett.*, A**12**, 300–5, 1987

5 Talvacchio J., Wagner G. R. and Talisa S. H. High T_c film development for electronic application, *Microwave Journal*, pp. 105–114, July, 1991
6 Braginsky V. B. and Panov V. I. Superconducting resonators on sapphire, *IEEE Trans. on Magnetics*, **MAG-15**(1), 3–33, 1979
7 Tellmann N., Klein N., Dähne U., Scholen A., Schulz H. and Chaloupka H. High Q $LaAlO_3$ dielectric resonator shielded by YBCO-films, *IEEE Trans. on Applied Superconductivity*, **4**(3), 143–8, 1994
8 Nuss M. C., Mankiewich P. M., Howard R. E., Straughn B. L., Harvey T. E., Brandle C. D., Berkstrasser G. W., Goossen K. W. and Smith P. R. Propagation of terahertz bandwidth electrical pulses on $YBa_2Cu_3O_7$ transmission lines on $LaAlO_3$ substrates, *Appl. Phys. Lett.*, **54**, 2265–7, 1989
9 Lyons W. G. Low loss substrates for high temperature superconductors, *MIT Lincoln Laboratory Quarterly Technical Report*, October, 1989
10 Sandstrom R. L., Giess E. A., Gallagher W. J., Cooper E. I., Chisholm M. F., Gupta A., Shinde S. and Laibowitz R. B. Lanthanum gallate substrates for epitaxial high T_c superconducting thin films, *Appl. Phys. Lett.*, **53**(19), 1874–6, 1988
11 Harvey A. F. *Microwave Engineering*, Academic Press, New York, 1963
12 Araki K., Iwasa I., Kobayshi Y., Nagata S. and Morisue M. Ultra broad band measurements on high T_c superconducting transmission lines, *IEEE Trans. on Magnetics*, **25**, 980–3, 1989
13 Gorshunov B. P., Kozlov G. V., Krasnosvobodstev S. I., Pechen E. V., Prokhorov A. M., Prokhorov A. S., Syrotnsky O. I. and Volkov A. A. Submillimetre properties of high T_c superconductors, *Physica*, C**153–5**, 667–70, 1988
14 Rupprecht G. and Bell R. O. Microwave losses in cubic strontium titanate above the phase transition *Phys. Rev.*, **123**(1), 97–8, 1961
15 Rupprecht G., Bell R. O. and Silverman B. D. Nonlinearity and microwave losses in cubic strontium titanate *Phys. Rev.*, **125**(6), 1915–20, 1962
16 Kobayashi Y., Sato Y. Y. and Yajima K. *Trans. IEICE*, **VE-72**, 290, 1989
17 Lanagan M. T., Yamamoto J. K., Bhalla A. and Sankar S. G. The dielectric properties of yttria stabilised zirconia, *Mater. Lett.*, **7**, 437, 1989
18 Anderson A. C., Tsaur B. Y., Stienbeck J. W. and Dilorio M. S. RF surface resistance of $YBa_2Cu_3O_{7-x}$ thin films, *MIT Lincoln Laboratory Quarterly Report*, March, 1988
19 Smith P. A. and Davis L. E. Dielectric loss tangent of yttria stabilised zirconia, *Electron. Lett.*, **28**(4), 424–5, 1992
20 Thorp J. S. and Enayati-Rad The dielectric behaviour of single crystal MgO, Fe/MgO, and Cr/MgO, *J. Mater. Sci.*, **16**, 255–60, 1981
21 Porch A., Lancaster M. J. and Humphreys R. G. The coplanar resonator technique for determining the surface impedance of $YBa_2Cu_3O_{7-\delta}$ thin films, *IEEE Trans. on Microwave Theory and Techniques*, **43**(2), 306–14, 1995
22 Berkowski M., Pajaczkowska A., Gierlowski P., Sobolewski R., Lweandowski S. J., Gorshunov B. P., Kozlov G. V., Lyudnirsky D. B., Sirotinsky O. I., Saltykov P. A., Soltner H., Poppe U., Cuchal C. and Lubig A. $CaNdAlO_4$ perovskite substrate for microwave and far-infrared applications of epitaxial high T_c superconducting thin films, *Appl. Phys. Lett.*, **57**, 632–4, 1990
23 Brown R., Pendrick V., Kalokitisis D. and Chai B. H. T. Low loss substrate for microwave application of high temperature superconductor films, *Appl. Phys. Lett.*, **57**, 1351–3, 1990
24 Young K. H., and Chai B. H. T. $CaYAlO_4$: New low loss substrate for epitaxial

growth of superconducting thin films, *Jap. J. Appl. Phys.*, **30**, L2116–L2119, 1992

25 Konaka T., Sato M., Asano H. and Kubo S. Relative permittivity and loss tangents of substrate materials for high T_c superconducting thin films, *J. Superconductivity*, **4**, 283–7, 1991

26 Konopka J. and Wolff I. Dielectric properties of substrates for deposition of high T_c films up to 40 GHz, *IEEE Trans. on Microwave Theory and Techniques*, **40**, 2418–23, 1992

27 Hollmann E. K, Vendik O. G., Zaitsev A. G. and Melekh B. T. Substrates for high T_c superconductor microwave integrated circuits, *Superconductivity Science and Technology*, **7**, 609–22, 1994

28 Rao K. V. and Smacula A. *J. Appl. Phys.*, **37**, 319, 1966

29 Smith P. A., Davis L. E., Button T. W. and Alford N. McN The dielectric loss tangent of liquid nitrogen, *Superconductor Science and Technology*, **4**, 128–9, 1991

Appendix 3
Some useful relations

A3.1 Introduction

Two tables are given in this appendix. Table A3.1.1 gives a summary of some of the expressions for various parameters of a plane wave in conductors, superconductors and dielectric materials, with the various approximations applied in their derivation. The derivation and description of these expressions are given in Chapter 1. Table A3.1.2 gives a summary of some useful fundamental physical constants.

Table A3.1.1. *The expressions for various parameters of a plane wave in different media*

	General superconductors	Normal conductors	Good dielectrics	Good superconductors
Approximations and assumptions	$\frac{\sigma}{\omega\varepsilon} \gg 1 \quad \sigma = \sigma_1 - j\sigma_2$	$\frac{\sigma}{\omega\varepsilon} \gg 1 \quad \sigma_2 = 0$	$\frac{\sigma}{\omega\varepsilon} \ll 1$	$\sigma_2 \gg \sigma_1$
Propagation constant	$\gamma = (1+j)\sqrt{\left(\frac{\omega\mu}{2}\right)}\sqrt{(\sigma_1 - j\sigma_2)}$	$\gamma = (1+j)\sqrt{\left(\frac{\omega\mu\sigma}{2}\right)}$	$\gamma = \frac{\sigma}{2}\sqrt{\left(\frac{\mu}{\varepsilon}\right)} + j\omega\sqrt{(\mu\varepsilon)}\left(1 + \frac{\sigma_2}{8\omega^2\varepsilon^2}\right)$	$\gamma = \sqrt{(\omega\mu\sigma_2)}\left(1 + j\frac{\sigma_1}{2\sigma_2}\right)$
Intrinsic impedance/ surface impedance	$Z_s = (1+j)\sqrt{\left(\frac{\omega\mu}{2}\right)}\sqrt{\left(\frac{\sigma_1 + j\sigma_2}{\sigma_1^2 + \sigma_2^2}\right)}$	$Z_s = (1+j)\sqrt{\left(\frac{\omega\mu}{2\sigma}\right)}$	$Z_s = \sqrt{\left(\frac{\mu}{\varepsilon}\right)}\left(1 + j\frac{\sigma}{2\omega\varepsilon} - \frac{3\sigma^2}{2\omega^2\varepsilon^2}\right)$	$Z_s = \sqrt{\left(\frac{\omega\mu}{\sigma_2}\right)}\left(\frac{\sigma_1}{2\sigma_2} + j\right)$
Surface resistance	$R_s = \frac{1}{2}\sqrt{(\omega\mu)}\,\mathrm{Re}\,(f_2(\sigma_1, \sigma_2))$	$R_s = \sqrt{\left(\frac{\omega\mu}{2\sigma}\right)}$	$R_s = \sqrt{\left(\frac{\mu}{\varepsilon}\right)}\left(1 - \frac{3\sigma^2}{2\omega^2\varepsilon^2}\right)$	$R_s = \frac{\omega^2\mu^2\sigma_1\lambda^3}{2}$
Surface reactance	$X_s = \frac{1}{2}\sqrt{(\omega\mu)}\,\mathrm{Im}\,(f_2(\sigma_1, \sigma_2))$	$X_s = \sqrt{\left(\frac{\omega\mu}{2\sigma}\right)}$	$X_s = \sqrt{\left(\frac{\mu}{\varepsilon}\right)}\left(\frac{\sigma}{2\omega\varepsilon}\right)$	$X_s = \omega\mu\lambda$
Velocity	$c = \frac{2\omega}{\sqrt{(\omega\mu)}\,\mathrm{Im}\,(f_1(\sigma_1, \sigma_2))}$	$c = \sqrt{\left(\frac{2\omega}{\mu\sigma}\right)}$	$c = \frac{1}{\sqrt{(\mu\varepsilon)}}\frac{1}{\left(1 + \frac{\sigma^2}{8\omega^2\varepsilon^2}\right)}$	$c = \sqrt{\left(\frac{4\sigma_2\omega}{\mu\sigma_1^2}\right)}$
Attenuation	$\alpha = \frac{1}{2}\sqrt{(\omega\mu)}\,\mathrm{Re}\,(f_1(\sigma_1, \sigma_2))$	$\alpha = \sqrt{\left(\frac{\omega\mu\sigma}{2}\right)}$	$\alpha = \frac{\sigma}{2}\sqrt{\left(\frac{\mu}{\varepsilon}\right)}$	$\alpha = \sqrt{(\omega\mu\sigma_2)}$
Penetration depth/skin depth	$\delta = \frac{2}{\sqrt{(\omega\mu)}\,\mathrm{Re}\,(f_1(\sigma_1, \sigma_2))}$	$\delta = \sqrt{\left(\frac{2}{\omega\mu\sigma}\right)}$	$\delta = \frac{2}{\sigma}\sqrt{\left(\frac{\varepsilon}{\mu}\right)}$	$\lambda = \frac{1}{\sqrt{(\omega\mu\sigma_2)}}$

General: $\gamma = \sqrt{(j\omega\mu(\sigma + j\omega\varepsilon))} \quad Z_s = \sqrt{\left(\frac{j\omega\mu}{\sigma + j\omega\varepsilon}\right)}$

$f_1(\sigma_1, \sigma_2) = \sqrt{p + \sigma_1} + \sqrt{p - \sigma_1} - j\sqrt{p - \sigma_1} + j\sqrt{p + \sigma_1}$

$f_2(\sigma_1, \sigma_2) = p^{-1}(\sqrt{p + \sigma_1} - \sqrt{p - \sigma_1} + j\sqrt{p - \sigma_1} + j\sqrt{p + \sigma_1}) \quad p = \sqrt{\sigma_1^2 + \sigma_2^2}$

Table A3.1.2. *Some useful fundamental constants*

Name	Symbol	Value	Unit
Avogadro's constant	N	6.02217×10^{23}	mol^{-1}
Boltzmann's constant	k_B	1.38062×10^{-23}	JK^{-1}
Electron charge	e	1.602192×10^{-19}	C
Electron mass	m	9.10956×10^{-31}	kg
Flux quantum	$\Phi_0 = h/2e$	2.06783×10^{-15}	Wb
Impedance of vacuum	η_0	376.7304	Ω
Light velocity	c	2.997925×10^{8}	ms^{-1}
Permittivity of vacuum	ε_0	8.854185×10^{-12}	Fm^{-1}
Permeability of vacuum	μ_0	$4\pi \times 10^{-7}$	Hm^{-1}
Planck's constant	h	6.6262×10^{-34}	Js
Planck's constant	$\hbar = h/2\pi$	1.054592×10^{-34}	Js

Index